T0258161

Advanced Topics in Electromagnetic Radiation

Advanced Topics in Electromagnetic Radiation

Edited by **Edgar Wilson**

New York

Published by NY Research Press,
23 West, 55th Street, Suite 816,
New York, NY 10019, USA
www.nyresearchpress.com

Advanced Topics in Electromagnetic Radiation
Edited by Edgar Wilson

International Standard Book Number: 978-1-63238-023-4 (Hardback)

Printed in the United States of America.

Contents

Preface

Every book is initially just a concept; it takes months of research and hard work to give it the final shape in which the readers receive it. In its early stages, this book also went through rigorous reviewing. The notable contributions made by experts from across the globe were first molded into patterned chapters and then arranged in a sensibly sequential manner to bring out the best results.

Electromagnetic radiation, a basic phenomenon of electromagnetism, has been described in this comprehensive book. Modern applications of electromagnetic radiation are an emerging technological practice. Experts have comprehensively dealt with two aspects of electromagnetic radiation, theory and application; together in the same book. It discusses numerous topics of practical relevance, including medical treatment, telecommunication systems and radiation effects. The book provides vivid description of the subjects along with contemporary examples which make this book an indispensable reference book for electromagnetic radiation applications.

It has been my immense pleasure to be a part of this project and to contribute my years of learning in such a meaningful form. I would like to take this opportunity to thank all the people who have been associated with the completion of this book at any step.

Editor

Generalized Orthogonality Relation
for Spherical Harmonics

A. Draux[1] and G. Gouesbet[2]

[1]Institut National des Sciences Appliquées (INSA) de Rouen, Département de Génie
Mathématique, Place Emile Blondel, BP08, 76131, Mont-Saint-Aignan Cédex
[2]Laboratoire d'Electromagnétisme des Systèmes Particulaires (LESP) Unité Mixte de
Recherche (UMR) 6614 du Centre National de la Recherche Scientifique (CNRS)
COmplexe de Recherche Interprofessionnel en Aérothermochimie (CORIA) Université de
Rouen et Institut National des Sciences Appliquées (INSA) de Rouen BP12,
Avenue de l'Université, Technopôle du Madrillet, 76801, Saint-Etienne-du Rouvray
France

1. Introduction

One of us (G.G), with collaborators, has been involved in the study of generalized Lorenz-Mie
theories (GLMTs) describing the interaction between electromagnetic arbitrary shaped beams
(typically laser beams) and a class of regular scatterers for which solutions to Maxwell's
equations can be found by using the method of separation of variables, e.g. Gouesbet &
Gréhan (2000a), J.A.Lock & Gouesbet (2009), Gouesbet (2009a), and references therein. It
has, at a certain time, been found interesting to examine whether the knowledge gained
in the effort of developing electromagnetic GLMTs could be, at least partially, adapted to
quantum mechanical problems. The examination of the issue of quantum scattering of
quantum arbitrary shaped beams produced several papers devoted to (i) the description
of quantum arbitrary shaped beams Gouesbet (2005), Gouesbet & J.A.Lock (2007) (ii) the
evaluation of cross-sections in the case of quantum arbitrary shaped beams interacting with
radial quantum potentials Gouesbet (2006a), Gouesbet (2007a) and (iii) the exhibition of
formal cross-sectional analogies between electromagnetic and quantum scatterings Gouesbet
(2004), Gouesbet (2006b), Gouesbet (2007b). During the development of this work, the
evaluation of two integrals, based on spherical harmonics, has been required. These integrals
may be obtained from a generalized orthogonality relation for spherical harmonics which is
established in this paper. Another recent application concerns the optical theorem and non
plane wave scattering in quantum mechanics Gouesbet (2009b).

Section II is devoted to the demonstrations used to reach the generalized orthogonality
relation mentioned above. Section III is a conclusion.

2. Demonstrations

2.1 Lemma 1

Let $X(b)$ denote the following expression :

$$X(b) = \sin a \sin \theta \cos(\varphi - b) + \cos a \cos \theta \tag{1}$$

Then :

$$\int_0^{2\pi} P_n(X(b))e^{-im(\varphi-b)}d\varphi = \int_0^{2\pi} P_n(X(0))\cos(|m|\ \varphi)d\varphi \tag{2}$$

where P_n is the Legendre polynomial of degree n, $i = \sqrt{-1}$ and $m \in \mathbf{Z}$.

As a remark for further use, let us insist on the fact that the r.h.s. of Eq.2 does not depend on b nor on the sign of m.

2.2 Proof of Lemma 1

The integrand $P_n(X(b))e^{-im(\varphi-b)}$ in the l.h.s. of Eq.2 is a linear combination of terms reading as :

$$[\cos(\varphi - b)]^j e^{-im(\varphi-b)}, j = 0...n \tag{3}$$

Therefore, to demonstrate Eq.2, it is sufficient to prove that :

$$\int_0^{2\pi} [\cos(\varphi - b)]^j\ e^{-im(\varphi-b)}d\varphi = \int_0^{2\pi} (\cos\varphi)^j \cos(|m|\ \varphi)d\varphi, \forall j \in \mathbf{N} \tag{4}$$

Let us introduce a symbol to denote the l.h.s. of Eq.4, setting :

$$K_{jm} = \int_0^{2\pi} [\cos(\varphi - b)]^j\ e^{-im(\varphi-b)}d\varphi \tag{5}$$

We then make a change of variables from $(\varphi - b)$ to Ψ, use the Leibniz formula, and establish:

$$K_{jm} = \int_{-b}^{-b+2\pi} (\cos\Psi)^j e^{-im\Psi}d\Psi \tag{6}$$

$$= \frac{1}{2^j} \sum_{k=0}^j \binom{j}{k} \int_{-b}^{-b+2\pi} e^{i(2k-j-m)\Psi}d\Psi$$

$$= \frac{2\pi}{2^j} \binom{j}{k} \delta_{0,2k-j-m}, k = 0...j$$

where δ denotes the Kronecker symbol.

Let us introduce :

$$\widehat{K}_{jm} = \int_0^{2\pi} (\cos\varphi)^j \cos(|m|\ \varphi)d\varphi \tag{7}$$

Converting cosines to exponentials and using again the Leibniz formula, it readily becomes :

$$\widehat{K}_{jm} = \frac{1}{2^{j+1}} \sum_{k=0}^j \binom{j}{k} \int_0^{2\pi} \left[e^{i(2k-j-|m|)\varphi} + e^{i(2k-j+|m|)\varphi} \right] d\varphi \tag{8}$$

leading to :

$$\widehat{K}_{jm} = \frac{2\pi}{2j+1} \binom{j}{k} \left[\delta_{0,2k-j-|m|} + \delta_{0,2k-j+|m|} \right] \tag{9}$$

We can then deduce the following results :

(i) If $(j - m)$ is odd, or $j < |m|$, then :

$$K_{jm} = \widehat{K}_{jm} = 0 \tag{10}$$

(ii) If $(j - m)$ is even, and $|m| \leq j$,

then :

$$K_{jm} = \widehat{K}_{jm} = \frac{2\pi}{2j+1} \left[\binom{j}{\frac{j+m}{2}} + \binom{j}{\frac{j-m}{2}} \right] \tag{11}$$

2.3 Corollary 2

By using Eqs.10 and 11, we obtain the following fairly obvious corollary : if $n < |m|$, then :

$$\int_0^{2\pi} P_n(X(b)) e^{-im(\varphi-b)} d\varphi = 0 \tag{12}$$

We are now going to establish two identities to be used in the sequel.

2.4 Lemma 3

For any integers r, s, n such that $s \leq r \leq [n/2]$, in which $[i]$ denotes the integer part of i, we have :

(i)

$$\frac{1}{2^n} \sum_{m=0}^{s} (-1)^m \frac{2^{2m}(2n-2m)!}{m!(n-m)!(s-m)!(r-m)!} = \frac{1}{r!s!} \prod_{j=1}^{n-s}(2j-1) \prod_{i=n-r-s+1}^{n-r}(2i-1) \tag{13}$$

with the convention that $\prod_{i=i_1}^{i_2} = 1$ if $i_2 < i_1$, and :

(ii)

$$\sum_{t=r}^{[n/2]} \frac{1}{2^{2t}(n-2t)!(t-r)!(t-s)!} = \frac{(2(n-r-s))!}{2^n(n-r-s)!(n-2r)!(n-2s)!} \tag{14}$$

with the convention that $\sum_{i=i_1}^{i_2} = 0$ if $i_2 < i_1$.

2.5 Proof of Lemma 3

(i)

Let us denote by S_1 the l.h.s. of Eq.13. But we have :

$$\overline{S_1} = \frac{(2n-2m)!}{2^{n-m}(n-m)!} = \prod_{j=1}^{n-m}(2j-1) \tag{15}$$

Hence, S_1 can be rewritten as :

$$S_1 = \prod_{j=1}^{n-s}(2j-1) \sum_{m=0}^{s}(-1)^m 2^m \frac{\prod_{j=n-s+1}^{n-m}(2j-1)}{m!(s-m)!(r-m)!} \quad (16)$$

$$= (-1)^s \frac{\prod_{j=1}^{n-s}(2j-1)}{r!s!} \sum_{m=0}^{s} 2^m m! \binom{s}{m}\binom{r}{m} \prod_{j=n-s+1}^{n-m}(1-2j)$$

From the r.h.s. of Eq.16, we introduce a polynomial $S_1(x)$ reading as :

$$S_1(x) = \sum_{m=0}^{s} 2^m m! \binom{s}{m}\binom{r}{m} \prod_{j=n-s+1}^{n-m}(x+1-2j) \quad (17)$$

We observe that $S_1(x)$ is a polynomial of degree s written in the Newton basis $\{N_i(x)\}_{i=0}^{s}$ with $N_0(x) = 1$ and $N_{i+1}(x) = (x-x_i)N_i(x)$ for $i \geqslant 0$, where $x_0 = 2n - 2s + 1, x_k = x_0 + kh$ and $h = 2$. Thus :

$$\prod_{j=n-s+1}^{n-m}(x+1-2j) = N_{s-m}(x) \quad (18)$$

Let us prove that :

$$S_1(x) = Q(x) = \prod_{i=n-r-s+1}^{n-r}(x-2i+1) \quad (19)$$

It is sufficient to verify that $S_1(x_k) = Q(x_k)$ for k=0...s. We have :

$$Q(x_k) = 2^s \frac{(r+k)!}{(r+k-s)!} \quad (20)$$

Since :

$$S_1(x) = \sum_{m=0}^{s} 2^{s-m} \frac{s!}{m!}\binom{r}{s-m} N_m(x) \quad (21)$$

and :

$$N_m(x_k) = 2^m \frac{k!}{(k-m)!} \quad (22)$$

we obtain :

$$S_1(x_k) = 2^s \frac{k!r!}{(r-s+k)!} \sum_{m=0}^{k} \binom{s}{m}\binom{r-s+k}{k-m} \quad (23)$$

in which the summation is originally found to range from $m = 0$ to s, but can afterward be reduced from $m = 0$ to k. By using the following identity (Abramowitz & Stegun (1964), p822) :

$$\sum_{m=0}^{n} \binom{r}{m}\binom{s}{n-m} = \binom{r+s}{n}, \forall (r+s) \geq n \quad (24)$$

we have :

$$S_1(x_k) = \frac{2^s k!r!}{(r-s+k)!}\binom{r+k}{k} = \frac{2^s(r+k)!}{(r+k-s)!} = Q(x_k) \quad (25)$$

But :

$$S_1 = \frac{(-1)^s}{r!s!} \prod_{j=1}^{n-s} (2j-1) S_1(0) \tag{26}$$

Hence, it is readily seen that the first identity holds.

(ii)

On one hand, we introduce the quantity B_2 according to :

$$(n-2r)!B_2 = \frac{1}{2^{r+s}(n-2s)!} \prod_{i=1}^{n-r-s} (2i-1) \tag{27}$$

$$= \frac{1}{2^{r+s}(n-2s)!} \prod_{i=1}^{n-[n/2]-s} (2i-1) \prod_{i=n-[n/2]-s+1}^{n-r-s} (2i-1)$$

Next, let $R(x)$ denote the polynomial of degree $[n/2] - r$, given by :

$$R(x) = \frac{1}{2^{r+s}(n-2s)!} \prod_{i=1}^{n-[n/2]-s} (2i-1) \prod_{i=n-[n/2]-s+1}^{n-r-s} (x+2i-1) \tag{28}$$

so that :

$$(n-2r)!B_2 = R(0) \tag{29}$$

On the other hand, we have :

$$(n-2r)!S_2 = \sum_{t=r}^{[n/2]} \frac{1}{2^{2t}} \frac{\prod_{j=r}^{t-1}(n-2j) \prod_{j=r}^{t-1}(n-2j-1)}{(t-r)!(t-s)!} \tag{30}$$

One of the two products only involves even integers while the other only involves odd integers. Thus, we can write them as :

$$\prod_{j=r}^{t-1} 2([n/2]-j) \prod_{j=r}^{t-1} (2n-2[n/2]-2j-1) \tag{31}$$

Therefore, with fairly obvious changes of variables :

$$(n-2r)!S_2 = \sum_{t=0}^{[n/2]-r} \frac{1}{2^{2t+2r}t!(t+r-s)!} \prod_{j=0}^{t-1} 2([n/2]-j-r) \prod_{j=0}^{t-1} [2(n-[n/2]-j-r)-1] \tag{32}$$

Let $\widetilde{S}_2(x)$ denote the polynomial :

$$\widetilde{S}_2(x) = \sum_{t=0}^{[n/2]-r} \frac{1}{2^{2t+2r}t!(t+r-s)!} \prod_{j=0}^{t-1} 2([n/2]-j-r) \prod_{j=0}^{t-1} [x+2(n-[n/2]-j-r)-1] \tag{33}$$

so that :

$$(n-2r)!S_2 = \widetilde{S}_2(0) \tag{34}$$

The polynomial $\widetilde{S}_2(x)$, which is of degree $([n/2] - r)$, is written in the Newton basis $\{N_m(x)\}_{m=0}^{[n/2]-r}$, with $x_0 = 1 - 2(n - [n/2] - r)$ and $x_m = x_0 + mh, m = 0...[n/2] - r, h = 2$.

Let us now prove that $\widetilde{S}_2(x)$ and $R(x)$ are identical, which is equivalent to verify that $\widetilde{S}_2(x_m) = R(x_m), m = 0...[n/2] - r$.

We have :

$$\widetilde{S}_2(x_m) = \sum_{t=0}^{m} \frac{1}{2^{2t+2r}} \frac{\prod_{j=0}^{t-1} 2([n/2] - j - r)}{t!(t+r-s)!} \frac{2^t m!}{(m-t)!} \tag{35}$$

$$= \frac{1}{2^{2r}} \sum_{t=0}^{m} \frac{([n/2] - r)!m!}{t!(t+r-s)!([n/2] - r - t)!(m-t)!}$$

$$= \frac{m!}{2^{2r}(m+r-s)!} \sum_{t=0}^{m} \binom{[n/2] - r}{t} \binom{m+r-s}{m-t}$$

By using Eq.24, this becomes :

$$\widetilde{S}_2(x_m) = \frac{m!}{2^{2r}(m+r-s)!} \binom{[n/2] + m - s}{m} = \frac{([n/2] + m - s)!}{2^{2r}(n+r-s)!([n/2] - s)!} \tag{36}$$

Next, we have :

$$R(x_m) = \frac{1}{2^{r+s}(n-2s)!} \prod_{i=1}^{n-[n/2]-s} (2i-1) \prod_{i=1}^{[n/2]-r} 2(r+m+i-s) \tag{37}$$

$$= \frac{1}{2^{2r+n-2[n/2]}} \frac{(2(n-[n/2]-s))!([n/2]+m-s)!}{(n-2s)!(n-[n/2]-s)!(r+m-s)!}$$

For any integer n, even or odd, we readily have :

$$(2(n - [n/2] - s))!([n/2] - s)! = 2^{n-2[n/2]}(n-2s)!(n - [n/2] - s)! \tag{38}$$

Therefore :

$$R(x_m) = \widetilde{S}_2(x_m), m = 0...[n/2] - r \tag{39}$$

Hence, because :

$$S_2 = \frac{\widetilde{S}_2(0)}{(n-2r)!} = \frac{R(0)}{(n-2r)!} \tag{40}$$

the second identity holds.

2.6 Theorem 4

We can now prove a first theorem reading as :

$$\int_0^{2\pi} P_n(\sin a \sin \theta \cos \varphi + \cos a \cos \theta) d\varphi = 2\pi P_n(\cos a) P_n(\cos \theta) \tag{41}$$

where P_n is the Legendre polynomial of degree n.

2.7 Proof of Theorem 4

By using an expansion of Legendre polynomials (e.g. G.B.Arfken & H.J.Weber (2005), p744), we obtain :

$$P_n(\sin a \sin \theta \cos \varphi + \cos a \cos \theta) \tag{42}$$

$$= \frac{1}{2^n} \sum_{m=0}^{[n/2]} (-1)^m \binom{n}{m} \binom{2n-2m}{n} (\sin a \sin \theta \cos \varphi + \cos a \cos \theta)^{n-2m}$$

But we have :

$$\int_0^{2\pi} (\sin a \sin \theta \cos \varphi + \cos a \cos \theta)^{n-2m} d\varphi \tag{43}$$

$$= \sum_{j=0}^{n-2m} \binom{n-2m}{j} \int_0^{2\pi} (\sin a \sin \theta \cos \varphi)^j (\cos a \cos \theta)^{n-2m-j} d\varphi$$

$$= \sum_{k=0}^{\left[\frac{n-2m}{2}\right]} \binom{n-2m}{2k} \int_0^{2\pi} (\sin a \sin \theta \cos \varphi)^{2k} (\cos a \cos \theta)^{n-2m-2k} d\varphi$$

in which we have used (see Eq.6) :

$$\int_0^{2\pi} (\cos \varphi)^j d\varphi = 0, \text{ for any odd integer } j \tag{44}$$

Using again Eq.6 for j even, we obtain :

$$\int_0^{2\pi} (\sin a \sin \theta \cos \varphi + \cos a \cos \theta)^{n-2m} d\varphi \tag{45}$$

$$= \sum_{k=0}^{\left[\frac{n-2m}{2}\right]} \frac{2\pi}{2^{2k}} \binom{n-2m}{2k} \binom{2k}{k} (1-\cos^2 a)^k (1-\cos^2 \theta)^k (\cos a \cos \theta)^{n-2m-2k}$$

Eq.45 implies that $\int_0^{2\pi} P_n(\sin a \sin \theta \cos \varphi + \cos a \cos \theta) d\varphi$ is a polynomial $q_n(\cos a, \cos \theta)$, symmetrical with respect to the two variables $\cos a$ and $\cos \theta$, of degree n with respect to each variable. We then invoke the Leibniz formula to develop $(1-\cos^2 a)^k$ and $(1-\cos^2 \theta)^k$ involved in Eq.45 leading to :

$$q_n(\cos a, \cos \theta) = \frac{2\pi}{2^n} \sum_{m=0}^{[n/2]} \sum_{k=0}^{\left[\frac{n-2m}{2}\right]} \sum_{j=0}^{k} \sum_{i=0}^{k} \frac{(-1)^{m-j-i}}{2^{2k}} \binom{n}{m} \binom{2n-2m}{n} \tag{46}$$

$$\binom{n-2m}{2k} \binom{2k}{k} \binom{k}{k-j} \binom{k}{k-i} (\cos a)^{n-2m-2i} (\cos \theta)^{n-2m-2j}$$

We now intend to identify Eq.46 and the expansion of $2\pi P_n(\cos a) P_n(\cos \theta)$ with respect to the variables $\cos a$ and $\cos \theta$.

For a fixed integer r such that $r = m + j$ and $r \leq [n/2]$, and a fixed integer s such that $s = m + i$ and $s \leq [n/2]$, the factor of $(\cos a)^{n-2s}(\cos\theta)^{n-2r}$ in Eq.46 can be obtained and compared with its counterpart in $2\pi P_n(\cos a)P_n(\cos\theta)$. This counterpart reads as :

$$\frac{2\pi}{2^{2n}}(-1)^{r+s}\binom{n}{r}\binom{n}{s}\binom{2n-2r}{n}\binom{2n-2s}{n} \tag{47}$$

Since q_n is a symmetrical polynomial, it is sufficient to carry out the comparison for $s \leq r$. To approach the aim, we interchange $k-$ and $j-$ summations in Eq.46, according to :

$$\sum_{k=0}^{\left[\frac{n-2m}{2}\right]}\sum_{j=0}^{k} = \sum_{j=0}^{\left[\frac{n-2m}{2}\right]}\sum_{k=j}^{\left[\frac{n-2m}{2}\right]} \tag{48}$$

Therefore, for $m + j = r$, the factor of $(\cos\theta)^{n-2r}$ is :

$$\frac{2\pi}{2^n}\sum_{m=0}^{r}\sum_{k=r-m}^{\left[\frac{n-2m}{2}\right]}\sum_{i=0}^{k} \tag{49}$$

$$\frac{(-1)^{r-i}}{2^{2k}}\binom{n}{m}\binom{2n-2m}{n}\binom{n-2m}{2k}\binom{2k}{k}\binom{k}{m+k-r}\binom{k}{k-i}$$

$$(\cos a)^{n-2m-2i}$$

Next, we interchange $k-$ and $i-$summations, according to :

$$\sum_{k=r-m}^{\left[\frac{n-2m}{2}\right]}\sum_{i=0}^{k} = \sum_{i=0}^{r-m}\sum_{k=r-m}^{\left[\frac{n-2m}{2}\right]} + \sum_{i=r-m+1}^{\left[\frac{n-2m}{2}\right]}\sum_{k=i}^{\left[\frac{n-2m}{2}\right]} \tag{50}$$

Thus, for $m + i = s \leq r$, the factor of $(\cos a)^{n-2s}(\cos\theta)^{n-2r}$ is :

$$\frac{2\pi}{2^n}\sum_{m=0}^{s}\sum_{k=r-m}^{\left[\frac{n-2m}{2}\right]}\frac{(-1)^{m+r+s}}{2^{2k}}\binom{n}{m}\binom{2n-2m}{n}\binom{n-2m}{2k} \tag{51}$$

$$\binom{2k}{k}\binom{k}{m+k-r}\binom{k}{m+k-s}$$

which is equal to :

$$(-1)^{r+s}\frac{2\pi}{2^n}\sum_{m=0}^{s}(-1)^m\frac{(2n-2m)!}{m!(n-m)!(s-m)!(r-m)!} \tag{52}$$

$$\sum_{k=r-m}^{[n/2]-m}\frac{1}{2^{2k}(n-2m-2k)!(m+k-r)!(m+k-s)!}$$

By using the two identities 13 and 14 from Lemma 3, this result may be rewritten as :

$$(-1)^{r+s}\frac{2\pi}{r!s!}\prod_{j=1}^{n-s}(2j-1)\prod_{i=n-r-s+1}^{n-r}(2i-1)\frac{(2(n-r-s))!}{2^n(n-r-s)!(n-2r)!(n-2s)!} \tag{53}$$

which is equal to the expression 47. Hence, the proof is done.

2.8 Corollary 5. Reproducing kernel

A new expression of the reproducing kernel of Legendre polynomials can be readily derived from Theorem 4. We then obtain a Corollary 5 as follows.

The reproducing kernel $K_n(x, t)$ of Legendre polynomials $P_i(x)$:

$$K_n(x, t) = \sum_{i=0}^{n} \frac{2i+1}{2} P_i(x) P_i(t) \tag{54}$$

has the following integral representation :

$$K_n(x, t) = \frac{1}{2\pi} \int_0^{2\pi} \sum_{i=0}^{n} \frac{2i+1}{2} P_i(xt + \cos\varphi \sqrt{1-x^2}\sqrt{1-t^2}) d\varphi \tag{55}$$

2.9 Jacobi polynomials

For further use, we now recall some results concerning Jacobi polynomials $P_n^{(\alpha,\beta)}(x), \alpha > -1$ and $\beta > -1$ Abramowitz & Stegun (1964), T.S.Chihara (1978), Szegö (1939).

Jacobi polynomials are orthogonal with respect to the linear functional :

$$\int_{-1}^{+1} .(1-x)^{\alpha}(1+x)^{\beta} dx, \ \alpha, \beta > -1 \tag{56}$$

They have the following L^2−norm :

$$\int_{-1}^{+1} (1-x)^{\alpha}(1+x)^{\beta} \left[P_n^{(\alpha,\beta)}(x)\right]^2 dx = \frac{2^{\alpha+\beta+1}}{2n+\alpha+\beta+1} \frac{\Gamma(n+\alpha+1)\Gamma(n+\beta+1)}{\Gamma(n+1)\Gamma(n+\alpha+\beta+1)} \tag{57}$$

where Γ is the Gamma function.

The derivative of $P_n^{(\alpha,\beta)}(x)$ is another Jacobi polynomial :

$$\frac{d}{dx} P_n^{(\alpha,\beta)}(x) = \frac{1}{2}(n+\alpha+\beta+1) P_{n-1}^{(\alpha+1,\beta+1)}(x) \tag{58}$$

We now provide relations valid for $\alpha = \beta > -1$.

Jacobi polynomials, for $\alpha = \beta > -1$, satisfy a three-term recurrence relation :

$$(n+\alpha+1)(2n+2\alpha+1)x P_n^{(\alpha,\alpha)}(x) \tag{59}$$
$$= (n+1)(n+2\alpha+1) P_{n+1}^{(\alpha,\alpha)}(x) + (n+\alpha)(n+\alpha+1) P_{n-1}^{(\alpha,\alpha)}(x)$$

with $P_0^{(\alpha,\alpha)} = 1$ and $P_{-1}^{(\alpha,\alpha)} = 0$.

Furthermore :

$$(2n + 2\alpha - 1)P_n^{(\alpha-1,\alpha-1)}(x) \tag{60}$$
$$= \frac{(n + 2\alpha - 1)(n + 2\alpha)}{2(n + \alpha)}P_n^{(\alpha,\alpha)}(x) - \frac{(n + \alpha - 1)}{2}P_{n-2}^{(\alpha,\alpha)}(x)$$

$$\frac{2n + 2\alpha + 3}{2}(1 - x^2)P_n^{(\alpha+1,\alpha+1)}(x) \tag{61}$$
$$= (n + \alpha + 1)P_n^{(\alpha,\alpha)}(x) - \frac{(n+1)(n+2)}{n + \alpha + 2}P_{n+2}^{(\alpha,\alpha)}(x)$$

$$(1 - x^2)\frac{d}{dx}P_n^{(\alpha,\alpha)}(x) = (1 - x^2)\frac{n + 2\alpha + 1}{2}P_{n-1}^{(\alpha+1,\alpha+1)}(x) \tag{62}$$
$$= -nxP_n^{(\alpha,\alpha)}(x) + (n + \alpha)P_{n-1}^{(\alpha,\alpha)}(x)$$

We now know enough to prove the main theorem.

2.10 Theorem 6

For any integer $m \in \mathbf{Z}$ and any integer $n \in \mathbf{N}$:

$$\int_0^{2\pi} P_n(\sin a \sin \theta \cos(\varphi - b) + \cos a \cos \theta)e^{-im(\varphi-b)}d\varphi \tag{63}$$
$$= 2\pi \frac{(n - |m|)!}{(n + |m|)!}P_n^{|m|}(\cos a)P_n^{|m|}(\cos \theta)$$

where $i = \sqrt{-1}$, P_n is the Legendre polynomial of degree n, and $P_n^{|m|}$ is the associated Legendre function of order $|m|$ defined by :

$$P_n^{|m|}(x) = (-1)^{|m|}(1 - x^2)^{\frac{|m|}{2}}\frac{d^{|m|}}{dx^{|m|}}P_n(x) \tag{64}$$

2.11 Proof of Theorem 6

We conveniently introduce, for further use, a specific notation for the l.h.s. of Eq.63 :

$$D_{nm} = \int_0^{2\pi} P_n(\sin a \sin \theta \cos(\varphi - b) + \cos a \cos \theta)e^{-im(\varphi-b)}d\varphi \tag{65}$$

Now, it happens that Eq.63 of Theorem 6 is already proved for $m = 0$ (Theorem 4) and for $n < |m|$ (Corollary 2). Moreover, by using Lemma 1, Eq.63 is equivalent to :

$$\int_0^{2\pi} P_n(\sin a \sin \theta \cos \varphi + \cos a \cos \theta)\cos(|m| \varphi)d\varphi = 2\pi \frac{(n - |m|)!}{(n + |m|)!}P_n^{|m|}(\cos a)P_n^{|m|}(\cos \theta) \tag{66}$$

But $P_j^{|m|}(x)$ possesses an expression in terms of the Jacobi polynomial $P_{j-|m|}^{(|m|,|m|)}(x)$. Indeed, by using Eqs.58 and 64 :

$$P_j^{|m|}(x) = (-1)^{|m|}(1-x^2)^{\frac{|m|}{2}} \frac{(j+1)_{|m|}}{2^{|m|}} P_{j-|m|}^{(|m|,|m|)}(x) \tag{67}$$

where $(c)_j$ is the Pochhammer symbol :

$$\left.\begin{array}{c} (c)_j = c(c+1)...(c+j-1), \forall j \in \mathbf{N} \\ (c)_0 = 1 \end{array}\right\} \tag{68}$$

We now prove Eq.63 for $m = 1$. From Eq.59 with $\alpha = 0$ and Eq.67, we have :

$$(2n+1)XP_n(X) \tag{69}$$
$$= (2n+1)\sin a \sin\theta \cos\varphi P_n(X) + (2n+1)\cos a \cos\theta P_n(X)$$
$$= (n+1)P_{n+1}(X) + nP_{n-1}(X)$$

in which we conveniently used $X = X(b = 0)$. Hence, by using Theorem 4, and remembering Lemma 1 and its associated remark :

$$(2n+1)\sin a \sin\theta D_{n1} \tag{70}$$
$$= (n+1)D_{n+1,0} - (2n+1)\cos a \cos\theta D_{n0} + nD_{n-1,0}$$
$$= 2\pi(n+1)P_{n+1}(\cos a)P_{n+1}(\cos\theta) - 2\pi(2n+1)\cos a \cos\theta P_n(\cos a)P_n(\cos\theta)$$
$$+2\pi nP_{n-1}(\cos a)P_{n-1}(\cos\theta)$$

With Eq.59 for $\alpha = 0$, this expression becomes :

$$\frac{2\pi}{n+1}[(2n+1)\cos aP_n(\cos a) - nP_{n-1}\cos(a)] \tag{71}$$
$$[(2n+1)\cos\theta P_n(\cos\theta) - nP_{n-1}(\cos\theta)]$$
$$-2\pi(2n+1)\cos a \cos\theta P_n(\cos a)P_n(\cos\theta) + 2\pi nP_{n-1}(\cos a)P_{n-1}(\cos\theta)$$

which can be factorized to :

$$2\pi\frac{(2n+1)}{n(n+1)}[-n\cos aP_n(\cos a) + nP_{n-1}(\cos a)][-n\cos\theta P_n(\cos\theta) + nP_{n-1}(\cos\theta)] \tag{72}$$

This expression is identical to :

$$2\pi(2n+1)\sin a \sin\theta\frac{(n-1)!}{(n+1)!}P_n^1(\cos a)P_n^1(\cos\theta) \tag{73}$$

This result is obtained by using Eq.62 with $\alpha = 0$ and Eq.64 for $m = 1$, namely :

$$(\sin a)^2\frac{d}{d\cos a}P_n(\cos a) = -n\cos aP_n(\cos a) + nP_{n-1}(\cos a) \tag{74}$$

$$P_n^1(\cos a) = -\sin a\frac{d}{d\cos a}P_n(\cos a) \tag{75}$$

Therefore, Eq.63 holds for $m = 1$. Now, we can complete the demonstration of Eq.63 by recurrence, assuming that it is satisfied by any integer $n \in \mathbf{N}$ up to $(m-1) \geq 1$ (we can assume that m is positive).

We have :

$$\cos(m\varphi) = 2\cos\varphi\cos((m-1)\varphi) - \cos((m-2)\varphi) \tag{76}$$

From Eqs.76 and 69, we obtain :

$$\sin a \sin\theta P_n(X)\cos(m\varphi) \tag{77}$$
$$= \frac{2}{2n+1}\cos((m-1)\varphi)\left[(n+1)P_{n+1}(X) - (2n+1)\cos a\cos\theta P_n(X) + nP_{n-1}(X)\right]$$
$$- \sin a \sin\theta P_n(X)\cos((m-2)\varphi)$$

Hence :

$$\sin a \sin\theta D_{nm} = \frac{2}{2n+1}\left[(n+1)D_{n+1,m-1} - (2n+1)\cos a\cos\theta D_{n,m-1} + nD_{n-1,m-1}\right] \tag{78}$$
$$- \sin a \sin\theta D_{n,m-2}$$

We now use the recurrence assumption, yielding, from Theorem 6 :

$$\sin a \sin\theta D_{nm} = \frac{4\pi}{2n+1}\left[(n+1)\frac{(n-m+2)!}{(n+m)!}P_{n+1}^{m-1}(\cos a)P_{n+1}^{m-1}(\cos\theta)\right. \tag{79}$$
$$-(2n+1)\frac{(n-m+1)!}{(n+m-1)!}\cos a\cos\theta P_n^{m-1}(\cos a)P_n^{m-1}(\cos\theta)$$
$$\left. +n\frac{(n-m)!}{(n+m-2)!}P_{n-1}^{m-1}(\cos a)P_{n-1}^{m-1}(\cos\theta)\right]$$
$$-2\pi\frac{(n-m+2)!}{(n+m-2)!}\sin a\sin\theta P_n^{m-2}(\cos a)P_n^{m-2}(\cos\theta)$$

Every P_j^{m-1} and P_j^{m-2} is replaced by using the expression 67, leading to :

$$\sin a\sin\theta D_{nm} = 2\pi\frac{(n-m)!}{(n+m)!}\frac{(\sin a\sin\theta)^{m-1}}{2^{2m-4}}((n+1)_m)^2\left\{\frac{1}{2(2n+1)}\right. \tag{80}$$
$$\left[\frac{(n-m+1)_2}{n+1}P_{n-m+2}^{(m-1,m-1)}(\cos a)P_{n-m+2}^{(m-1,m-1)}(\cos\theta)\right.$$
$$-(2n+1)\frac{n-m+1}{n+m}\cos a\cos\theta P_{n-m+1}^{(m-1,m-1)}(\cos a)P_{n-m+1}^{(m-1,m-1)}(\cos\theta)$$
$$\left.+\frac{n^3}{(n+m-1)_2}P_{n-m}^{(m-1,m-1)}(\cos a)P_{n-m}^{(m-1,m-1)}(\cos\theta)\right]$$
$$\left.-\frac{(n-m+1)_2}{(n+m-1)_2}P_{n-m+2}^{(m-2,m-2)}(\cos a)P_{n-m+2}^{(m-2,m-2)}(\cos\theta)\right\}$$

We afterward substitute $P_{n-m+2}^{(m-2,m-2)}$ for $P_j^{(m-1,m-1)}$ by using Eq.60 and $\cos a P_{n-m+1}^{(m-1,m-1)}(\cos a)$ for $P_j^{(m-1,m-1)}$ by using Eq.59. We then obtain :

$$\sin a \sin\theta D_{nm} = 2\pi \frac{(n-m)!}{(n+m)!} \frac{(\sin a \sin\theta)^{m-1}}{2^{2m-4}} ((n+1)_m)^2 \{ \frac{1}{2(2n+1)} \tag{81}$$

$$[\frac{(n-m+1)_2}{n+1} P_{n-m+2}^{(m-1,m-1)}(\cos a) P_{n-m+2}^{(m-1,m-1)}(\cos\theta)$$

$$-\frac{n-m+1}{(n+m)(n+1)^2(2n+1)}((n+m)(n-m+2)P_{n-m+2}^{(m-1,m-1)}(\cos a)$$

$$+n(n+1)P_{n-m}^{(m-1,m-1)}(\cos a))((n+m)(n-m+2)P_{n-m+2}^{(m-1,m-1)}(\cos\theta)$$

$$+n(n+1)P_{n-m}^{(m-1,m-1)}(\cos\theta)) +$$

$$\frac{n^3}{(n+m-1)_2}P_{n-m}^{(m-1,m-1)}(\cos a)P_{n-m}^{(m-1,m-1)}(\cos\theta)]$$

$$-\frac{(n-m+1)_2}{(2n+1)^2(n+m-1)_2}(\frac{(n+m-1)_2}{2(n+1)}P_{n-m+2}^{(m-1,m-1)}(\cos a)$$

$$-\frac{n}{2}P_{n-m}^{(m-1,m-1)}(\cos a))$$

$$(\frac{(n+m-1)_2}{2(n+1)}P_{n-m+2}^{(m-1,m-1)}(\cos\theta) - \frac{n}{2}P_{n-m}^{(m-1,m-1)}(\cos\theta))\}$$

becoming :

$$\sin a \sin\theta D_{nm} = 2\pi \frac{(n-m)!}{(n+m)!} \frac{(\sin a \sin\theta)^{m-1}}{2^{2m-4}} ((n+1)_m)^2 \tag{82}$$

$$[\frac{((n-m+1)_2)^2}{4(n+1)^2(2n+1)^2} P_{n-m+2}^{(m-1,m-1)}(\cos a) P_{n-m+2}^{(m-1,m-1)}(\cos\theta)$$

$$-\frac{n(n-m+1)_2}{4(n+1)(2n+1)^2}(P_{n-m+2}^{(m-1,m-1)}(\cos a) P_{n-m}^{(m-1,m-1)}(\cos\theta)$$

$$+P_{n-m+2}^{(m-1,m-1)}(\cos\theta) P_{n-m}^{(m-1,m-1)}(\cos a))$$

$$+\frac{n^2}{4(2n+1)^2}P_{n-m}^{(m-1,m-1)}(\cos a)P_{n-m}^{(m-1,m-1)}(\cos\theta)]$$

which factorizes to :

$$\sin a \sin\theta D_{nm} = 2\pi \frac{(n-m)!}{(n+m)!} \frac{(\sin a \sin\theta)^{m-1}}{2^{2m-2}} \frac{((n+1)_m)^2}{(2n+1)^2} \tag{83}$$

$$[nP_{n-m}^{(m-1,m-1)}(\cos a) - \frac{(n-m+1)_2}{n+1}P_{n-m+2}^{(m-1,m-1)}(\cos a)]$$

$$[nP_{n-m}^{(m-1,m-1)}(\cos\theta) - \frac{(n-m+1)_2}{n+1}P_{n-m+2}^{(m-1,m-1)}(\cos\theta)]$$

We then invoke Eq.61, with $\alpha = m-1, n \mapsto n-m$, to obtain :

$$\sin a \sin\theta D_{nm} = 2\pi \frac{(n-m)!}{(n+m)!} \frac{(\sin a \sin\theta)^{m+1}}{2^{2m}} ((n+1)_m)^2 P_{n-m}^{(m,m)}(\cos a) P_{n-m}^{(m,m)}(\cos\theta) \tag{84}$$

Therefore, recalling the definition of D_{nm}, we find that Eq.63, i.e. Theorem 6, holds for any integer $m \in \mathbf{Z}$.

2.12 A consequence

An important consequence of Theorem 6 concerns the spherical harmonics $Y_j^m(\theta, \varphi)$ which are defined as :

$$Y_j^m(\theta, \varphi) = \sqrt{\frac{(2j+1)}{4\pi} \frac{(j-m)!}{(j+m)!}} P_j^m(\cos\theta) e^{im\varphi} \tag{85}$$

When $m < 0$, $P_j^m(\cos\theta)$ is defined as :

$$P_j^m(\cos\theta) = (-1)^m \frac{(j+m)!}{(j-m)!} P_j^{-m}(\cos\theta) \tag{86}$$

Therefore, we may uniquely define $P_j^m(\cos\theta)$ for any integer $m \in \mathbf{Z}$, according to :

$$P_j^m(\cos\theta) = (-1)^{\frac{m-|m|}{2}} \frac{(j-|m|)!}{(j-m)!} P_j^{|m|}(\cos\theta) \tag{87}$$

in which $P_j^{|m|}(x)$ is defined by Eq.64.

Eq.85 can then be given an unique form for any integer $m \in \mathbf{Z}$, reading as :

$$Y_j^m(\theta, \varphi) = (-1)^{\frac{m-|m|}{2}} \sqrt{\frac{2j+1}{4\pi}} \sqrt{\frac{(j-m)!}{(j+m)!} \frac{(j-|m|)!}{(j-m)!}} P_j^{|m|}(\cos\theta) e^{im\varphi} \tag{88}$$

simplifying to :

$$Y_j^m(\theta, \varphi) = (-1)^{\frac{m-|m|}{2}} \sqrt{\frac{2j+1}{4\pi}} \sqrt{\frac{(j-|m|)!}{(j+|m|)!}} P_j^{|m|}(\cos\theta) e^{im\varphi} \tag{89}$$

From these equations, the complex conjuguate of $Y_j^m(\theta, \varphi)$ is :

$$\overline{Y_j^m(\theta, \varphi)} = (-1)^m Y_j^{-m}(\theta, \varphi), \forall m \in \mathbf{Z} \tag{90}$$

2.13 Corollary 7

For any integers $n, j \in \mathbf{N}$, and any integer $m \in \mathbf{Z}$:

$$\int_0^\pi \left[\int_0^{2\pi} P_n(\sin a \sin\theta \cos(\varphi - b) + \cos a \cos\theta) e^{-im(\varphi - b)} d\varphi \right] \tag{91}$$

$$P_j^m(\cos\theta) \sin\theta d\theta$$

$$= \frac{4\pi}{2j+1} P_j^m(\cos a)\delta_{nj}$$

2.14 Proof of Corollary 7

>From Theorem 6 and Eq.87, the l.h.s. of Eq.91 is :

$$LHS = 2\pi(-1)^{\frac{m-|m|}{2}}\frac{(n-|m|)!}{(n+|m|)!}\frac{(j-|m|)!}{(j-m)!}P_n^{|m|}(\cos a)\int_0^\pi P_n^{|m|}(\cos\theta)P_j^{|m|}(\cos\theta)\sin\theta d\theta \quad (92)$$

But, using Eqs.67 and 57 :

$$\int_0^\pi P_n^{|m|}(\cos\theta)P_j^{|m|}(\cos\theta)\sin\theta d\theta \quad (93)$$

$$= \frac{1}{2^{2|m|}}(n+1)_{|m|}(j+1)_{|m|}\int_{-1}^{+1}P_{n-|m|}^{(|m|,|m|)}(x)P_{j-|m|}^{(|m|,|m|)}(x)(1-x^2)^{|m|}dx$$

$$= \frac{1}{2^{2|m|}}((n+1)_{|m|})^2\frac{2^{2|m|+1}}{2n+1}\frac{(\Gamma(n+1))^2}{\Gamma(n-|m|+1)\Gamma(n+|m|+1)}\delta_{nj} = \frac{2}{2n+1}\frac{(n+|m|)!}{(n-|m|)!}\delta_{nj}$$

Therefore, Eq.92 becomes :

$$LHS = \frac{4\pi}{2n+1}(-1)^{\frac{m-|m|}{2}}\frac{(n-|m|)!}{(n-m)!}P_n^{|m|}(\cos a)\delta_{nj} = \frac{4\pi}{2j+1}P_j^m(\cos a)\delta_{nj} \quad (94)$$

in which we invoked Eq.87. This ends the proof.

2.15 Corollary 8

For any integers $n, j \in \mathbf{N}$ and any integer $m \in \mathbf{Z}$:

$$\int_0^\pi\int_0^{2\pi}\overline{Y_j^m(\theta,\varphi)}P_n(\sin a\sin\theta\cos(\varphi-b)+\cos a\cos\theta)\sin\theta d\theta d\varphi \quad (95)$$

$$= (-1)^m\frac{4\pi}{2j+1}Y_j^{-m}(a,b)\delta_{nj} = \frac{4\pi}{2j+1}\overline{Y_j^m(a,b)}\delta_{nj}$$

2.16 Proof of Corollary 8

Corollary 8 is a simple consequence of Corollary 7. Invoking also Eqs.88-90, we indeed have :

$$\int_0^\pi\int_0^{2\pi}\overline{Y_j^m(\theta,\varphi)}P_n(\sin a\sin\theta\cos(\varphi-b)+\cos a\cos\theta)\sin\theta d\theta d\varphi$$

$$= (-1)^{\frac{m-|m|}{2}}\sqrt{\frac{2j+1}{4\pi}}\sqrt{\frac{(j-|m|)!}{(j+|m|)!}}$$

$$\int_0^\pi \int_0^{2\pi} P_j^{|m|}(\cos\theta)e^{-im\varphi}P_n(\sin a \sin\theta \cos(\varphi - b) + \cos a \cos\theta)\sin\theta d\theta d\varphi$$

$$= (-1)^{\frac{m-|m|}{2}}e^{-imb}P_j^{|m|}(\cos a)\sqrt{\frac{4\pi}{2j+1}}\sqrt{\frac{(j-|m|)!}{(j+|m|)!}}\delta_{nj} = (-1)^m \frac{4\pi}{2j+1}Y_j^{-m}(a,b)\delta_{nj} \quad (96)$$

2.17 Additional remarks

In the series of papers Gouesbet (2006a), Gouesbet (2006b), Gouesbet (2007b), Gouesbet (2007a), one of the integrals required for use in the considered physical issues is Eq.95 of Corollary 8, with however $b = 0$. The second integral required in the same series of papers was given under the following form :

$$I_{ij} = \int_0^\pi \int_0^{2\pi} Y_i^0(\sin a \sin\theta \cos\varphi + \cos a \cos\theta)Y_j^0(\theta)\sin\theta d\theta d\varphi = \left[\sum_{k=0}^{(i-\epsilon)/2}C_{ki}^\epsilon(\cos a)^{2k+\epsilon}\right]\delta_{ij} \quad (97)$$

in which :

$$C_{ki}^\epsilon = \frac{(-1)^{(i-\epsilon)/2}}{2^i}(-1)^k \frac{(i+2k+\epsilon)!}{(\frac{i-\epsilon}{2}-k)!(2k+\epsilon)!(\frac{i+\epsilon}{2}+k)!} \quad (98)$$

where $\epsilon = 0, 1$ for i even, odd, respectively.

We again use the expansion of Legendre polynomials already invoked at the beginning of the proof of Theorem 4 (Eq.42), and establish that Eq.97 becomes :

$$I_{ij}(a) = P_i(\cos a)\delta_{ij} \quad (99)$$

Then, by using :

$$Y_l^0(\theta) = \sqrt{\frac{2l+1}{4\pi}}P_l(\cos\theta) \quad (100)$$

it is easily established that Eq.100 is a special case of Eq.98 (Corollary 8) for $m = 0$.

2.18 Summary

In the framework of a study examining analogies between electromagnetic and quantum scatterings, the evaluation of two integrals were required. One integral is given by Corollary 8, for $b = 0$, namely :

$$\int_0^\pi \int_0^{2\pi} \overline{Y_j^m(\theta,\varphi)}P_n(\sin a \sin\theta \cos\varphi + \cos a \cos\theta)\sin\theta d\theta d\varphi \quad (101)$$

$$= (-1)^m \frac{4\pi}{2j+1}Y_j^{-m}(a,b)\delta_{nj} = \frac{4\pi}{2j+1}\overline{Y_j^m(a,b)}\delta_{nj}$$

The second integral can be obtained as a special case of this result, and can be written as :

$$\int_0^\pi \int_0^{2\pi} \overline{Y_i^0}(\sin a \sin\theta \cos\varphi + \cos a \cos\theta)Y_j^0(\theta)\sin\theta d\theta d\varphi = P_i(\cos a)\delta_{ij} \quad (102)$$

3. Conclusion

Legendre polynomials, associated Legendre polynomials and associated Legendre functions, are widely used in physics, and particularly in light scattering. One of the most famous occurrences of associated Legendre functions is to be found in the Lorenz-Mie theory describing the interaction between an illuminating electromagnetic plane wave and a spherical particle defined by its diameter and its complex refractive index Mie (1908). However, in this theory, only associated Legendre functions $P_n^{\pm 1}$ (or P_n^1) do appear. In a generalized Lorenz-Mie theory describing the interaction between an electromagnetic arbitrary shaped beam and (again) a spherical particle defined by its diameter and its complex refractive index, all P_n^m's (or $P_n^{|m|}$'s) may appear, e.g. Gouesbet et al. (1988), Maheu et al. (1988).They actually appear in several places, first of all in the basis functions on which the electromagnetic fields are expanded. Second, expansion coefficients (called beam shape coefficients) can be evaluated by numerical integrations involving the expressions of the electromagnetic fields and associated Legendre functions, e.g. Gouesbet, Lock & Gréhan (2011). Associated Legendre functions also appear in many expressions generated by the theory, for the evaluation in particular of various cross-sections, under the form of yet other quadratures which, however, may be analytically performed, e.g. appendices in Gouesbet & Gréhan (2011). Homogeneous spheres defined by a diameter and a complex refractive index are not the only cases of light scattering theories in which associated Legendre functions are involved. They are actually involved whenever the symmetries of the scattering particle require the use of spherical coordinates, such as for multilayered spheres Onofri et al. (1995), assemblies of spheres and aggregates Gouesbet & Gréhan (1999), or for a spherical particle with an eccentric host sphere Gouesbet & Gréhan (2000b). May be more surprisingly, associated Legendre functions also play an important role in spheroidal coordinates insofar as, at the present time, beam shape coefficients in spheroidal coordinates are best expressed in terms of beam shape coefficients in spherical coordinates Gouesbet, Xu & Han (2011). Motivated by the successes of generalized Lorenz-Mie theories and by their numerous applications, an effort has then been devoted to the examination of analogies between electromagnetic arbitrary shaped beams and quantum arbitrary shaped beams, somehow culminating in a generalized optical theorem for non plane wave scattering in quantum mechanics Gouesbet (2009b). During this effort, one of us (G.G) encountered quadratures involving associated Legendre functions which he never encountered before in the framework of generalized Lorenz-Mie theories. This paper provides an analytical evaluation of these quadratures. The reader might be interested in playing with the obtained results, using a symbolic computation software like Maple, as we did to extensively check our derivations.

4. References

Abramowitz, M. & Stegun, I. (1964). *Handbook of mathematical functions*, Dover Publications, New-York.

G.B.Arfken & H.J.Weber (2005). *Mathematical methods for physicists, sixth edition*, Elsevier Academic Press.

Gouesbet, G. (2004). Cross-sections in Lorenz-Mie theory and quantum scattering : formal analogies, *Optics Communications* 231: 9–15.

Gouesbet, G. (2005). Expansion in free space of arbitrary quantum wavepackets, quantum laser beams, and Gaussian quantum laser beams (on-axis and off-axis) in terms of free spherical waves, *Particle and Particle Systems Characterization* 22: 38–44.

Gouesbet, G. (2006a). Asymptotic quantum elastic generalized Lorenz-Mie theory, *Optics Communications* 266: 704–709.

Gouesbet, G. (2006b). A transparent macroscopic sphere is cross-sectionnally equivalent to a superposition of two quantum-like radial potentials, *Optics Communications* 266: 710–715.

Gouesbet, G. (2007a). Asymptotically quantum inelastic generalized Lorenz-Mie theory, *Optics Communications* 278: 215–220.

Gouesbet, G. (2007b). Electromagnetic scattering by an absorbing macroscopic sphere is cross-sectionnally equivalent to a superposition of two effective quantum processes, *Journal of Optics A : Pure and Applied Optics* 9: 369–375.

Gouesbet, G. (2009a). Generalized Lorenz-Mie theories, the third decade : a perspective, an invited review paper, *Journal of Quantitative Spectroscopy and Radiative Transfer*, 110, 1223-1238 .

Gouesbet, G. (2009b). On the optical theorem and non plane wave scattering in quantum mechanics, *Journal of Mathematical Physics* 50: paper 112302, 1–5.

Gouesbet, G. & Gréhan, G. (1999). Generalized Lorenz-Mie theory for assemblies of spheres and aggregates, *Journal of Optics A : Pure and Applied Optics* 1: 706–712.

Gouesbet, G. & Gréhan, G. (2000a). Generalized Lorenz-Mie theories : from past to future, *Atomization and Sprays* 10 (3-5): 277–333.

Gouesbet, G. & Gréhan, G. (2000b). Generalized Lorenz-Mie theory for a sphere with an eccentrically located spherical inclusion, *Journal of Modern Physics* 47,5: 821–837.

Gouesbet, G. & Gréhan, G. (2011). *Generalized Lorenz-Mie theories*, Springer.

Gouesbet, G. & J.A.Lock (2007). Quantum arbitrary shaped beams revisited, *Optics Communications* 273: 296–305.

Gouesbet, G., Lock, J. & Gréhan, G. (2011). Generalized Lorenz-Mie theories and description of electromagnetic arbitrary shaped beams : localized approximations and localized beam models, *Journal of Quantitative Spectroscopy and Radiative Transfer* 112: 1–27.

Gouesbet, G., Maheu, B. & Grehan, G. (1988). Light scattering from a sphere arbitrarily located in a Gaussian beam, using a Bromwich formulation, *Journal of the Optical Society of America A* 5,9: 1427–1443.

Gouesbet, G., Xu, F. & Han, Y. (2011). Expanded description of electromagnetic arbitrary shaped beams in spheroidal coordinates, for use in light scattering theories, a review, *Journal of Quantitative Spectroscopy and Radiative Transfer* 112: 2249–2267.

J.A.Lock & Gouesbet, G. (2009). Generalized Lorenz-Mie theory and applications, an invited review paper, *Journal of Quantitative Spectroscopy and Radiative Transfer*, 110, 800-807 .

Maheu, B., Gouesbet, G. & Gréhan, G. (1988). A concise presentation of the generalized Lorenz-Mie theory for arbitrary location of the scatterer in an arbitrary incident profile, *Journal of Optics, Paris* 19,2: 59–67.

Mie, G. (1908). Beiträge zur Optik trüben Medien speziell kolloidaler Metalösungen, *Annalen der Physik* 25: 377–452.

Onofri, F., Gréhan, G. & Gouesbet, G. (1995). Electromagnetic scattering from a multilayered sphere located in an arbitrary beam, *Applied Optics* 34,30: 7113–7124.

Szegö, G. (1939). Orthogonal polynomials, *A.M.S. colloquium publications, Vol XXIII, American Mathematical Society, Providence, RI*.

T.S.Chihara (1978). *An introduction to orthogonal polynomials*, Gordon and Breach, New-York.

Ampère's Law Proved Not to Be Compatible with Grassmann's Force Law

Jan Olof Jonson
Stockholm University
Sweden

1. Introduction

Efforts have frequently been made to create links between different approaches within electromagnetism. Names like Ampère, Coulomb, Lorentz, Grassmann, Maxwell et.al. are all linked to efforts to create a comprehensive understanding of electromagnetism. In this paper the very focus is on breaking the alleged links between Ampère's law and the Grassmann-Lorentz force. Thanks to extensive mathematical efforts it appears to be possible to disprove earlier assumed links. That will tend to lead the further investigation of the subject effectively forwards

2. The fallacious derivation of Lorentz's law made by Grassmann

2.1 Finding the conceptual roots of the Lorentz force

Graneau discusses the assumption that the Ampère and the Lorentz forces are mathematically equal, claiming that this is not true [1]. He further makes the statement that the magnetic component of the Lorentz force was first proposed by Grassmann [2],[3]. This author shows that when Grassmann makes the derivation beginning with Ampère's law, he commits faults, which finally results in a term that is similar to the Lorentz force.

2.2 Grassmann on the electromagnetic force between currents

Grassmann himself discusses the conceptual problems that arise when studying Ampère's law [2],[3]. He says that the complicated form of Ampère's law arouses suspicion. [4] Among others he complains over the fact that the formula in no way resembles that for gravitational attraction, indicating thereby the lack of analogy between the two kinds of forces

$$(2\cos\varepsilon - r\cos\alpha.\cos\beta).ab / r^2 \qquad (1)$$

It ought to be mentioned that in the time of Grassmann and Ampère the electron has not yet been discovered and hence, the concept of what constituted a current must have been rather vague. It is therefore understandable that neither of them were able to apply Coulomb's law on the problem a law that by form fulfills the requirement to resemble the gravitational force. This author has been successful in doing so, beginning in his first paper on the subject in 1997 [5].

Applying Ampère's law on a configuration [6] consisting of two current carrying conductors, with the elements a and b respectively, an angle ε between the currents, α and β are the angles formed by the elements a and b respectively with the line drawn between the two mid-points, the current from the attracting element being i and its length ds, l being the perpendicular from the midpoints of the attracted element on the circuit element b on to the line of the attracting one, or

$$l = r \sin \alpha \tag{2}$$

gives the force between the two elements

$$-(ids.b / r^2)\cos \beta (2\cos \varepsilon - 3\cos \alpha.\cos \beta) \tag{3}$$

which develops into

$$ds = -d(I \cot \alpha) = l.d\alpha / \sin^2 \alpha = r^2.d\alpha / l \tag{4}$$

Where

$$\varepsilon = \alpha - \beta \tag{5}$$

$$\text{and } d\beta = d\alpha \tag{6}$$

This leads the development of eq. (4) into:

$$-(ib / l)(\cos^2 \beta.\cos \alpha.d\alpha - 2\sin \alpha.\sin \beta.\cos \beta.d\beta) \tag{7}$$

After he has arbitrarily chosen to put $d\beta$ at the end of the second term, due to the statement (6) above.

He thereafter integrates over the whole attracting line, thereby getting *nothing from the second term*. This is not mathematics. If there were originally undertaken an incremental step $d\alpha$ attached to both terms, the integration along α can of course not be avoided by stating that $d\beta = d\alpha$ and choose to change $d\alpha$ into $d\beta$ on the second term.

He continues the treatment of the two currents by deriving the force perpendicular to the attracted element b due to the attracting one

$$-(ib_l / l)(\cos \varepsilon.\cos \beta + \cos \alpha.\sin^2 \beta) \tag{8}$$

This time, when performing the integral, he does it not with respect to α this time; instead he does it with respect to β. He further claims the values of α and β now to be those of the 'initial point' of the line. This indicates that he has not treated the necessary rest of the closed circuit.

$$(ib_l / l)(1 + \cos \varepsilon.\cos \beta + \cos \alpha.\sin^2 \beta) \tag{9}$$

Now it has appeared a new term '1' at the very beginning. At infinity he states that α becomes 180°, and β becomes 180°- ε

Hence, the expression (9) may after recalling that

$$\varepsilon = \alpha - \beta \qquad (10)$$

be rewritten

$$(ib_l\ /\ l)(\cot(\tfrac{1}{2}\alpha) + \sin\beta.\cos\beta) \qquad (11)$$

Thereafter he neglects the second term of eq. (11), while leaving no indication of the reason for doing so and attains after using eq. (2)

$$(ib_l\ /\ r)(\cot(\tfrac{1}{2}\alpha) \qquad (12)$$

In the following a passage is cited from his paper in order to make it easier to understand his way of thinking [8]

"From this expression he now at once attains the mutual interaction of two current elements, as he prefers to express it. It happens, since he is regarding the attracting current element *i.ds* as the combination of the two lines through which the current is passing, "these possessing the direction and intensity (i) of this element, and one of them having its current flowing in the same direction as that in the element, and the other in the opposite direction, while the first of them has its starting point in the initial point of the element , and the second has its starting point in the end of the element. We then obtain $(ab_l\ /\ r^2).\sin\alpha$ as the effect exerted by a current element a on another b, distant r from it"

This term proportional to $\sin\alpha$ is apparently that which has impelled people to name him as the first to define "Biôt-Savart's law".

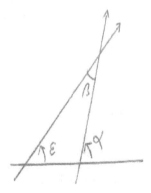

Fig. 1.

2.3 Some comments on the method of Grassmann

If following the derivation Grassmann has made, at several occasions there are clear cases of confusion concerning what he is doing and why. The mathematical steps recall more an 'ad-hoc' way of using mathematical terminology than a real logical way of working. Hence, his results seem to be of no value with respect to factual electrodynamics.

It seems to be totally irrelevant to use the theoretical treatment by Grassmann in order to construct any links between Ampère's law and Biôt-Savart's law and related Lorentz force. They must be kept apart from each other.

3. Analysis of the results by Assis and Bueno in comparing Lorentz's force law and Ampère's law

Assis and Bueno have written a paper [9] in which he claims Ampère's law to be consistent with Grassmann's force law. They derive expressions for the force between the support and the bridge within a set of Ampère's bridge, using both laws. Conceding that the laws are not equal at every point of the circuit, they claim that the result for a whole closed circuit is equal. One special point of observation that they make is that Biôt-Savart's law and the related Lorentz force do not obey Newton's third law, whereas Ampère's law does [10] The set of Ampére's bridge is described by detail in two consecutive figures, with respect to the definitions of the integration domains [11]. The width of the conductor is small relative to the lengths of the branches. The bridge is described by two variables, the laminar thickness is being ignored. The shape is rectangular, the branches being of length l_3 (along the x direction) and l_2 (along the y direction). The branches along the y axis are cut off at the distance l_1 the segments thus attained have been numerated from one to six counterclockwise, beginning with the part of the support being situated along the x axis, where presumably the supporting battery is practically applied. Thus, the cuts appear between segment 2-3 and 5-6 respectively. In this paper it is especially being focused on the force that appears from segment 5 onto segment 4, since the Lorentz force from segment 6 to segment 5 (and similarly between segment 2 and 3) is zero, whereas Ampère's law is not. Interestingly, the Lorentz force from segment 5 onto segment 4 differs from zero, contrary to the force in the opposite direction. The authors claim that this net force **within** this branch (the bridge) is able to account for the force that otherwise Ampère's law produces from segment 6 to 5. Normally, according to Newtons's third law, there cannot be any net force, if there is no acceleration, since during conditions of balance, no net force is active. Biôt-Svart's law and the Lorentz force law implies a single-directed force, that if it were to be real would immediately blow up the circuit, which does not happen. The force that Ampère's law produces between the two segments corresponds to an internal **tension** that is made visible when cutting of the bridge, thereby creating two equal forces, but of opposite direction, that exactly cancel.

To conclude, this discussion seems to verify that the Lorentz force is by nature unphysical, whereas Ampère's law is physical. Furtheron, the derivation that will follow in the next chapter, gains momentum to this conclusion, since it will appear that the derivation by Assis and Bueno that constitutes a fundamental basis for their claim, has been fallaciously performed.

3.1 Basic formulas for the derivation

Assis and Bueno defines the Grassmann force as follows:

$$d^2\vec{F}_{ji}{}^G = I_i d\vec{l}_i \times d\vec{B}_j^{B=S} \stackrel{\cong}{=} I_i dl_i \times (\frac{\mu_0}{4\pi} I_j \frac{d\vec{l}_j \times \hat{r}}{r^2}) = -\frac{\mu_0}{4\pi} \frac{I_i I_j}{r^2} \left[(d\vec{l}_i \bullet d\vec{l}_j)\hat{r} - (d\vec{l}_i \bullet \hat{r})d\vec{l}_j \right] \quad (13)$$

Since the integration involves four variables, a change of the name of the left hand variable will be done according to: $d^4 \bar{F}_{ji}{}^G$

It must be mentioned that the Assis and Bueno mention the law as "Grassmann's", without showing the reference. The reference that this author has found is that given by Peter Graneau [13], [14], [15]. It would be preferred to be used the name "Lorentz' force law (based on Biôt-Savart's law)". For the reader's convenience, also Ampère's law is given here:

$$d^2 \bar{F}_{ji}{}^A = -\frac{\mu_0}{4\pi} I_i I_j \frac{\hat{r}}{r^2} (2(d\bar{l}_i \bullet d\bar{l}_j) - 3(\hat{r} \bullet d\bar{l}_i)(\hat{r} \bullet d\bar{l}_j)) \tag{14}$$

3.2 Result of the calculations with respect to segment 5-4

The integrations can usually be performed straightforwardly, thereby using the normal rules for integrations. At some points, however, it appears necessary to make approximations, when a term is extremely small as compared to the others, for example terms $\propto w$ in the numerator. That has been dome at *every* actual occasion in this work. This is the precondition for attaining a closed expression at the end.

3.3 The result according to Assis and Bueno

Assis and Bueno have been using two ways to approximate the integrals that have to be performed. In the first case they assume the circuit to be divided into rectangles (21), in the second case they let a diagonal line at the corners define the border between two segments (22). In the first case they claim that the result of the calculation of Grassmann's force on segment 4 due to segment 5 is:

$$(F^G{}_{54})_y \cong \frac{\mu_0}{4\pi} (\ln \frac{l_2 - l_1}{w} - \ln \frac{(l_2 - l_1) + ((l_2 - l_1)^2 + l_3^2)^{1/2}}{l_3} + \ln 2 - \frac{3}{2} \ln(1 + \sqrt{2}) + \frac{\sqrt{2}}{2} + \frac{1}{2}) \tag{15}$$

In the second case their result is

$$(F^G{}_{54})_y \cong \frac{\mu_0}{2\pi} (\ln \frac{l_2}{w} - \ln \frac{l_2 + (l_2^2 + l_3^2)^{1/2}}{l_3} + \frac{(l_2^2 + l_3^2)^{1/2}}{l_2} + \ln 2 + \frac{1}{2}) \tag{16}$$

a result that they identify as equal to that which Ampère's law gives rise to.

3.4 The result according to the analysis of this author, first approach

However, the intention with this paper is to judge the claims by Assis and Bueno. In order to attain that goal, the Grassmann force they have been using will be used in this paper in an independent derivation, by this author.

The first step in the calculation procedure is to give the problem a strict formulation in the shape of an integral, thereby identifying as well the variables of the integrations as the boarders. Applying Eq.(13) above to the segments 5-4 will give rise to the following integral equation:

$$(F^G{}_{54})_y \cong \frac{\mu_0 I^2}{4\pi w^2} \int_0^w dx_5 \int_0^{l_3} dx_4 \int_{l_1}^{l_2-w} dy_5 \int_{l_2-w}^{l_2} dy_4 \frac{x_4 - x_5}{((x_4 - x_5)^2 + (y_4 - y_5)^2)^{3/2}} \tag{17}$$

3.5 The first step: Integration with respect to x_4

Integrating first with respect to x_4, gives the result:

$$(F^G{}_{54})_y = \frac{\mu_0 I^2}{4\pi w^2} \int_{l_1}^{l_2-w} dy_5 \int_{l_2-w}^{l_2} dy_4 \int_0^w dx_5 \left(-\frac{1}{((l_3 - x_5)^2 + (y_4 - y_5)^2)^{1/2}} + \frac{1}{(x_5^2 + (y_4 - y_5)^2)^{1/2}}\right) \tag{18}$$

3.6 The second step: Integration with respect to y_5

The subsequent integration with respect to y_5 gives rise to the following expression:

$$(F^G{}_{54})_y = \frac{\mu_0 I^2}{4\pi w^2} \int_0^w dx_5 \int_{l_2-w}^{l_2} dy_4 (\ln(y_4 - l_2 + w + \sqrt{(y_4 - l_2 + w)^2 + (l_3 - x_5)^2})$$

$$-\ln(y_4 - l_1 + \sqrt{(y_4 - l_1)^2 + (l_3 - x_5)^2}) - \ln(y_4 - l_2 + w + \sqrt{(y_4 - l_2 + w)^2 + x_5^2}) \tag{19}$$

$$+\ln(y_4 - l_1 + \sqrt{(y_4 - l_1)^2 + x_5^2})$$

For convenience, the four terms may be named (19a), (19b), (19c) and (19d) in consecutive order.

3.7 The third step: Integration with respect to y_4 and x_5

3.7.1 The first term of Eq.(19) above treated, (19a)

In order to solve the integration, some integration formulas must be used:

$$\int \ln(z + (z^2 \pm 1)^{\frac{1}{2}}) dz = z \ln(z + (z^2 \pm 1)^{\frac{1}{2}}) - (z^2 \pm 1) \tag{20}$$

$$\int (x^2 + a^2)^{\frac{1}{2}} dx = \frac{x}{2}(x^2 + a^2)^{\frac{1}{2}} \pm \frac{a^2}{2} \ln(x + (x^2 \pm a^2)^{\frac{1}{2}}) \tag{21}$$

$$\int \ln z \, dz = z \ln z - z \tag{22}$$

In order to solve Eq. (19a), the substitution

$$\frac{y_4 - l_2 + w}{l_3 - x_5} = z \tag{23}$$

will favorably be used. This makes eq. (19a) transform into:

$$(F^G{}_{54})_y = \frac{\mu_0 I^2}{4\pi w^2} \int_0^w dx_5 (l_3 - x_5) \int_{z=0}^{\frac{w}{l_3 - x_5}} (-dz)(\ln(l_3 - x_5) + \ln(z + \sqrt{z^2 + 1})) \tag{24}$$

This expression contains two terms: The first term becomes after integration:

$$(F^G{}_{54})_{y,1} = \frac{\mu_0 I^2}{4\pi w} \ln l_3 \tag{25}$$

The second term

$$(F^G{}_{54})_{y,2} = \frac{\mu_0 I^2}{4\pi w^2} \int_0^w dx_5 (l_3 - x_5) \int_{z=0}^{\frac{w}{l_3 - x_5}} (-dz)\ln(z + \sqrt{z^2 + 1}) \tag{26}$$

Solving with respect to z, thereby using formula [10] gives:

$$(F^G{}_{54})_{y,2} = \frac{\mu_0 I^2}{4\pi w^2} \int_0^w dx_5 (l_3 - x_5)(-1)(z\ln(z + \sqrt{z^2 + 1}) - \sqrt{z^2 + 1}) \tag{27}$$

Straightforward integration accordingly gives, provided also the approximation

$$w << l_3 - x_5 \tag{28}$$

is used in the final stage, both terms of the integral equal zero in the limit. The second term of those also requires the usage of an integration formula, Eq. [20], before the null result can be achieved. Hence,

$$(F^G{}_{54})^y \langle (19a) \rangle = \frac{\mu_0 I^2}{4\pi} (\ln l_3) \tag{29}$$

3.8 The second term of Eq.(19) above treated, (19b)

In order to solve Eq. (19b), it is at first reasonable to use the substitution

$$t = y_4 - l_1 \tag{30}$$

which gives rise to

$$(F^G{}_{54})^y \langle (19b) \rangle = \frac{\mu_0 I^2}{4\pi} (-\int_{x_5=0}^w dx_5 \int_{t=l_2-w-l_1}^{l_2-l_1} dt \ln(t + \sqrt{t^2 + (l_3 - x_5)^2})) \tag{31}$$

Preparing to perform the substitution

$$z = \frac{t}{l_3 - x_5} \tag{32}$$

an intermediate step will be to write

$$(F^G{}_{54})^y \langle\langle (19b)\rangle\rangle = \frac{\mu_0 I^2}{4\pi}\{-\int\limits_{x_5=0}^{w} dx_5 \int\limits_{t=l_2-w-l_1}^{l_2-l_1} dt \ln((l_3-x_5)(\frac{t}{l_3-x_5}+\sqrt{(\frac{t}{l_3-x_5})^2+1}))\} \qquad (33)$$

Here it appears to be evident that this expression might simply be divided into two, according to the logarithm product law (Eq. (61) later in the text):

$$(F^G{}_{54})^y \langle\langle (19b)\rangle\rangle = \frac{\mu_0 I^2}{4\pi}\{-\int\limits_{x_5=0}^{w} dx_5(l_3-x_5)\cdot \ln(l_3-x_5)\int\limits_{z=\frac{l_2-l_1-w}{l_3-x_5}}^{\frac{l_2-l_1}{l_3-x_5}} dz -$$

$$-\int\limits_{x_5=0}^{w} dx_5(l_3-x_5)\int\limits_{z=\frac{l_2-l_1-w}{l_3-x_5}}^{\frac{l_2-l_1}{l_3-x_5}} dz\ln(z+\sqrt{z^2+1})\} \qquad (34)$$

These terms will hereafter be denoted $(F^G{}_{54})^y \langle\langle (19b)\rangle\rangle_1$ and $(F^G{}_{54})^y \langle\langle (19b)\rangle\rangle_2$ respectively.

Solving straightforwardly, and using series expansions of the ln function, thereby neglecting terms of higher order of w, gives for the first term of the expression (34) above:

$$(F^G{}_{54})^y \langle\langle (19b)\rangle\rangle_1 = \frac{\mu_0 I^2}{4\pi}\{-\ln l_3\} \qquad (35)$$

The second part will demand some more computational work, as will appear below:

Evaluating now the integral with respect to z, leads directly to the following, rather complicated expression:

$$(F^G{}_{54})^y \langle\langle (19b)\rangle\rangle_2 = \frac{\mu_0 I^2}{4\pi w^2}(-\int\limits_{x_5=0}^{w} dx_5((l_2-l_1)\ln(\frac{l_2-l_1}{l_3-x_5}+\sqrt{(\frac{l_2-l_1}{l_3-x_5})^2+1})-$$

$$-(l_2-l_1-w)\ln(\frac{l_2-l_1-w}{l_3-x_5}+\sqrt{(\frac{l_2-l_1-w}{l_3-x_5})^2+1}))+$$

$$+\frac{\mu_0 I^2}{4\pi w^2}(-\int\limits_{x_5=0}^{w} dx_5(l_3-x_5)(\sqrt{(\frac{l_2-l_1-w}{l_3-x_5})^2+1}-\sqrt{(\frac{l_2-l_1}{l_3-x_5})^2+1})) \qquad (36)$$

The three terms on the right hand expression will be called (36a), (36b) and (36c) respectively.

Now, instead of evaluating the expressions straightforwardly, it appears to be favorable to find suitable combinations of terms that would be able to make the solution simpler.

3.9 The sum of the first two terms of Eq. (36)

In order to solve the integral (36) it will be favorable to make the following variable substitution:

$$g = \frac{l_2 - l_1}{l_3 - x_5} \tag{37}$$

Applying this on the first term above (36a) gives:

$$\frac{\mu_0 I^2}{4\pi w^2}(-\int_{g=\frac{l_2-l_1}{l_3}}^{\frac{l_2-l_1}{l_3-w}} dg(l_2 - l_1)^2 \frac{1}{g^2}\ln(g + \sqrt{g^2+1})) \tag{38}$$

Applying a similar variable substitution on the second term above (36b) gives:

$$\frac{\mu_0 I^2}{4\pi w^2}\int_{g=\frac{l_2-l_1-w}{l_3}}^{\frac{l_2-l_1-w}{l_3-w}} dg(l_2 - l_1 - w)^2 \frac{1}{g^2}\ln(g + \sqrt{g^2+1}) \tag{39}$$

The treatment of the third term (36c) will be postponed, until the first two have been developed and simplified. Doing so, leads to the result with respect to the first term (36a)

$$\frac{\mu_0 I^2}{4\pi w^2}\{(l_2 - l_1)\{(l_3 - w)\ln(\frac{1}{l_3}(1 + \frac{w}{l_3})(l_2 - l_1 + \sqrt{(l_2-l_1)^2 + (l_3-w)^2})) -$$

$$-l_3\ln(\frac{1}{l_3}(l_2 - l_1 + \sqrt{(l_2-l_1)^2 + l_3^2})) + (l_2 - l_1)\ln(\frac{l_3 - w}{l_2 - l_1}(1 + \sqrt{(\frac{l_2-l_1}{l_3-w})^2 + 1})) - \tag{40}$$

$$-(l_2 - l_1)\ln(\frac{l_3}{l_2-l_1}(1 + \sqrt{(\frac{l_2-l_1}{l_3})^2 + 1}))\}\}$$

and for (39):

$$\frac{\mu_0 I^2}{4\pi w^2}\{-(l_2 - l_1 - w)\{(l_3 - w)\ln(\frac{1}{l_3}(1 + \frac{w}{l_3})(l_2 - l_1 - w + \sqrt{(l_2-l_1-w)^2 + (l_3-w)^2})) +$$

$$+l_3\ln(\frac{1}{l_3}(l_2 - l_1 - w + \sqrt{(l_2-l_1-w)^2 + l_3^2})) - (l_2 - l_1 - w)\ln(\frac{l_3 - w}{l_2-l_1-w}(1 + \sqrt{(\frac{l_2-l_1-w}{l_3-w})^2 + 1})) + \tag{41}$$

$$+(l_2 - l_1 - w)\ln(\frac{l_3}{l_2-l_1-w}(1 + \sqrt{(\frac{l_2-l_1-w}{l_3})^2 + 1}))\}\}$$

In the next step the two results above, (40) and (41) will be added. The result is:

$$\frac{\mu_0 I^2}{4\pi w^2}\{(l_2-l_1)(l_3-w)(\frac{w}{l_3}+\ln(\frac{l_2-l_1}{l_3}+\sqrt{(\frac{l_2-l_1}{l_3})^2+(\frac{l_3-w}{l_3})^2}))-$$

$$-(l_2-l_1-w)(l_3-w)(\frac{w}{l_3}+\ln(\frac{l_2-l_1-w}{l_3}+\sqrt{(\frac{l_2-l_1-w}{l_3})^2+(\frac{l_3-w}{l_3})^2}))-$$

$$-(l_2-l_1)l_3\ln(\frac{l_2-l_1}{l_3}+\sqrt{(\frac{l_2-l_1}{l_3})^2+1}))+$$

$$+(l_2-l_1-w)l_3\ln(\frac{l_2-l_1-w}{l_3}+\sqrt{(\frac{l_2-l_1-w}{l_3})^2+1}))+$$

$$+(l_2-l_1)^2\ln(\frac{l_3-w}{l_2-l_1}(1+\sqrt{(\frac{l_2-l_1}{l_3-w})^2+1}))-$$

$$-(l_2-l_1-w)^2\ln(\frac{l_3-w}{l_2-l_1-w}(1+\sqrt{(\frac{l_2-l_1-w}{l_3-w})^2+1}))-$$

$$-(l_2-l_1)^2\ln(\frac{l_3}{l_2-l_1}(1+\sqrt{(\frac{l_2-l_1}{l_3})^2+1}))+$$

$$+(l_2-l_1-w)^2\ln(\frac{l_3}{l_2-l_1-w}(1+\sqrt{(\frac{l_2-l_1-w}{l_3})^2+1}))\}$$

(42)

It has been used the series expansion of $\ln(1+x)\cong x$ for $0\le x\le1$ [22]

Since the 'x' is very small, only the first term in the expansion has been taken into account.

The terms within expression (42) will for practical reasons be denoted (42:1) until (42:8).

The next step will be to simplify the expression (42). This must be done in a deliberate way in order to succeed. A practical method is to separate out the dominant terms first, thereafter put the terms of first order in w thereafter those of second order, neglecting the terms of even higher order. The reason for this is that the numerator in the very first term of the expression contains w^2 in the numerator.

In the following this integral formula will again be usable:

$$\int\ln(z+(z^2\pm1)^{\frac{1}{2}})dz=z\ln(z+(z^2\pm1)^{\frac{1}{2}})-(z^2\pm1)^{\frac{1}{2}}$$

(43)

Following this procedure gives for the terms of expression (29) respectively:

$$\frac{\mu_0 I^2}{4\pi w^2}\{(l_2-l_1)(l_3-w)(\{\ln(\frac{l_2-l_1}{l_3}+\sqrt{(\frac{l_2-l_1}{l_3})^2+1}))-$$

$$-\frac{w}{l_3}\cdot\frac{1}{\frac{l_2-l_1}{l_3}+\sqrt{(\frac{l_2-l_1}{l_3})^2+1}}\cdot\frac{1}{\sqrt{(\frac{l_2-l_1}{l_3})^2+1}}+\frac{w}{l_3}\}$$

(42:1)

$$\frac{\mu_0 I^2}{4\pi w^2}\{(-(l_2-l_1-w)(l_3-w)(\ln(\frac{l_2-l_1}{l_3}+\sqrt{(\frac{l_2-l_1}{l_3})^2+1})-$$

$$-\frac{w}{l_3^2}\cdot\frac{l_2-l_1+l_3(1+\sqrt{(\frac{l_2-l_1}{l_3})^2+1})}{\frac{l_2-l_1}{l_3}+\sqrt{(\frac{l_2-l_1}{l_3})^2+1}}\cdot\frac{1}{\sqrt{(\frac{l_2-l_1}{l_3})^2+1}}+\frac{w}{l_3})\}$$

(42:2)

$$\frac{\mu_0 I^2}{4\pi w^2}\{-(l_2-l_1)l_3\cdot\ln(\frac{l_2-l_1}{l_3}+\sqrt{(\frac{l_2-l_1}{l_3})^2+1})\}$$

(42:3)

$$\frac{\mu_0 I^2}{4\pi w^2}\{(l_2-l_1-w)l_3\cdot\{\ln(\frac{l_2-l_1}{l_3}+\sqrt{(\frac{l_2-l_1}{l_3})^2+1})-$$

$$-\frac{w}{l_3^2}\cdot\frac{l_2-l_1+l_3\sqrt{(\frac{l_2-l_1}{l_3})^2+1}}{\frac{l_2-l_1}{l_3}+\sqrt{(\frac{l_2-l_1}{l_3})^2+1}}\cdot\frac{1}{\sqrt{(\frac{l_2-l_1}{l_3})^2+1}}\}\}$$

(42:4)

$$\frac{\mu_0 I^2}{4\pi w^2}\{(l_2-l_1)^2\cdot\{\ln(\frac{l_3}{l_2-l_1}+\sqrt{(\frac{l_3}{l_2-l_1})^2+1})-$$

$$+$$

$$-\frac{w}{(l_2-l_1)^2}\cdot\frac{l_3+(l_2-l_1)\sqrt{(\frac{l_3}{l_2-l_1})^2+1}}{\frac{l_3}{l_2-l_1}+\sqrt{(\frac{l_3}{l_2-l_1})^2+1}}\cdot\frac{1}{\sqrt{(\frac{l_3}{l_2-l_1})^2+1}}\}\}$$

(42:5)

$$\frac{\mu_0 I^2}{4\pi w^2}\{-(l_2-l_1-w)^2\cdot\{\ln(\frac{l_3}{l_2-l_1}+\sqrt{(\frac{l_3}{l_2-l_1})^2+1})-$$

$$-\frac{w}{(l_2-l_1)^3}\cdot\frac{l_3(l_2-l_1-l_3)+(l_2-l_1-l_3)(l_2-l_1)\sqrt{(\frac{l_3}{l_2-l_1})^2+1}}{\frac{l_3}{l_2-l_1}+\sqrt{(\frac{l_3}{l_2-l_1})^2+1}}\cdot\frac{1}{\sqrt{(\frac{l_3}{l_2-l_1})^2+1}}\}\}$$

(42:6)

$$\frac{\mu_0 I^2}{4\pi w^2}\{-(l_2-l_1)^2\cdot\ln(\frac{l_3}{l_2-l_1}+\sqrt{(\frac{l_3}{l_2-l_1})^2+1})\}$$

(42:7)

$$\frac{\mu_0 I^2}{4\pi w^2}\{(l_2-l_1-w)^2\cdot\{\ln(\frac{l_3}{l_2-l_1}+\sqrt{(\frac{l_3}{l_2-l_1})^2+1})+$$

$$+\frac{w}{(l_2-l_1)^3}\cdot\frac{l_3^2+l_3(l_2-l_1)\sqrt{(\frac{l_3}{l_2-l_1})^2+1}}{\frac{l_3}{l_2-l_1}+\sqrt{(\frac{l_3}{l_2-l_1})^2+1}}\cdot\frac{1}{\sqrt{(\frac{l_3}{l_2-l_1})^2+1}}\}\} \tag{42:8}$$

The sum of these eight expressions is equal to expression (42). Now, it will be easier to perform the addition of the terms of Eq. (42), by adding the eight separate terms above to each other. An important key to success has been to approximate every ln and root expression with a series expansion for every case when a 'small' term is appearing besides the big ones. By using this method it has been possible to restrict the complicated ln and root expression to the 'standard' ones, without 'small' terms. In doing so it is possible to gather similar terms from the different expression, even when there have been 'small' terms added inside the ln and root expressions.

The result is:

$$\frac{\mu_0 I^2}{4\pi}\{-\ln l_3-\ln(\frac{l_2-l_1}{l_3}+\sqrt{(\frac{l_2-l_1}{l_3})^2+1})\}+1-$$

$$-\frac{(l_2-l_1)^2+l_3^2+(l_2-l_1)\sqrt{(\frac{l_2-l_1}{l_3})^2+1}}{l_3^2}\cdot\frac{1}{\frac{l_2-l_1}{l_3}+\sqrt{(\frac{l_2-l_1}{l_3})^2+1}}\cdot\frac{1}{\sqrt{(\frac{l_2-l_1}{l_3})^2+1}}+$$

$$+\frac{1}{(\frac{l_3}{l_2-l_1}+\sqrt{(\frac{l_3}{l_2-l_1})^2+1})\cdot\sqrt{(\frac{l_3}{l_2-l_1})^2+1}}\cdot\{\frac{1}{2}-\frac{1}{4}\cdot\frac{l_3^2}{l_3^2+(l_2-l_1)^2}+$$

$$+\frac{1}{2}\cdot\frac{(l_1-l_2+l_3)^2(l_2-l_1+2l_3)}{(l_2-l_1)^3}+$$

$$+\frac{3}{2}\cdot\frac{l_3^2(l_2-l_1)}{(l_2-l_1)^3}-\frac{1}{4}\cdot\frac{l_3^4(l_2-l_1)}{(l_2-l_1)^5}+\sqrt{(\frac{l_3}{l_2-l_1})^2+1}\}+ \tag{44}$$

$$+\frac{1}{(\frac{l_3}{l_2-l_1}+\sqrt{(\frac{l_3}{l_2-l_1})^2+1})^2\cdot((\frac{l_3}{l_2-l_1})^2+1)}\cdot\{\frac{1}{(l_2-l_1)^4}((2l_3^2+(l_2-l_1)^2)(l_2-l_1)^2+$$

$$\frac{1}{2}\cdot(l_1-l_2+l_3)^2(l_3^2+(l_2-l_1)^2)-$$

$$-\frac{1}{2}\cdot l_3^2(l_3^2+(l_2-l_1)^2)-\frac{1}{2}\cdot l_3^4)-\frac{l_3}{l_2-l_1}\cdot\sqrt{(\frac{l_3}{l_2-l_1})^2+1}-\frac{l_3^3}{(l_2-l_1)^3}\cdot\sqrt{(\frac{l_3}{l_2-l_1})^2+1}\}$$

Comment: The four last rows above (in Eq.(44)) arise due to the "other terms" of Eq. (42.5) to (42.8). To be

3.10 The third term of Eq.(36)

The third term of Eq. (36) was denoted (36c) is repeated here for convenience.

$$(F^G_{54})^y \langle (19b) \rangle_3 = \frac{\mu_0 I^2}{4\pi w^2}\left(-\int_{x_5=0}^{w} dx_5 (l_3 - x_5)\left(\sqrt{(\frac{l_2 - l_1 - w}{l_3 - x_5})^2 + 1} - \sqrt{(\frac{l_2 - l_1}{l_3 - x_5})^2 + 1}\right)\right) \quad (36c)$$

It can be simplified easily by recognizing that the integrand may be written approximately

$$\frac{\mu_0 I^2}{4\pi w^2}\sqrt{(l_2 - l_1)^2 + l_3^2} \cdot \left(\frac{w(l_2 - l_1)}{(l_2 - l_1)^2 + l_3^2}\right) \quad (45)$$

after having neglected terms of higher order than one in w, since the integration implies a multiplication with w.

The result is

$$\frac{\mu_0 I^2}{4\pi}\left(\frac{l_2 - l_1}{\sqrt{(l_2 - l_1)^2 + l_3^2}}\right) \quad (46)$$

3.11 The sum of all terms due to the second term of Eq.(19) (i.e. 19b)

Now all the partial results due to Eq.(19b) have been attained and the task remains only to sum them together. In order to do so, Eq. (35), (44) and (46) must be added. The result is accordingly:

$$\frac{\mu_0 I^2}{4\pi}\{-\ln l_3 - \ln(\frac{l_2 - l_1}{l_3} + \sqrt{(\frac{l_2 - l_1}{l_3})^2 + 1}) + 1 -$$

$$-\frac{(l_2 - l_1)^2 + l_3^2 + (l_2 - l_1)l_3\sqrt{(\frac{l_2 - l_1}{l_3})^2 + 1}}{l_3^2} \cdot \frac{1}{\frac{l_2 - l_1}{l_3} + \sqrt{(\frac{l_2 - l_1}{l_3})^2 + 1}} \cdot \frac{1}{\sqrt{(\frac{l_2 - l_1}{l_3})^2 + 1}} +$$

$$+\frac{1}{(\frac{l_3}{l_2 - l_1} + \sqrt{(\frac{l_3}{l_2 - l_1})^2 + 1}) \cdot \sqrt{(\frac{l_3}{l_2 - l_1})^2 + 1}} \cdot (\frac{1}{2} - \frac{1}{4}\frac{l_3^2}{l_3^2 + (l_2 - l_1)^2} +$$

$$+\frac{1}{2}\frac{(l_1 - l_2 + l_3)^2(l_2 - l_1 + 2l_3)}{(l_2 - l_1)^3} + \frac{3}{2}\frac{l_3^2(l_2 - l_1)}{(l_2 - l_1)^3} - \frac{1}{4}\frac{l_3^4(l_2 - l_1)}{(l_2 - l_1)^5} + \sqrt{(\frac{l_3}{l_2 - l_1})^2 + 1}) + \quad (47)$$

$$+\frac{1}{(\frac{l_3}{l_2 - l_1} + \sqrt{(\frac{l_3}{l_2 - l_1})^2 + 1})^2 \cdot ((\frac{l_3}{l_2 - l_1})^2 + 1)} \cdot (\frac{1}{(l_2 - l_1)^4}((2l_3^2 + (l_2 - l_1)^2)(l_2 - l_1)^2 +$$

$$+\frac{1}{2} \cdot (l_1 - l_2 + l_3)^2(l_3^2 + (l_2 - l_1)^2) - \frac{1}{2} \cdot l_3^2(l_3^2 + (l_2 - l_1)^2) - \frac{1}{2} \cdot l_3^4) - \frac{l_3}{l_2 - l_1} \cdot \sqrt{(\frac{l_3}{l_2 - l_1})^2 + 1} -$$

$$-\frac{l_3^3}{(l_2 - l_1)^3} \cdot \sqrt{(\frac{l_3}{l_2 - l_1})^2 + 1} + \frac{l_2 - l_1}{\sqrt{(l_2 - l_1)^2 + l_3^2}})\}$$

3.12 The third term of Eq.(19) above treated, (19c)

Straightforward integration leads again to a result, requiring first a long chain of partial integrations:

In order to solve the third term, Eq. (19c), it is at first reasonable to use the substitution

$$\frac{y_4 - l_2 + w}{x_5} = z \tag{48}$$

This makes it possible to write the integral (19c):

$$\frac{\mu_0 I^2}{4\pi w^2} \{ \int_{x_5=0}^{w} dx_5 x_5 \int_{z=0}^{\frac{w}{x_5}} dz \ln(x_5(z + \sqrt{z^2 + 1})) \} \tag{49}$$

This expression may be dissolved into two terms:

$$\frac{\mu_0 I^2}{4\pi w^2} \{ \int_{x_5=0}^{w} dx_5 x_5 \int_{z=0}^{\frac{w}{x_5}} dz \ln(x_5 + \ln(z + \sqrt{z^2 + 1})) \} \tag{49a and 49b}$$

The first term simply becomes

$$\frac{\mu_0 I^2}{4\pi}(-\ln w + 1) \tag{50}$$

The second term can be rewritten:

$$\frac{\mu_0 I^2}{4\pi w^2} (\int_{x_5=0}^{w} dx_5 w(\ln(w + \sqrt{w^2 + x_5^2}) - \ln x_5) - \int_{x_5=0}^{w} dx_5 \sqrt{w^2 + x_5^2} + \int_{x_5=0}^{w} dx_5 x_5) \tag{51}$$

These three terms will now be denoted (51a), (51b) and (51c) respectively. In order to solve these integrals, it appears practical to make the variable substitution

$$x_5 = \frac{w}{u} \tag{52}$$

Having done this, Eq.(65) may be rewritten:

$$\frac{\mu_0 I^2}{4\pi w^2} \{ \int_{u \to +\infty}^{0} du(-\frac{1}{u^2}(\ln(u + \sqrt{u^2 + 1})) + \int_{u \to +\infty}^{0} du \frac{1}{u^3}\sqrt{u^2 + 1} - \int_{u \to 0+\infty}^{0} du \frac{1}{u^3} \} \tag{53}$$

Well, straightforwardly performing variable substitution, and, using the integral formula:

$$\int \frac{dx}{x(x^2 + a^2)^{\frac{1}{2}}} = -\frac{1}{a}\ln \left| \frac{a + (x^2 + a^2)^{\frac{1}{2}}}{x} \right| \tag{54}$$

finally leads to the result: Term (53a) becomes

$$\frac{\mu_0 I^2}{4\pi}(2\ln(1+\sqrt{2})) \tag{55}$$

The same procedure is repeated with the next term, which leads to:

Term (53b) becomes

$$\frac{\mu_0 I^2}{4\pi}(-\frac{\sqrt{2}}{2}-\frac{1}{2}\cdot\ln(1+\sqrt{2})) \tag{56}$$

Finally, term (53c) becomes

$$\frac{\mu_0 I^2}{4\pi}(\frac{1}{2}) \tag{57}$$

The total result of term (6c) then will be attained, if adding (50), (55), (56) and (57) to each other.

The result is:

$$(F^G_{54})_y \langle(19c)\rangle = \frac{\mu_0 I^2}{4\pi}(-\ln w + \frac{3}{2} - \frac{\sqrt{2}}{2} + \frac{3}{2}\ln(1+\sqrt{2})) \tag{58}$$

3.13 The fourth term of Eq.(19) above treated, (19d)

Straightforward integration leads again to a result. Many partial integration to be done, but in a rather straightforward way. In order to solve the fourth term, Eq. (19c), it is at first reasonable to use the substitution

$$\frac{y_4 - l_1}{x_5} = z \tag{59}$$

This makes it possible to write the integral (19d) :

$$\frac{\mu_0 I^2}{4\pi w^2}\{\int_{x_5=0}^{w} dx_5 x_5 \int_{z=\frac{l_2-l_1-w}{x_5}}^{\frac{l_2-l_1}{x_5}} dz \ln(x_5 z + \sqrt{(x_5 z)^2 + x_5^2})\} \tag{60}$$

This expression may be dissolved into two terms, using the logarithmic product formula:

$$\ln\{A \cdot B\} = \ln A + \ln B \tag{61}$$

The first term simply becomes

$$\frac{\mu_0 I^2}{4\pi}(-\ln w + 1) \tag{62}$$

The second term can be rewritten:

$$\frac{\mu_0 I^2}{4\pi w^2}\{\int\limits_{x_5=0}^{w} dx_5((l_2-l_1)(\ln(\frac{l_2-l_1}{x_5}+\sqrt{(\frac{l_2-l_1}{x_5})^2+1})-\sqrt{(l_2-l_1)^2+x_5^2}-$$

$$-(l_2-l_1-w)\ln(\frac{l_2-l_1-w}{x_5}+\sqrt{(\frac{l_2-l_1-w}{x_5})^2+1})+\sqrt{(l_2-l_1-w)^2+x_5^2}\}$$

(63)

In order to perform the evaluation successfully, the integral formulas [20] and [25] will be needed. Further series expansions have to be done with respect to root and logarithmic expressions that will arise during the work, namely [28], [29] and [30]. It is the hope that these advices will lead the reader successfully to the result, which is

$$(F^G{}_{54})_y\langle(19d)\rangle=\frac{\mu_0 I^2}{4\pi}\{-\ln w+\frac{1}{2}+\ln(l_2-l_1)+\ln 2\}$$

(64)

3.14 The sum of all contributions

In order to attain an expression for the total sum of the integrations above, it would be favorable to gather them in consecutive order below:

$$(F^G{}_{54})^y\langle(19a)\rangle=\frac{\mu_0 I^2}{4\pi}(\ln l_3)$$

(29)

$$(F^G{}_{54})^y\langle(19b)\rangle=\frac{\mu_0 I^2}{4\pi}\{-\ln l_3+\ln(\frac{l_2-l_1}{l_3}+\sqrt{(\frac{l_2-l_1}{l_3})^2+1})+1-$$

$$-\frac{(l_2-l_1)^2+l_3^2+(l_2-l_1)l_3\sqrt{(\frac{l_2-l_1}{l_3})^2+1}}{l_3^2}\cdot\frac{1}{\frac{l_2-l_1}{l_3}+\sqrt{(\frac{l_2-l_1}{l_3})^2+1}}\cdot\frac{1}{\sqrt{(\frac{l_2-l_1}{l_3})^2+1}}+$$

$$+\frac{1}{(\frac{l_3}{l_2-l_1}+\sqrt{(\frac{l_3}{l_2-l_1})^2+1})\cdot\sqrt{(\frac{l_3}{l_2-l_1})^2+1}}\cdot(\frac{1}{2}-\frac{1}{4}\cdot\frac{l_3^2}{l_3^2+(l_2-l_1)^2}+$$

$$+\frac{1}{2}\cdot\frac{(l_1-l_2+l_3)^2(l_2-l_1+2l_3)}{(l_2-l_1)^3}+\frac{3}{2}\cdot\frac{l_3^2(l_2-l_1)}{(l_2-l_1)^3}-\frac{1}{4}\cdot\frac{l_3^4(l_2-l_1)}{(l_2-l_1)^5}+\sqrt{(\frac{l_3}{l_2-l_1})^2+1})+$$

$$+\frac{1}{(\frac{l_3}{l_2-l_1}+\sqrt{(\frac{l_3}{l_2-l_1})^2+1})^2\cdot((\frac{l_3}{l_2-l_1})^2+1)}\cdot(\frac{1}{(l_2-l_1)^4}((2l_3^2+(l_2-l_1)^2)(l_2-l_1)^2+$$

$$+\frac{1}{2}\cdot(l_1-l_2+l_3)^2(l_3^2+(l_2-l_1)^2)-\frac{1}{2}\cdot l_3^2(l_3^2+(l_2-l_1)^2)-\frac{1}{2}\cdot l_3^4)-$$

$$-\frac{l_3}{l_2-l_1}\cdot\sqrt{(\frac{l_3}{l_2-l_1})^2+1}-\frac{l_3^3}{(l_2-l_1)^3}\cdot\sqrt{(\frac{l_3}{l_2-l_1})^2+1}+\frac{l_2-l_1}{\sqrt{(l_2-l_1)^2+l_3^2}})\}$$

(47)

$$(F^G{}_{54})_y\langle(19c)\rangle=\frac{\mu_0 I^2}{4\pi}(-\ln w+\frac{3}{2}-\frac{\sqrt{2}}{2}+\frac{3}{2}\ln(1+\sqrt{2}))$$

(58)

$$(F^G{}_{54})_y \langle (19d) \rangle = \frac{\mu_0 I^2}{4\pi} \{-\ln w + \frac{1}{2} + \ln(l_2 - l_1) + \ln 2\} \tag{64}$$

The sum of all these four terms (omitted here) (65)

3.14.1 The sum of all contributions, simplified expression

After having performed the summation, some simplification occurs and gives rise to the following result:

$$(F^G{}_{54})^y \langle (19) \rangle = \frac{\mu_0 I^2}{4\pi} \{-\ln(\frac{l_2 - l_1}{l_3} + \sqrt{(\frac{l_2 - l_1}{l_3})^2 + 1}) + 1 -$$

$$-\frac{(l_2 - l_1)^2 + l_3^2 + (l_2 - l_1)\sqrt{(\frac{l_2 - l_1}{l_3})^2 + 1}}{l_3^2} \cdot \frac{1}{\frac{l_2 - l_1}{l_3} + \sqrt{(\frac{l_2 - l_1}{l_3})^2 + 1}} \cdot \frac{1}{\sqrt{(\frac{l_2 - l_1}{l_3})^2 + 1}} +$$

$$+\frac{1}{(\frac{l_3}{l_2 - l_1} + \sqrt{(\frac{l_3}{l_2 - l_1})^2 + 1}) \cdot \sqrt{(\frac{l_3}{l_2 - l_1})^2 + 1}} \cdot \{\frac{1}{2} - \frac{1}{4} \cdot \frac{l_3^2}{l_3^2 + (l_2 - l_1)^2} +$$

$$+\frac{1}{2} \cdot \frac{(l_1 - l_2 + l_3)^2 (l_2 - l_1 + 2l_3)}{(l_2 - l_1)^3} + \frac{3}{2} \cdot \frac{l_3^2(l_2 - l_1)}{(l_2 - l_1)^3} - \frac{1}{4} \cdot \frac{l_3^4(l_2 - l_1)}{(l_2 - l_1)^5} + \sqrt{(\frac{l_3}{l_2 - l_1})^2 + 1}\} +$$

$$+\frac{1}{(\frac{l_3}{l_2 - l_1} + \sqrt{(\frac{l_3}{l_2 - l_1})^2 + 1})^2 \cdot ((\frac{l_3}{l_2 - l_1})^2 + 1)} \cdot (\frac{1}{(l_2 - l_1)^4}((2l_3^2 + (l_2 - l_1)^2)(l_2 - l_1)^2 +$$

$$+\frac{1}{2} \cdot (l_1 - l_2 + l_3)^2 (l_3^2 + (l_2 - l_1)^2) - \frac{1}{2} \cdot l_3^2 (l_3^2 + (l_2 - l_1)^2) - \frac{1}{2} \cdot l_3^4) - \frac{l_3}{l_2 - l_1} \cdot \sqrt{(\frac{l_3}{l_2 - l_1})^2 + 1}$$

$$-\frac{l_3^3}{(l_2 - l_1)^3} \cdot \sqrt{(\frac{l_3}{l_2 - l_1})^2 + 1} + \frac{l_2 - l_1}{\sqrt{(l_2 - l_1)^2 + l_3^2}}) - 2\ln w + 2 - \frac{\sqrt{2}}{2} +$$

$$\frac{3}{2}\ln(1 + \sqrt{2}) + \ln(l_2 - l_1) + \ln 2\} \tag{66}$$

4. The result according to the analysis of this author, second approach

The first step in the calculation procedure is to give the problem a strict formulation in the shape of an integral, thereby identifying as well the variables of the integrations as the boarders. Applying Eq.(13) above to the segments 5-4 will give rise to the following integral equation, now in this case with a diagonal line at the corner, defining the border between two segments. The change appears at the upper border in the integral over y_5: $l_2 - x_5$ instead of $l_2 - w$ of Eq. (4) above.

$$(F^G{}_{54})_y \cong \frac{\mu_0 I^2}{4\pi w^2} \int_0^w dx_5 \int_0^{l_3} dx_4 \int_{l_1}^{l_2 - x_5} dy_5 \int_{l_2 - w}^{l_2} dy_4 \frac{x_4 - x_5}{((x_4 - x_5)^2 + (y_4 - y_5)^2)^{3/2}} \tag{67}$$

4.1 The first step: Integration with respect to x_4

Integrating first with respect to x_4, gives the result:

$$(F^G{}_{54})_y = \frac{\mu_0 I^2}{4\pi w^2} \int\limits_{l_1}^{l_2-x_5} dy_5 \int\limits_{l_2-w}^{l_2} dy_4 \int\limits_0^w dx_5 \left(-\frac{1}{((l_3-x_5)^2+(y_4-y_5)^2)^{1/2}} + \frac{1}{(x_5^2+(y_4-y_5)^2)^{1/2}}\right) \quad (68)$$

This corresponds to the procedure in section 4.5 and the integration result here (Eq.(68)) is equal to that (Eq.(18)). The border line at the corner has namely not yet been involved.

4.2 The second step: Integration with respect to y_5

The subsequent integration with respect to y_5 gives rise to the following expression:

$$(F^G{}_{54})_y = \frac{\mu_0 I^2}{4\pi w^2} \int\limits_0^w dx_5 \int\limits_{l_2-w}^{l_2} dy_4 \left(\ln(y_4-l_2+x_5+\sqrt{(y_4-l_2+x_5)^2+(l_3-x_5)^2})\right.$$
$$-\ln(y_4-l_1+\sqrt{(y_4-l_1)^2+(l_3-x_5)^2}) - \ln(y_4-l_2+x_5+\sqrt{(y_4-l_2+x_5)^2+x_5^2}) \quad (69)$$
$$\left.+\ln(y_4-l_1+\sqrt{(y_4-l_1)^2+x_5^2})\right)$$

Here, at this level, the border line has begun to affect the integration result, now when integrating with respect to y_5.

The second and the fourth term of Eq.(69) are different than the corresponding terms of Eq.(19). For convenience, it might appear suitable to denote the terms of Eq.(69) (69a) etc until (69d). The terms which must be further integrated here are accordingly (69a) and (69c). Eq.(69b) and (69d) are equal to Eq.(19b) and (19d) respectively.

4.3 The third step: Integration with respect to y_4 and x_5

4.3.1 The first term of Eq.(69) above treated, (69a)

In order to solve Eq. (69a), the substitution

$$\frac{y_4-l_2+x_5}{l_3-x_5} = z \quad (70)$$

will favorably be used. This makes eq. (69a) to transform into:

$$(F^G{}_{54})_y = \frac{\mu_0 I^2}{4\pi w^2} \int\limits_0^w dx_5 (l_3-x_5) \int\limits_{z=\frac{x_5-w}{l_3-x_5}}^{\frac{x_5}{l_3-x_5}2} (-dz)(\ln(z(l_3-x_5)+\sqrt{z^2(l_3-x_5)^2+1})) \quad (71)$$

This expression must be separated into several separate terms in order to be solved.

The first term will be:

$$\frac{\mu_0 I^2}{4\pi w^2} \int_0^w dx_5 \ln(l_3 - x_5) \cdot w \tag{72}$$

Integration and, finally, series expansion of the ln function gives the result

$$\frac{\mu_0 I^2}{4\pi w}\{\ln l_3\} \tag{73}$$

The second term will accordingly be:

$$\frac{\mu_0 I^2}{4\pi w^2} \int_0^w dx_5 (l_3 - x_5) \int_{z=\frac{x_5-w}{l_3-x_5}}^{\frac{x_5}{l_3-x_5}} (-dz)\ln(z+\sqrt{z^2+1}) \tag{74}$$

The second terms appears to give a zero result, hence the total result may be written

$$(F^G{}_{54})^y\langle(69a)\rangle = \frac{\mu_0 I^2}{4\pi}\{\ln l_3\} \tag{75}$$

4.4 The second term of Eq.(69) above treated, (69b)

As discussed above, the integration of the second term of Eq.(69) must be equal to the result of the integration of Eq.(19b) in section. 3.12, i.e. Eq.(47). Hence,

$$(F^G{}_{54})^y\langle(69b)\rangle = \frac{\mu_0 I^2}{4\pi}\{-\ln l_3 - \ln(\frac{l_2-l_1}{l_3} + \sqrt{(\frac{l_2-l_1}{l_3})^2+1})\}+1-$$

$$-\frac{(l_2-l_1)^2+l_3{}^2+(l_2-l_1)\sqrt{(\frac{l_2-l_1}{l_3})^2+1}}{l_3{}^2} \cdot \frac{1}{\frac{l_2-l_1}{l_3}+\sqrt{(\frac{l_2-l_1}{l_3})^2+1}} \cdot \frac{1}{\sqrt{(\frac{l_2-l_1}{l_3})^2+1}}+$$

$$+\frac{1}{(\frac{l_3}{l_2-l_1}+\sqrt{(\frac{l_3}{l_2-l_1})^2+1})\cdot\sqrt{(\frac{l_3}{l_2-l_1})^2+1}} \cdot \{\frac{1}{2}-\frac{1}{4}\cdot\frac{l_3{}^2}{l_3{}^2+(l_2-l_1)^2}+$$

$$+\frac{1}{2}\cdot\frac{(l_1-l_2+l_3)^2(l_2-l_1+2l_3)}{(l_2-l_1)^3}+\frac{3}{2}\cdot\frac{l_3{}^2(l_2-l_1)}{(l_2-l_1)^3}-\frac{1}{4}\cdot\frac{l_3{}^4(l_2-l_1)}{(l_2-l_1)^5}+\sqrt{(\frac{l_3}{l_2-l_1})^2+1}\}+ \tag{76}$$

$$+\frac{1}{(\frac{l_3}{l_2-l_1}+\sqrt{(\frac{l_3}{l_2-l_1})^2+1})^2\cdot((\frac{l_3}{l_2-l_1})^2+1)} \cdot \{\frac{1}{(l_2-l_1)^4}((2l_3{}^2+(l_2-l_1)^2)(l_2-l_1)^2+$$

$$\frac{1}{2}\cdot(l_1-l_2+l_3)^2(l_3{}^2+(l_2-l_1)^2)-\frac{1}{2}\cdot l_3{}^2(l_3{}^2+(l_2-l_1)^2)-\frac{1}{2}\cdot l_3{}^4)-$$

$$\frac{l_3}{l_2-l_1}\cdot\sqrt{(\frac{l_3}{l_2-l_1})^2+1}-\frac{l_3{}^3}{(l_2-l_1)^3}\cdot\sqrt{(\frac{l_3}{l_2-l_1})^2+1}+\frac{l_2-l_1}{\sqrt{(l_2-l_1)^2+l_3{}^2}})\}$$

4.5 The third term of Eq.(69) above treated, (69c)

Due to the rather complicated expressions that will follow, the integral that has to be solved will be repeated here for convenience (from section 5.3.):

$$(F^G{}_{54})_y \langle(69c)\rangle = \frac{\mu_0 I^2}{4\pi w^2} \int_0^w dx_5 \{-\ln(y_4 - l_2 + x_5 + \sqrt{(y_4 - l_2 + x_5)^2 + x_5^2}\} \tag{69c}$$

In order to solve Eq. (69c), the variable substitution

$$\frac{y_4 - l_2 + x_5}{x_5} = z \tag{77}$$

may appear suitable

In doing so, the following two terms will result:

$$\frac{\mu_0 I^2}{4\pi w^2} \{ \int_{x_5=0}^w dx_5 x_5 \int_{z=\frac{x_5-w}{x_5}}^1 dz(\ln x_5 + \ln(z + \sqrt{z^2 + 1}))\} \tag{78}$$

The first term gives

$$\frac{\mu_0 I^2}{4\pi w^2} \{-\ln w + 1\} \tag{79}$$

Using the integral equation [19] the second term develops to

$$\frac{\mu_0 I^2}{4\pi w^2} \int_0^w dx_5 \cdot x_5 \{-\ln(1 + \sqrt{2}) + \sqrt{2} + \frac{x_5 - w}{x_5} \ln(\frac{x_5 - w}{x_5} + \sqrt{(\frac{x_5 - w}{x_5})^2 + 1}) - \sqrt{(\frac{x_5 - w}{x_5}) + 1}\} \tag{80}$$

This integral apparently consists of three separate terms, which may be treated separately from each other.

The first term is only a constant term, which has to be integrated with x_5 .over the variable x_5 . Hence, the result is easily written

$$\frac{\mu_0 I^2}{4\pi w^2} \{-\frac{\ln(1 + \sqrt{2})}{2} + \frac{\sqrt{2}}{2}\} \tag{81}$$

The second term of Eq. (80) causes a real difficulty when trying to integrate the expression, since no approximations in using seres expansions are allowed. That is so, since the terms are all of the same order. Instead, it is possible to estimate the limits, between which the result must lay. The method is given by the expression

$$m(b-a) \le \int_a^b f(x)dx \le M(b-a) \text{ [23]} \tag{82}$$

By analyzing the asymptotic behavior of the integrand, it may thus be stated that the result has to be within the interval

$$(0, \frac{\mu_0 I^2}{4\pi w} \int_{x_5=0}^{w} dx_5 \{\ln x_5 + \ln(-w + \sqrt{w^2 + x_5^2}\} \ ) \tag{83}$$

which may be simplified to

$$(0, \frac{\mu_0 I^2}{4\pi}(-(1 + \ln(1 + \sqrt{2})) \ \} \tag{84}$$

The last term of Eq.(80) implies a straightforward integration of a square root expression with the variable squared. A suitable table solution may be used [24]

The result may be written

$$\frac{\mu_0 I^2}{4\pi} \{\frac{1}{2} + \frac{1}{4\sqrt{2}} \ln(\sqrt{2} + 1) - \ln(\sqrt{2} - 1\} \tag{85}$$

Now it is possible to write the total result of the integral of Eq.(69c):

$$(F^G{}_{54})^y \langle(69c)\rangle = \frac{\mu_0 I^2}{4\pi}\{-\ln w + \frac{3}{2} + \frac{\sqrt{2}}{2} + (\frac{1}{4\sqrt{2}} - \frac{1}{2})\ln(\sqrt{2} + 1) +$$
$$+ \frac{1}{4\sqrt{2}}\ln(\sqrt{2} - 1) + (0, -(1 + \ln(1 + \sqrt{2})\} \tag{86}$$

4.6 The fourth term of Eq.(69) above treated, (69d)

As discussed above (in section 4.3), the integration of the fourth term of Eq (69) must be equal to the result of the integration of (19d), i.e. Eq.(64) in section 3.14. Hence,

$$(F^G{}_{54})_y \langle(69d)\rangle = \frac{\mu_0 I^2}{4\pi}\{-\ln w + \frac{1}{2} + \ln(l_2 - l_1) + \ln 2\} \tag{87}$$

4.7 The sum of all contributions

$$(F^G{}_{54})^y \langle(69a)\rangle = \frac{\mu_0 I^2}{4\pi}\{\ln l_3\} \tag{75}$$

$$(F^G{}_{54})^y \langle\langle(69b)\rangle\rangle = \frac{\mu_0 I^2}{4\pi}\{-\ln l_3 - \ln(\frac{l_2-l_1}{l_3} + \sqrt{(\frac{l_2-l_1}{l_3})^2+1}) + 1 -$$

$$-\frac{(l_2-l_1)^2 + l_3^2 + (l_2-l_1)l_3\sqrt{(\frac{l_2-l_1}{l_3})^2+1}}{l_3^2} \cdot \frac{1}{\frac{l_2-l_1}{l_3}+\sqrt{(\frac{l_2-l_1}{l_3})^2+1}} \cdot \frac{1}{\sqrt{(\frac{l_2-l_1}{l_3})^2+1}} +$$

$$+\frac{1}{(\frac{l_3}{l_2-l_1}+\sqrt{(\frac{l_3}{l_2-l_1})^2+1})\cdot\sqrt{(\frac{l_3}{l_2-l_1})^2+1}} \cdot (\frac{1}{2}-\frac{1}{4}\cdot\frac{l_3^2}{l_3^2+(l_2-l_1)^2} +$$

$$+\frac{1}{2}\cdot\frac{(l_1-l_2+l_3)^2(l_2-l_1+2l_3)}{(l_2-l_1)^3} + \frac{3}{2}\cdot\frac{l_3^2(l_2-l_1)}{(l_2-l_1)^3} - \frac{1}{4}\cdot\frac{l_3^4(l_2-l_1)}{(l_2-l_1)^5} + \sqrt{(\frac{l_3}{l_2-l_1})^2+1}) +$$

$$+\frac{1}{(\frac{l_3}{l_2-l_1}+\sqrt{(\frac{l_3}{l_2-l_1})^2+1})^2\cdot((\frac{l_3}{l_2-l_1})^2+1)}\cdot(\frac{1}{(l_2-l_1)^4}((2l_3^2+(l_2-l_1)^2)(l_2-l_1)^2+$$

$$\frac{1}{2}\cdot(l_1-l_2+l_3)^2(l_3^2+(l_2-l_1)^2) - \frac{1}{2}\cdot l_3^2(l_3^2+(l_2-l_1)^2) - \frac{1}{2}\cdot l_3^4) - \frac{l_3}{l_2-l_1}\cdot\sqrt{(\frac{l_3}{l_2-l_1})^2+1} -$$

$$\frac{l_3^3}{(l_2-l_1)^3}\cdot\sqrt{(\frac{l_3}{l_2-l_1})^2+1} + \frac{l_2-l_1}{\sqrt{(l_2-l_1)^2+l_3^2}})\}$$

$$\tag{76}$$

$$(F^G{}_{54})^y \langle\langle(69c)\rangle\rangle = \frac{\mu_0 I^2}{4\pi}\{-\ln w + \frac{3}{2} + \frac{\sqrt{2}}{2} + (\frac{1}{4\sqrt{2}}-\frac{1}{2})\ln(\sqrt{2}+1) + \frac{1}{4\sqrt{2}}\ln(\sqrt{2}-1) +$$
$$(0,-(1+\ln(1+\sqrt{2}))\}$$

$$\tag{86}$$

$$(F^G{}_{54})_y \langle\langle(69d)\rangle\rangle = \frac{\mu_0 I^2}{4\pi}\{-\ln w + \frac{1}{2} + \ln(l_2-l_1) + \ln 2\}$$

$$\tag{87}$$

The sum of all these four terms (omitted here) $\tag{88}$

4.8 Numerical check with the Assis/Bueno result necessary

4.8.1 The first approach

By reasons of comparison the result by Assis and Bueno (Eq. (3) above) is repeated here. Obviously, they have not attained the same result as this author (66). Some features are similar, but the differences are also apparent. Therefore it is adequate to claim that the laws are not equal.

$$(F^G{}_{54})_y \cong \frac{\mu_0}{4\pi}(\ln\frac{l_2-l_1}{w} - \ln\frac{(l_2-l_1)+((l_2-l_1)^2+l_3^2)^{1/2}}{l_3} + \ln 2 - \frac{3}{2}\ln(1+\sqrt{2}) + \frac{\sqrt{2}}{2} + \frac{1}{2})$$

$$\tag{3}$$

A numerical check for one case shows also that not even an integral gives rise to equal result. It was chosen a wire width $w = 3mm$, $l_1 = 0.7m$, $l_2 = 1.2m$ and, finally, $l_3 = 0.7m$

The result thus attained is

$$(F^G{}_{54})_y \cong \frac{\mu_0}{4\pi} \cdot 10.8 \tag{89}$$

Using, however, the calculation by this author, based on Eq. (76) above, the result will be:

$$(F^G{}_{54})_y \cong \frac{\mu_0}{4\pi} \cdot 12.6 \tag{90}$$

4.8.2 The second approach

Since it was not possible to attain a solution on closed form, a numerical check for one case can be sufficient, provided the result is not in accordance with the formula. It was chosen a wire width $w = 3mm$, $l_1 = 0.7m$, $l_2 = 1.2m$ and, finally, $l_3 = 0.7m$

The result thus attained is

$$(F^G{}_{54})_y \cong \frac{\mu_0}{4\pi} (8.4, 10.3) \tag{91}$$

whereas the formula used by Assis and Bueno gives

$$(F^G{}_{54})_y \cong \frac{\mu_0}{4\pi} \cdot 12.8 \tag{92}$$

Hence, Assis and Bueno are wrong in their claim that the Grassmann force gives the same result as Ampère's law [1]. To conclude, it seemed to be rather wise to reject the claim of coincidence between these two laws, as they did not coincide before integrating. To be stated again,

5. Discussion and conclusions

From the rigorous analysis that has been performed above it is evident that Grassmann's law is not equal to Ampère's law. It is also evident that the very roots of the idea of their equality by Grassmann is false, too. However, it does not exist any need to regard them as equal. It was only one idea among many invented by science in its search for a better understanding of physics. In this respect it constitutes a progress for physics that it has been possible to reject one of the speculative ideas that must inevitably arise in the search for the truth. More seriously, it is of course grave for the proponents of the Lorentz force (identified as the Grassmann force), if it cannot be justified by referring to preceding established ideas but must stand as a mere ad-hoc invention.

6. List of variables

6.1 Variables specified in chapter 3

a, b infinitesimal elements of respective current

r the distance between any two points of two respective electric currents

l the perpendicular from the midpoints of the attracted element on the circuit element b on to the line of the attracting one, or

b_l cosine component of the b element on the perpendicular

ε the angle between two electric currents

α, β the angles formed by the elements a and b respectively with the line drawn between the two mid-points

6.2 Variables specified from chapter 4 onwards

$d^2 \vec{F}_{ji}{}^G$ the electromagnetic force between two infinitesimal elements of two electric currents (Assis and Bueno: Grassmann's force)

$d^4 \vec{F}_{ji}{}^G$ the same force as $d^2 \vec{F}_{ji}{}^G$ above, but with a more adequate notation, indicating the four variables being differentiated

$(F^G{}_{ji})_y$ the y component of the Grassmann force due to two segments

$(F^G{}_{ji})_y \langle ... \rangle$ the y component of the Grassmann force due to two segments, for a specific term

$d^2 \vec{F}_{ji}{}^A$ the force between two electric currents according to Ampère's law

$\hat{r}$ unit vector along the line connecting the two conductors

I_i, I_j the currents of respective conductor

$d\vec{l}_i, d\vec{l}_j$ length elements of respective conductor

l_1 the position of the cut-off point at the y branch

l_2 length of Ampère's bridge along the y axis

l_3 length of Ampère's bridge along the x axis

w the width of branches of Ampère's bridge

7. References

[1] Peter Graneau, 'Ampere and Lorentz forces', Physics Letters, VOL. 107 A, Number 5, 4 February 1985, 235.

[2] H.G. Grassmann, Poggendorffs Ann. Phys. Chemie 64 (1845), 1

[3] R.A.R. Tricker, 'Early Electrodynamics, The First Law of Circulation', Pergamon Press, 1965,201. This is the English translation of ref. [2]

[4] ibid p. 203.

[5] Jonson, J. O., 'The Magnetic Force between Two Currents Explained Using Only Coulomb's Law', Chinese Journal of Physics, VOL. 35, NO. 2, April 1997, p. 139-149,
http://psroc.phys.ntu.edu.tw/cjp/find_content.php?year=1997&vol=35&no=2

[6] R.A.R. Tricker, 'Early Electrodynamics. The First Law of Circulation', Pergamon Press, 1965, 206, Figure 60

[7] R.A.R. Tricker, 'Early Electrodynamics. The First Law of Circulation', Pergamon Press, 1965, 205. The equations on this page

[8] R.A.R. Tricker, 'Early Electrodynamics. The First Law of Circulation', Pergamon Press, 1965, 207.

[9] Assis, A.K.T., and Bueno, A., 'Equivalence between Ampère and Grassmann's Forces', IEEE Transactions on Magnetics, VOL. 32, NO. 2, March 1996, 431-436

[10] ibid. p. 431

[11] ibid. p. 432 and p. 433

[12] ibid. p. 432, Eq. (4)

[13] ibid p. 433, Eq. (10)

[14] Peter Graneau, 'Ampere and Lorentz forces', Physics Letters, VOL. 107 A, Number 5, 4 February 1985, 235

[15] G. Grassmann, ,Poggendorffs Ann. Phys. Chemie 64 (1845), 1

[16] R.A.R. Tricker, 'Early Electrodynamics. The First Law of Circulation', Pergamon Press, 1965,201. This is the English translation of ref. [14]

[17] Assis, A.K.T., and Bueno, A., 'Equivalence between Ampère and Grassmann's Forces', IEEE Transactions on Magnetics, VOL. 32, NO. 2, March 1996, p. 432, Eq. (2).

[18] ibid. p. 433 (Eq. 12)

[19] ibid. p. 434, Eq. (15)

[20] Abramowitz, M. and Stegun A., 'Handbook of Mathematical Functions', Dover Publications, Inc., New York, 1970, p. 69, Formula 4.1.53

[21] ibid. p. 13, 3.3.41

[22] ibid. p. 69, 4.1.49

[23] Abramowitz, M. and Stegun, A., 'Handbook of Mathematical Functions', Dover Publications, Inc., New York, 1970, p. 69, Formula 4.1.44

[24] Ullemar, Leo, 'Funktioner av en variabel', Studentlitteratur, Lund, Sweden, 1972, p. 96

[25] Abramowitz, M. and Stegun A., 'Handbook of Mathematical Functions', Dover Publications, Inc., New York, 1970, p. 13, formula 3.3.37

[26] ibid. p. 13, 3.3.42

[27] ibid. p. 67, 4.1.6

[28] Assis, A.K.T., and Bueno, A. Bueno, 'Equivalence between Ampère and Grassmann's Forces', IEEE Transactions on Magnetics, VOL. 32, NO. 2, March 1996, p. 433, Eq. (10)

[29] Abramowitz, M. and Stegun A., 'Handbook of Mathematical Functions', Dover Publications, Inc., New York, 1970, p. 15, 3.6.11

[30] ibid p. 15, 3.6.9
[31] ibid. p. 68, 4.1.24

3

Vortices in Electric Dipole Radiation

X. Li[1], D.M. Pierce[2] and H.F. Arnoldus[2]
[1]California Polytechnic State University,
[2]Mississippi State University,
USA

1. Introduction

In the geometrical optics limit of light propagation, light travels from a source to an observer along straight lines, known as optical rays. On the other hand, the energy in an electromagnetic field flows along the field lines of the Poynting vector. It can be shown (Born & Wolf, 1980) that in the geometrical optics limit, where variations in the radiation field on the scale of a wavelength are neglected, the optical rays coincide with the field lines of the Poynting vector, and both are straight lines. In nanophotonics and near-field optics, where sub-wavelength resolution of the energy transport is of interest, the optical rays lose their significance. Energy flows along the field lines of the Poynting vector, and these field lines are in general curves. When sub-wavelength resolution is required, we need the exact solution of Maxwell's equations. In order to study the fundamental aspects of energy propagation, we consider the simplest and most important source of radiation, which is the electric dipole. When a small object, like an atom, molecule or nanoparticle, is placed in an external electromagnetic field (usually a laser beam), oscillating with angular frequency ω, a current will be induced in the particle, and this gives the particle an electric dipole moment of the form

$$\mathbf{d}(t) = d_0 \operatorname{Re}(\mathbf{u}\, e^{-i\omega t}), \tag{1}$$

with d_0 an overall (real) constant, and $\mathbf{u}$ a complex-valued unit vector, normalized as $\mathbf{u} \cdot \mathbf{u}^* = 1$. The oscillating dipole moment emits electromagnetic radiation. The electric field will have the form

$$\mathbf{E}(\mathbf{r}, t) = \operatorname{Re}[\mathbf{E}(\mathbf{r}) e^{-i\omega t}], \tag{2}$$

with $\mathbf{E}(\mathbf{r})$ the complex amplitude, and the magnetic field $\mathbf{B}(\mathbf{r}, t)$ has a similar form. We shall allow for the possibility that the dipole is embedded in a linear isotropic homogeneous medium with relative permittivity ε_r and relative permeability μ_r, and both are complex in general. The imaginary parts of ε_r and μ_r are non-negative, as can be shown from causality arguments. The index of refraction n of the medium is a solution of $n^2 = \varepsilon_r \mu_r$, and we take the solution with $\operatorname{Im} n \geq 0$. When ε_r and μ_r are both positive or both negative, we have $n^2 > 0$, and this leaves the sign of n ambiguous. It can be shown with a limit procedure

(McCall et al., 2002) that we should take $n > 0$ when ε_r and μ_r are both positive and $n < 0$ when ε_r and μ_r are both negative. The time-averaged Poynting vector in such a medium is (Jackson, 1998)

$$S(r) = \frac{1}{2\mu_o} \text{Re}\left[\frac{1}{\mu_r} E(r)^* \times B(r) \right] \tag{3}$$

2. Electric dipole radiation

Let the dipole be located at the origin of coordinates. The complex amplitudes of the electric and magnetic fields are then found to be (Li et al., 2011b)

$$E(r) = \frac{\mu_r d_o k_o^2}{4\pi\varepsilon_o}\left\{ u - (\hat{r} \cdot u)\hat{r} + [u - 3(\hat{r} \cdot u)\hat{r}]\frac{i}{nk_o r}\left(1 + \frac{i}{nk_o r} \right) \right\}\frac{e^{ink_o r}}{r} , \tag{4}$$

$$B(r) = \frac{n\mu_r d_o k_o^2}{4\pi\varepsilon_o c}(\hat{r} \times u)\left(1 + \frac{i}{nk_o r} \right)\frac{e^{ink_o r}}{r} , \tag{5}$$

for $r \neq 0$. Here, $k_o = \omega / c$ is the wavenumber of the radiation in free space and $\hat{r} = r / r$ is the unit vector in the radially outward direction. With these expressions for $E(r)$ and $B(r)$, the Poynting vector from (3) can be worked out. First we introduce

$$P_o = \frac{cd_o^2 k_o^4}{12\pi\varepsilon_o} , \tag{6}$$

which equals the power that would be emitted by the same dipole in free space. As we shall see, the field lines of the Poynting vector scale with k_o, so we introduce $q = k_o r$ as the dimensionless position vector of a field point. Similarly, we set $\bar{x} = k_o x$, etc., for the dimensionless Cartesian coordinates of a field point. Therefore, in dimensionless coordinates, a distance of 2π corresponds to one optical free-space wavelength. Then, the field lines of a vector field are determined by the directions of the vectors at each point in space, but not by their magnitudes. So when we set

$$S(r) = \frac{3P_o}{8\pi r^2} | \mu_r |^2 e^{-2q\text{Im}\,n}\sigma(q) , \tag{7}$$

then the field lines of $\sigma(q)$ are the same as the field lines of $S(r)$, since the overall factor that is split off in (7) is positive. We shall simply refer to $\sigma(q)$ as the Poynting vector. We find

$$\sigma(q) = [1 - (\hat{r} \cdot u)(\hat{r} \cdot u^*)]\text{Re}\left[\frac{n}{\mu_r}\left(1 + \frac{i}{nq} \right) \right]\hat{r}$$

$$+ \frac{1}{q|n|^2}\left| 1 + \frac{i}{nq} \right|^2 \{[1 - 3(\hat{r} \cdot u)(\hat{r} \cdot u^*)]\text{Im}(\varepsilon_r)\hat{r} + 2\text{Im}[\varepsilon_r(\hat{r} \cdot u^*)u]\} \tag{8}$$

Vector $\boldsymbol{\sigma}(\mathbf{q})$ is dimensionless, and only depends on the field point through its dimensionless representation $\mathbf{q}$. It furthermore depends on the polarization vector $\mathbf{u}$ of the dipole moment, and it depends in a complicated way on the material parameters ε_r and μ_r.

3. Field lines of the poynting vector

A field line of the vector field $\boldsymbol{\sigma}(\mathbf{q})$ can be parametrized as $\mathbf{q}(u)$, with u a dummy variable. Since at any point $\mathbf{q}$ on a field line the vector $\boldsymbol{\sigma}(\mathbf{q})$ is on the tangent line, the field lines are a solution of

$$\frac{d\mathbf{q}}{du} = \boldsymbol{\sigma}(\mathbf{q}) \tag{9}$$

Given an initial point $\mathbf{q}_i$, equation (9) determines the field line through this point. The dimensionless position vector $\mathbf{q}$ has Cartesian coordinates $(\overline{x}, \overline{y}, \overline{z})$, in terms of which (9) becomes

$$\frac{d\overline{x}}{du} = \sigma_{\overline{x}}(\overline{x},\overline{y},\overline{z}) \ , \quad \frac{d\overline{y}}{du} = \sigma_{\overline{y}}(\overline{x},\overline{y},\overline{z}) \ , \quad \frac{d\overline{z}}{du} = \sigma_{\overline{z}}(\overline{x},\overline{y},\overline{z}) \tag{10}$$

Here, $\sigma_x(\overline{x},\overline{y},\overline{z})$ is the x component of $\boldsymbol{\sigma}(\mathbf{q})$, expressed in the variables $\overline{x}$, $\overline{y}$ and $\overline{z}$. The field line pictures in this chapter are made by numerically integrating the set (10). For further analysis it is useful to express (9) in spherical coordinates (q,θ,ϕ). This gives

$$\frac{dq}{du} = \boldsymbol{\sigma}(\mathbf{q}) \cdot \hat{\mathbf{r}} \ , \tag{11}$$

$$q\frac{d\theta}{du} = \boldsymbol{\sigma}(\mathbf{q}) \cdot \mathbf{e}_{\theta} \ , \tag{12}$$

$$q\sin\theta\frac{d\phi}{du} = \boldsymbol{\sigma}(\mathbf{q}) \cdot \mathbf{e}_{\phi} \tag{13}$$

At a large distance from the dipole, compared to a wavelength, we have $q \gg 1$. Then the Poynting vector (8) simplifies to

$$\boldsymbol{\sigma}(\mathbf{q}) \approx [1 - (\hat{\mathbf{r}} \cdot \mathbf{u})(\hat{\mathbf{r}} \cdot \mathbf{u}^*)]\mathrm{Re}\left(\frac{n}{\mu_r}\right)\hat{\mathbf{r}} \tag{14}$$

It can be shown (McCall et al., 2002) that $\mathrm{Re}(n / \mu_r)$ is positive, and therefore the Poynting vector is approximately in the radially outward direction. We shall not consider the limiting case where n / μ_r is imaginary. Therefore, at a large distance from the dipole the field lines are approximately straight lines, running away from the dipole. Conversely, any curving of the field lines can only occur close to the dipole, e.g., within a distance of about a wavelength. All terms in (8) are proportional to $\hat{\mathbf{r}}$, except for the term containing the factor $\mathrm{Im}[\varepsilon_r(\hat{\mathbf{r}} \cdot \mathbf{u}^*)\mathbf{u}]$. Therefore, any curving of the field lines is due to this term. This can happen, for instance, when **u** is complex, or when ε_r has an imaginary part due to damping in the material.

4. Dipole in free space

Let us first consider a dipole in free space, so that $\varepsilon_r = \mu_r = n = 1$. The simplest case is when the incident field is linearly polarized, say along the z axis. Then the dipole moment will oscillate along the z axis, and we have $\mathbf{u} = \mathbf{e}_z$. With (1) we have

$$\mathbf{d}(t) = d_o \mathbf{e}_z \cos(\omega t) \tag{15}$$

The Poynting vector (8) becomes

$$\boldsymbol{\sigma}(\mathbf{q}) = \hat{\mathbf{r}} \sin^2 \theta , \tag{16}$$

which is in the radially outward direction at all points. Therefore, the field lines are straight lines coming out of the dipole. Figure 1(a) shows the field line pattern.

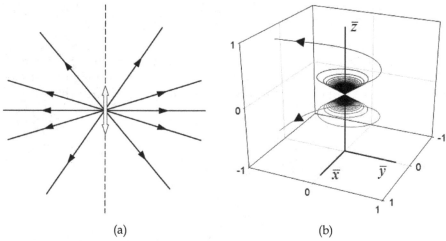

(a)   (b)

Fig. 1. The figure on the left shows the field lines of the Poynting vector for a dipole which oscillates linearly in the direction of the arrow. In the figure on the right two field lines of the Poynting vector are shown for a dipole which rotates in the xy plane. The field lines wind around the z axis in the same direction as the direction of rotation of the dipole moment. The x and y axes in the figure have been lowered for clarity of drawing. The constant θ_o is $\pi / 4$ for the upper field line, and this angle is the angle of the cone around the z axis on which the field line runs. For the lower field line this angle is $3\pi / 4$.

When the incident field is circularly polarized, with the electric field vector rotating in the xy plane, the unit vector **u** is

$$\mathbf{u} = -\frac{1}{\sqrt{2}}(\mathbf{e}_x + i\mathbf{e}_y) , \tag{17}$$

and the dipole moment becomes

$$\mathbf{d}(t) = -\frac{d_\circ}{\sqrt{2}}[\mathbf{e}_x \cos(\omega t) + \mathbf{e}_y \sin(\omega t)] \tag{18}$$

This dipole moment rotates in the xy plane, and the direction of rotation is counterclockwise when viewed down the positive z axis. The Poynting vector becomes

$$\boldsymbol{\sigma}(\mathbf{q}) = (1 - \tfrac{1}{2}\sin^2\theta)\hat{\mathbf{r}} + \frac{1}{q}\left(1 + \frac{1}{q^2}\right)\sin\theta\,\mathbf{e}_\phi \tag{19}$$

The first term on the right-hand side is in the radially outward direction, and the second term is proportional to $\mathbf{e}_\phi$. For q large, this second term vanishes, so at large distances, the field lines run radially outward. For small q, this second term dominates, and the field lines run approximately in the $\mathbf{e}_\phi$ direction. Since ϕ is the angle around the z axis, we expect the field lines to rotate around the z axis. Equations (11)-(13) for the field lines become

$$\frac{dq}{du} = 1 - \tfrac{1}{2}\sin^2\theta \ , \tag{20}$$

$$\frac{d\theta}{du} = 0 \ , \tag{21}$$

$$\frac{d\phi}{du} = \frac{1}{q^2}\left(1 + \frac{1}{q^2}\right) \tag{22}$$

From (21) we see that θ is constant along a field line, and we shall indicate this constant by $\theta_\circ$. Therefore, a field line lies on a cone with its axis as the z axis. Field lines run into the direction of increasing u, and since the right-hand side of (22) is positive, angle ϕ increases along a field line. Therefore, the field lines wind around the z axis in a counterclockwise direction when viewed down the positive z axis. From (22) and (20), and with $\theta = \theta_\circ$, we have

$$\frac{d\phi}{dq} = Z(\theta_\circ)\frac{1}{q^2}\left(1 + \frac{1}{q^2}\right) \ , \tag{23}$$

where we have set

$$Z(\theta_\circ) = \frac{1}{1 - \tfrac{1}{2}\sin^2\theta_\circ} \tag{24}$$

We now see q as the independent variable, and then the solution of (23) is

$$\phi(q) = \phi_\circ - Z(\theta_\circ)\frac{1}{q}\left(1 + \frac{1}{3q^2}\right) \ , \tag{25}$$

where ϕ_o is the integration constant. A field line starts at the location of the dipole, so at $q = 0$. The function $\phi(q)$ increases with q, and for $q \to \infty$ it reaches the asymptotic value of ϕ_o. The field line spirals around the z axis in the counterclockwise direction, when viewed down the positive z axis, and this spiral lies on the cone $\theta = \theta_o$. Two field lines are shown in Fig. 1(b). The resulting field line picture has a vortex structure, and this was called '*the dipole vortex*' (Arnoldus & Foley, 2004). The spatial extent of this vortex is less than a wavelength, as can be seen from the figure. On this scale, a distance of 2π corresponds to one wavelength.

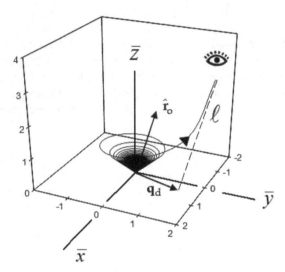

Fig. 2. A field line of the Poynting vector approaches asymptotically a line ℓ at a large distance from the source. This field line appears to come from a point in the xy plane with position vector $\mathbf{q}_d$. Therefore, the dipole seems to be displaced over vector $\mathbf{q}_d$, when observed from the far field.

5. Virtual displacement of the source

For the circular dipole in free space, the field lines form a vortex structure, as illustrated in Fig. 1(b). Close to the source, the field lines wind around the z axis numerous times, and then they run away to the far field, while remaining on a cone. In the far field, a field line approaches asymptotically a straight line, and this line is the optical ray from geometrical optics. In geometrical optics, a sub-wavelength spatial structure like the vortex is not resolved, and it would appear as if the optical ray comes from the location of the dipole. However, when sub-wavelength resolution is of interest, this is not the case anymore. Due to the rotation of a field line near the source, it appears as if the ray comes from a point in the xy plane which is displaced with respect to the position of the dipole, as shown in Fig. 2. The line ℓ is the asymptote of the field line, and this is the optical ray. This line intersects the xy plane at a location indicated by the displacement vector $\mathbf{q}_d$, so this vector represents the apparent location of the source.

The dimensionless Cartesian coordinates $(\bar{x},\bar{y},\bar{z})$ for a point on the field line are parametrized as

$$\bar{x} = q\sin\theta_o \cos\phi(q) \, , \tag{26}$$

$$\bar{y} = q\sin\theta_o \sin\phi(q) \, , \tag{27}$$

$$\bar{z} = q\cos\theta_o \tag{28}$$

Here, θ_o is the angle of the cone on which the field line lies, and $\phi(q)$ is given by (25). The free parameter is q. In order to obtain a representation of the line ℓ, we expand the right-hand sides of (26)-(28) in an asymptotic series in $1/q$. Due to the overall factors of q, we need to expand $\cos\phi(q)$ and $\sin\phi(q)$ up to order $1/q$, so that the combination yields a constant. From (25) we have

$$\phi(q) = \phi_o - \frac{1}{q}Z(\theta_o) + \dots \, , \tag{29}$$

and this gives

$$\cos\phi(q) = \cos\phi_o + \frac{1}{q}Z(\theta_o)\sin\phi_o + \dots \, , \tag{30}$$

$$\sin\phi(q) = \sin\phi_o - \frac{1}{q}Z(\theta_o)\cos\phi_o + \dots \tag{31}$$

Then we substitute the right-hand sides of (30) and (31) in (26) and (27), respectively, and omit the ellipses. We then obtain

$$\bar{x} = \sin\theta_o[t\cos\phi_o + Z(\theta_o)\sin\phi_o] \, , \tag{32}$$

$$\bar{y} = \sin\theta_o[t\sin\phi_o - Z(\theta_o)\cos\phi_o] \, , \tag{33}$$

$$\bar{z} = t\cos\theta_o \tag{34}$$

Here we have replace q by t, since this parameter does not represent the distance to the origin anymore. Equations (32)-(34) are the parameter equations for the line ℓ. This line is the asymptote of the field line that runs into the observation direction (θ_o,ϕ_o), and this direction is indicated by the eye in Fig. 2.

The unit vector in the observation direction (θ_o,ϕ_o) is

$$\hat{\mathbf{r}}_o = \sin\theta_o(\cos\phi_o\mathbf{e}_x + \sin\phi_o\mathbf{e}_y) + \cos\theta_o\mathbf{e}_z \, , \tag{35}$$

and the position vector of a point on the line is $\mathbf{q} = \bar{x}\mathbf{e}_x + \bar{y}\mathbf{e}_y + \bar{z}\mathbf{e}_z$. The parameter equations (32)-(34) can then be written in vector form as

$$\ell: \mathbf{q} = t\hat{\mathbf{r}}_o + \mathbf{q}_d \; , \tag{36}$$

where we have set

$$\mathbf{q}_d = \sin\theta_o Z(\theta_o)(\mathbf{e}_x \sin\phi_o - \mathbf{e}_y \cos\phi_o) \tag{37}$$

It follows from (34) that $\overline{z} = 0$ for $t = 0$, so the line ℓ intersects the xy plane for $t = 0$. From (36) we see that for $t = 0$ we have $\mathbf{q} = \mathbf{q}_d$ so vector $\mathbf{q}_d$ is the displacement vector from Fig. 2. From (35) and (37) it follows that $\mathbf{q}_d \cdot \hat{\mathbf{r}}_o = 0$, and therefore the displacement is perpendicular to the observation direction. The magnitude of the displacement vector is

$$q_d = \frac{2\sin\theta_o}{1 + \cos^2\theta_o} \tag{38}$$

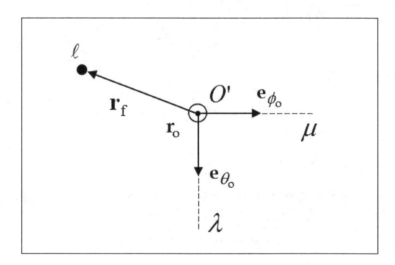

Fig. 3. Illustration of the image plane and the coordinate system in this plane.

The displacement is zero for observation along the z axis, and has its maximum for $\theta_o = \pi / 2$, e.g., for observation in the xy plane. The maximum displacement is $q_d = 2$. Since 2π corresponds to one wavelength, this displacement equals λ / π, with $\lambda = 2\pi / k_o$ the wavelength of the radiation. For visible light, this is of the order of about 150 nm. In nanophotonics, where structures with dimensions of a few nanometers are studied, this displacement is not negligible.

6. Displacement in the far field

In the far field, the line ℓ is the asymptote of the field line that runs into the observation direction (θ_o, ϕ_o). The virtual displacement of the source in the xy plane is then inferred from an extrapolation of the line ℓ from the far field to its intersection with the xy plane. We shall now consider this from a different point of view (Shu et al., 2008). For a given vector

$\mathbf{r}_o$, we define the observation plane, or image plane, as the plane which is perpendicular to $\mathbf{r}_o$, and which contains the point $\mathbf{r}_o$. We shall take the direction of $\mathbf{r}_o$ as specified by (θ_o, ϕ_o), so that $\hat{\mathbf{r}}_o$ is given by (35). Therefore, we can view the image plane as the tangent plane of a sphere with radius r_o around the dipole, and located at the spherical position (θ_o, ϕ_o). So the position of the image plane is specified by its angular location (θ_o, ϕ_o), and by its distance r_o to the origin of coordinates. The point $\mathbf{r}_o$ is taken as the origin O' in the plane, as illustrated in Fig. 3. The spherical-coordinate unit vectors $\mathbf{e}_{\theta_o}$ and $\mathbf{e}_{\phi_o}$ lie in the image plane, and are given explicitly by

$$\mathbf{e}_{\theta_o} = (\mathbf{e}_x \cos\phi_o + \mathbf{e}_y \sin\phi_o)\cos\theta_o - \mathbf{e}_z \sin\theta_o , \tag{39}$$

$$\mathbf{e}_{\phi_o} = -\mathbf{e}_x \sin\phi_o + \mathbf{e}_y \cos\phi_o \tag{40}$$

We then introduce a λ and a μ axis in the image plane, as shown in Fig. 3, so that a point in the image plane is represented by the Cartesian coordinates (λ, μ) in the plane. A point $\mathbf{r}$ in the image plane can therefore be written as

$$\mathbf{r} = \mathbf{r}_o + \lambda \mathbf{e}_{\theta_o} + \mu \mathbf{e}_{\phi_o} \tag{41}$$

A field line of the Poynting vector for the radiation emitted by a circular dipole in free space is parametrized by the angles θ_o and ϕ_o, which are the asymptotic values of θ and ϕ along the field line. We now consider an image plane which is located at the angular location (θ_o, ϕ_o), and we assume that the image plane is located in the far field, so r_o is much larger than a wavelength of the radiation. Then the field line is approximately along the asymptote ℓ. This line intersects the image plane at the location of the black dot in the figure. This point is represented by vector $\mathbf{r}_f$ in the image plane, as shown, or by the dimensionless vector $\mathbf{q}_f = k_o \mathbf{r}_f$. The parameter equation for ℓ is given by (36), and $\mathbf{q}_d$ can be written as

$$\mathbf{q}_d = -\sin\theta_o Z(\theta_o)\mathbf{e}_{\phi_o} \tag{42}$$

Since this is a vector in the image plane, we see immediately that

$$\mathbf{q}_f = \mathbf{q}_d \tag{43}$$

If the field lines were straight, then the field line in the (θ_o, ϕ_o) direction would intersect the image plane at O'. Due to the rotation of the field lines, a field line intersects the image plane at $\mathbf{q}_f$, so $\mathbf{q}_f$ is the displacement of the field line in the far field. Apparently, this displacement is the same as the virtual displacement of the source in the xy plane. This conclusion holds for a circular dipole in free space, but not in general, as we shall see below. The dimensionless coordinates in the image plane are $\bar{\lambda} = k_o \lambda$ and $\bar{\mu} = k_o \mu$. We then see from (42) that the intersection point of ℓ and the image plane has coordinates

$$\bar{\lambda}_f = 0 , \quad \bar{\mu}_f = -\sin\theta_o Z(\theta_o) \tag{44}$$

7. Image of the dipole

A single field line may not be directly observable by a detector. Usually, an image is formed on an image plane, and the intensity distribution over the plane is determined by a bundle of field lines that hit the plane. We shall take the plane of Fig. 3 as the image plane, and we consider the image formed on the plane by the radiation emitted by the circular dipole. Since $\hat{\mathbf{r}}_0$ is the unit normal on the plane, the intensity (power per unit area) at a point $\mathbf{r}$ on the plane is

$$I(\mathbf{r}_0;\lambda,\mu) = \mathbf{S}(\mathbf{r})\cdot\hat{\mathbf{r}}_0 \tag{45}$$

The Poynting vector $\mathbf{S}(\mathbf{r})$ is evaluated at point $\mathbf{r}$ on the plane, with $\mathbf{r}$ written as in (41). The intensity depends on the position of the image plane, represented by $\mathbf{r}_0$. This position is determined by its angular location (θ_0,ϕ_0) and by its distance r_0 to the dipole. We shall write $q_0 = k_0 r_0$ for the dimensionless representation of this distance. The dependence on (λ,μ) then gives the intensity distribution over the plane.

With the Poynting vector given by (7) and (19), the intensity distribution can be evaluated immediately. We obtain (Shu et al., 2009)

$$I(\mathbf{r}_0;\lambda,\mu) = I_0\left(\frac{q_0}{q}\right)^3\left\{1-\frac{1}{2q^2}[(q_0\sin\theta_0 + \bar{\lambda}\cos\theta_0)^2 + \bar{\mu}^2] - \frac{1}{q_0 q}\left(1+\frac{1}{q^2}\right)\bar{\mu}\sin\theta_0\right\}, \tag{46}$$

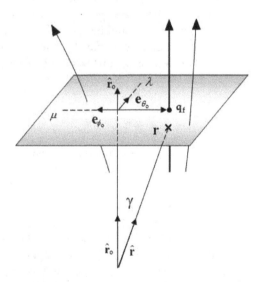

Fig. 4. The figure shows a 3D view of the image plane, the coordinate system, and several field lines. The bold field line is the field line that runs into the direction (θ_0,ϕ_0) and it intersects the image plane under 90° if the image plane is located in the far field. Angle γ is the angle under which the point $\mathbf{r}$ on the image plane is seen from the location of the dipole.

where we have set

$$q = \sqrt{q_o^2 + \overline{\lambda}^2 + \overline{\mu}^2} \, , \tag{47}$$

for the dimensionless distance between the point **r** in the image plane and the position of the dipole, and

$$I_o = \frac{3P_o}{8\pi r_o^2} \tag{48}$$

The overall factor $(q_o / q)^3$ has two contributions. A factor of $(q_o / q)^2$ comes from the fact that $\mathbf{S(r)}$ is proportional to $1/r^2$ and a factor of q_o / q comes from the dot product of $\mathbf{S(r)}$ with $\hat{\mathbf{r}}_o$ in (45). So this comes from the projection of $\mathbf{S(r)}$ onto the normal direction, and therefore this factor accounts for the fact that a field line crosses the image plane under an angle other than 90°. We see form Fig. 4 that $q_o / q = \cos\gamma$, with γ the angle under which the point **r** on the image plane is seen from the location of the dipole. An intensity distribution $I_o(q_o / q)^3$ would be a single peak at the origin of the image plane, and this peak is rotational symmetric around the normal vector $\hat{\mathbf{r}}_o$. The angular half-width at half maximum of the peak, as seen from the location of the dipole, would be 37°. The expression in braces in (46) accounts for the angular dependence of the emitted power and for the rotation of the field lines near the site of the dipole.

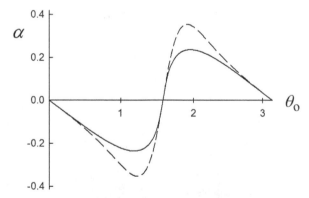

Fig. 5. The location of the peak of the intensity distribution is $(\overline{\lambda}_p, \overline{\mu}_p)$, and $\overline{\lambda}_p = \alpha q_o$. The solid line in the figure shows the dependence of α on θ_o, and the dashed line is an approximation (Shu et al., 2009).

It can be shown that the intensity distribution (46) has a single peak in the (λ, μ) plane, when the image plane is located in the far field ($q_o \gg 1$). Let $(\overline{\lambda}_p, \overline{\mu}_p)$ be the dimensionless coordinates of the position of the peak. If we write $\overline{\lambda}_p = \alpha q_o$, then α depends on the observation angle θ_o, and can be computed numerically. The result is shown in Fig. 5. Since $\overline{\lambda}_p$ scales with q_o, the location of the peak in the $\overline{\lambda}$ direction is a result of the angular distribution of the radiated power, and not of the rotation of the field lines near the dipole. More interesting is the location of the peak in the $\overline{\mu}$ direction. We obtain

$$\bar{\mu}_p = -(1+\alpha^2)^{3/2} \frac{2\sin\theta_o}{8(1+\alpha^2) - 5(\sin\theta_o + \alpha\cos\theta_o)^2} \tag{49}$$

This shift depends on θ_o, and this dependence is shown in Fig. 6. It is independent of q_o, and it is a direct result of the rotation of the field lines near the source. The shift is maximum for $\theta_o = \pi/2$, and the maximum shift is 2/3. Figure 7 shows the intensity distribution for this case.

The field line that runs into the (θ_o, ϕ_o) direction intersects the image plane at $\bar{\lambda}_f = 0$ and

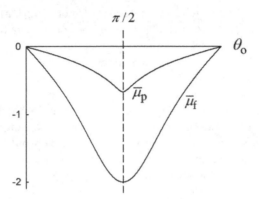

Fig. 6. The figure shows the dependence of $\bar{\mu}_p$ and $\bar{\mu}_f$ on θ_o.

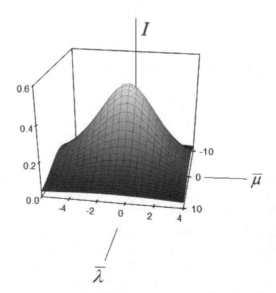

Fig. 7. The figure shows the intensity distribution over the image plane for $\theta_o = \pi/2$. The coordinates of the peak are $(\bar{\lambda}_p, \bar{\mu}_p) = (0, -2/3)$, and this represents the largest possible shift from the origin due to the rotation of the field lines.

$$\bar{\mu}_f = -\frac{2\sin\theta_o}{1+\cos^2\theta_o} \, , \tag{50}$$

according to (44). The value of $\bar{\mu}_f$ is also shown in Fig. 6, and we see that it does not coincide with the location of the maximum of the intensity distribution. The displacement of the field line is larger than the shift of the peak of the profile. So, the displacement is a good indicator of the position of the image, but due to subtle effects the position of the image is not exactly at the same location where the field line for this (θ_o,ϕ_o) direction intersects the image plane. Figure 8 illustrates how a bundle of field lines forms the image, rather than the field line for this direction only.

8. The difference profile

The shift of the peak in Fig. 7 is of the same order of magnitude as the spatial extent of the dipole vortex, and it is independent of the distance between the dipole and the observation plane, provided the image plane is in the far field. An experimental observation of this shift would confirm the existence of the vortex near the source. In this fashion, a property of the near field is observable through a measurement in the far field. However, the magnitude of the shift is less than a wavelength, and in the visible region of the spectrum, this is extremely small. Any observation of this shift would also require a very precise calibration of the experimental setup, since the shift is measured with respect to the origin O' of the image plane. Furthermore, the profile has a large background, as can be seen from Fig. 7, and the shape of the background depends on the observation angle θ_o .

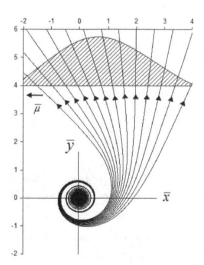

Fig. 8. The figure shows several field lines for a dipole that rotates counterclockwise in the xy plane. The image plane here is located at $q_o = 4$, and at $\theta_o = \phi_o = \pi / 2$, so perpendicular to the y axis. The field line in the (θ_o,ϕ_o) direction intersects the image plane at $\bar{x} = 2$ and the maximum of the intensity profile is located at $\bar{x} = 2/3$.

We see from Fig. 8 that the shift is due to the spiraling of the field lines. If we would reverse the rotation direction of the dipole, by reversing the helicity of the driving laser, the peak in the profile would move to $\bar{x} < 0$ in Fig. 8, and the background would remain approximately the same. When changing the direction of rotation of the dipole, expression (46) for the intensity remains the same, except that the term with $\bar{\mu}\sin\theta_o$ picks up a minus sign. We therefore introduce the difference profile ΔI as the intensity from (46) minus the same intensity for the radiation emitted by a dipole which rotates in the reverse direction. We then find (Li & Arnoldus, 2010a)

$$\Delta I(\mathbf{r}_o; \lambda, \mu) = -\frac{\zeta}{q^4}\left(1 + \frac{1}{q^2}\right)\bar{\mu}\sin\theta_o , \tag{51}$$

where

$$\zeta = \frac{3P_o k_o^2}{4\pi} \tag{52}$$

In (51), q is given by (47), so we see that ΔI depends on q_o, $\bar{\lambda}$ and $\bar{\mu}$. The dependence on the observation angle θ_o only enters through the overall factor $\sin\theta_o$. For a given θ_o, the

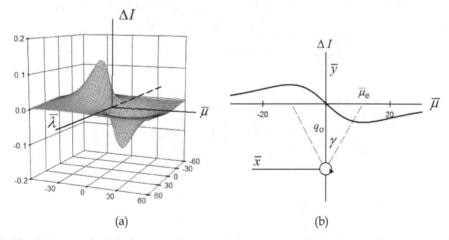

(a)   (b)

Fig. 9. The figure on the left shows a 3D view of the intensity difference profile for $q_o = 20$, and the figure on the right shows the profile as a function of $\bar{\mu}$. The extrema have an angular location of $\gamma = 30°$.

difference profile in the image plane is a function of $\bar{\lambda}$ and $\bar{\mu}$, with q_o fixed. It is easily verified from (51) that ΔI has two extrema on the $\bar{\mu}$ axis, and since ΔI is antisymmetric in $\bar{\mu}$, there is a maximum and a minimum. This difference profile is shown in Fig. 9 for $q_o = 20$. The coordinates of the extrema are

$$\bar{\mu}_e = \pm\frac{q_o}{\sqrt{3}} , \tag{53}$$

for $q_o \gg 1$, e.g., in the far field. The location of the peak in Fig. 7 is at $\bar{\mu} = -2/3$, and this is a displacement of about one-tenth of a wavelength with respect to O'. The extrema of the difference profile are proportional to q_o, and therefore these extrema are located at macroscopic distances from O'. As viewed from the dipole, they appear under an angle of $\gamma = 30°$, as follows from (53). Therefore, the extrema in the difference profile are a macroscopic feature of the intensity distribution. They appear in the far field due to the rotation of the field lines in the near field. Even though the vortex is of microscopic dimension, the location of the peaks in ΔI is macroscopic. This opens the possibility to observe the dipole vortex through a measurement in the far field. Such an experiment was performed recently (Haefner et al., 2009). The small particles were polystyrene spheres with a diameter of 4.6 μm, the laser light had a wavelength of 532 nm and the observation angle was $\theta_o = 90°$. The results of the experiment were in good agreement with Fig. 9(b).

9. Linear dipole in a medium

So far we have considered the radiation emitted by a dipole in free space. We shall now consider the effect of an embedding medium with relative permittivity ε_r and relative permeability μ_r, both of which are complex in general. When a plane electromagnetic wave travels through a material, say in the positive z direction, then the magnitude of the Poynting vector decays exponentially along the direction of propagation. This magnitude is proportional to $\exp(-2k_o z \operatorname{Im} n)$, so energy is dissipated when the imaginary part of the index of refraction is finite. Since the field lines of the Poynting vector are determined by the direction of $\mathbf{S}(\mathbf{r})$, and not its magnitude, the damping by the material does not affect the field lines of the Poynting vector (which are straight lines, parallel to the z axis, for this case). Let us now consider a linear dipole, oscillating along the z axis. In free space, the Poynting vector is given by (16), and the field lines are straight lines, coming out of the dipole, as shown in Fig. 1(a). One may expect that the result of damping by the medium is a diminished power flow in the radially outward direction, but with a field line picture that is the same as in Fig. 1(a). This reduced power flow was already split off in (7) as the factor $\exp(-2q \operatorname{Im} n)$. We shall now show for dipole radiation the result of the damping is far more dramatic.

When we set $\mathbf{u} = \mathbf{e}_z$ in (8), we obtain for the Poynting vector

$$\sigma(\mathbf{q}) = \hat{\mathbf{r}} \sin^2 \theta \operatorname{Re}\left[\frac{n}{\mu_r}\left(1 + \frac{i}{nq}\right)\right]$$

$$+ \frac{1}{|n|^2 q}\left|1 + \frac{i}{nq}\right|^2 [\hat{\mathbf{r}}(1 - 3\cos^2 \theta) + 2\mathbf{e}_z \cos\theta]\operatorname{Im}\varepsilon_r \qquad (54)$$

For a dipole in free space, this simplifies to $\sigma(\mathbf{q}) = \hat{\mathbf{r}} \sin^2 \theta$, giving field lines that run radially outward. When the dipole is embedded in a medium, this gets multiplied by $\operatorname{Re}[...]$. It can be shown that this factor is positive (unless n/μ_r is imaginary, which we shall not consider here), and therefore this would still give field lines that run radially outward. Furthermore, in the far field the second line in (54) vanishes, and we have $\sigma(\mathbf{q}) \approx \hat{\mathbf{r}} \sin^2 \theta \operatorname{Re}(n / \mu_r)$. So, in the far field, the field lines run approximately radially outward. When the imaginary part of ε_r is non-zero, the second line of (54) contributes to the Poynting vector. This term dominates in the near field, where q is small, and it has a part which contains $\mathbf{e}_z$. This part is responsible for a deviation of the field lines from the radially outgoing pattern. Since this part only contributes when $\operatorname{Im} \varepsilon_r \neq 0$, any deviation from the radial pattern is due to damping.

The vector field $\sigma(\mathbf{q})$ from (54) is rotationally symmetric around the z axis and reflection symmetric in the xy plane. Therefore we only need to consider the field lines in the first quadrant of the yz plane. In this quadrant, $\cos\theta$ is positive, and therefore the part of $\sigma(\mathbf{q})$ containing $\mathbf{e}_z$ is in the positive z direction. As a result, the field lines will bend away from the radial direction, and upward. The factor in square brackets in (54) is

$$\hat{\mathbf{r}}(1 - 3\cos^2 \theta) + 2\mathbf{e}_z \cos\theta = \sin\theta[\mathbf{e}_y (1 - 3\cos^2 \theta) + 3\mathbf{e}_z \sin\theta \cos\theta] \ , \qquad (55)$$

from which we see that the y component vanishes for $\cos\theta = 1 / \sqrt{3}$, so $\theta = 54.7°$. Therefore, when θ equals 54.7°, the part of the Poynting vector that contains $\operatorname{Im} \varepsilon_r$ is in the positive z direction. Consequently, the field lines in the near field cross the line $\theta = 54.7°$ in a vertical, upward direction. Furthermore, for field points with a smaller angle θ, the y component of $\sigma(\mathbf{q})$ is negative, and this means that the field line through such a point is headed towards the z axis. At the z axis we have $\theta = 0$, and it follows from (55) that the z component vanishes, relative to the y component, and therefore each field line in the near field approaches the z axis under 90°. The resulting field line pattern is illustrated in Fig. 10.

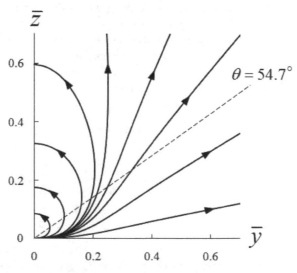

Fig. 10. The figure shows field lines of the Poynting vector for a dipole oscillating along the z axis and embedded in a material with $\varepsilon_r = 1.7 + 0.06i$ and $\mu_r = 1$. These are the values for water at 3 μm. The index of refraction is $n = 1.3 + 0.023i$.

Interestingly, we see that very close to the dipole some field lines form semiloops. The energy that flows along such a semiloop comes out of the dipole and is then entirely dissipated by the material. The field line ends at the z axis. A more careful analysis (Li et al., 2011a) shows that all field lines start off horizontally, so along the xy plane. Therefore, all energy is initially emitted along the xy plane. Some field lines form semiloops and some run to the far field where they eventually run approximately in the radial direction. This situation is in sharp contrast to the emission in free space, where the energy is emitted in all directions (except along the z axis), as shown in Fig. 1(a). Another remarkable difference is the near the z axis all field lines approach the z axis under $90°$, whereas for emission in free space the field lines near the z axis run parallel to the z axis.

When a linear dipole is embedded in a medium, the field lines of energy flow are curves, rather than straight lines, when the imaginary part of the relative permittivity is finite. Because of the damping, the energy flow is redistributed in the material. The effect of the dissipation is not only a weakening of the power transported along a field line, as for a plane wave, but the absorption during propagation results in a dramatic change in the direction of power flow in the near field.

10. Circular dipole in a medium

Let us now consider a rotating dipole moment, embedded in a medium. With $\mathbf{u}$ given by (17), the Poynting vector (8) becomes

$$\boldsymbol{\sigma}(\mathbf{q}) = \left(1 - \tfrac{1}{2}\sin^2\theta\right) \mathrm{Re}\left[\frac{n}{\mu_r}\left(1 + \frac{i}{nq}\right)\right]\hat{\mathbf{r}}$$

$$+ \frac{1}{|n|^2 q}\left|1 + \frac{i}{nq}\right|^2 \left\{\left[\left(1 - \tfrac{1}{2}\sin^2\theta\right)\hat{\mathbf{r}} + \tfrac{1}{2}\sin(2\theta)\mathbf{e}_\theta\right]\mathrm{Im}\,\varepsilon_r + \mathbf{e}_\phi(\sin\theta)\mathrm{Re}\,\varepsilon_r\right\} \qquad (56)$$

The right-hand side of the first line of (56) is proportional to $\hat{\mathbf{r}}$. This is the leading term in the far field, so far from the dipole the field lines run approximately in the radially outward direction. The term on the second line of (56) has a part proportional to $\mathbf{e}_\phi(\sin\theta)\mathrm{Re}\,\varepsilon_r$, and if the imaginary part of ε_r is zero, this is the only additional term. In that case, the Poynting vector has no $\mathbf{e}_\theta$ component, and therefore θ is constant along a field line, according to (12). Consequently, a field line lies on a cone around the z axis, as in Fig. 1(b) for $\varepsilon_r = 1$. When there is damping in the material due to the imaginary part of ε_r, the vector $\boldsymbol{\sigma}(\mathbf{q})$ has an $\mathbf{e}_\theta$ component. This will lead to a redirection of the field lines, and hence the field lines will not lie on a cone anymore. In other words, the flow of energy will be redistributed due to the damping, just like in the previous section.

The field line equations (11)-(13) become with (56)

$$\frac{dq}{du} = g(q,\theta) \ , \qquad (57)$$

$$\frac{d\theta}{du} = \frac{1}{2|n|^2 q^2}\left|1 + \frac{i}{nq}\right|^2 \sin(2\theta)\mathrm{Im}\,\varepsilon_r \ , \qquad (58)$$

$$\frac{d\phi}{du} = \frac{1}{|n|^2 q^2}\left|1+\frac{i}{nq}\right|^2 \operatorname{Re}\varepsilon_r ,$$ (59)

with

$$g(q,\theta) = \left(1-\tfrac{1}{2}\sin^2\theta\right)\left\{\operatorname{Re}\left[\frac{n}{\mu_r}\left(1+\frac{i}{nq}\right)\right]+\frac{1}{|n|^2 q}\left|1+\frac{i}{nq}\right|^2\operatorname{Im}\varepsilon_r\right\}$$ (60)

A field line runs into the direction of increasing u. Since $\operatorname{Im}\varepsilon_r \ge 0$, the function $g(q,\theta)$ is positive, and it then follows from (57) that q increases without bounds along a field line. Therefore, field lines start at the dipole, and run away to the far field. For the linear dipole, some field lines form semiloops, as seen in Fig. 10, but for a circular dipole this can not happen. For q large, the right-hand sides of (58) and (59) vanish, and therefore angles θ and ϕ for points on a field line approach the constant values θ_o and ϕ_o, just like in section 4 for the circular dipole in free space. Therefore, field lines approach asymptotically a straight line ℓ, as illustrated in Fig. 11.

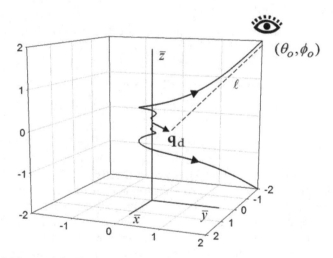

Fig. 11. Two field lines of the Poynting vector for a rotating dipole moment, embedded in a material with $\varepsilon_r =1+0.2i$ and $\mu_r =1$. Field lines approach asymptotically a straight line, and this gives rise to a virtual displacement $\mathbf{q}_d$ of the source.

10.1 Rotation of the field lines

Angle ϕ is the angle around the z axis, and we see from (59) that $d\phi/du$ is positive when $\operatorname{Re}\varepsilon_r$ is positive. Then ϕ increases along a field line, and the field line swirls around the z axis in the counterclockwise direction when viewed down the z axis. However, when $\operatorname{Re}\varepsilon_r$ is negative, angle ϕ decreases along a field line, and the field line rotates clockwise around the z axis. In this case, the direction of rotation of the field lines is opposite to the direction of rotation of the dipole moment. A material with $\operatorname{Re}\varepsilon_r <0$ (and $\mu_r =1$) is metallic, and the index of refraction is approximately imaginary. We see from (7) that this gives a very strong damping in the medium, and therefore hardly any radiation will be emitted. However, for materials known as double negative metamaterials the real part of the permeability is also negative, and then the index of refraction becomes approximately real (and negative). In that case the material is transparent, and the dipole radiation can propagate away from the site of the source with little damping. Among the many peculiar features of these metamaterials, this reversal of rotation of the dipole vortex is certainly noteworthy (Li & Arnoldus, 2010c).

Because q increases monotonically along a field line, we can consider q as the independent variable rather than u. We then find from (57)-(59)

$$\frac{d\theta}{dq} = \frac{1}{g(q,\theta)} \frac{1}{2|n|^2 q^2} \left|1 + \frac{i}{nq}\right|^2 \sin(2\theta) \operatorname{Im}\varepsilon_r \,, \tag{61}$$

$$\frac{d\phi}{dq} = \frac{1}{g(q,\theta)} \frac{1}{|n|^2 q^2} \left|1 + \frac{i}{nq}\right|^2 \operatorname{Re}\varepsilon_r \tag{62}$$

Let us now consider the solution for q small. For $\operatorname{Im}\varepsilon_r = 0$ we see from (61) that θ is constant, and so a field line lies on a cone. Then we expand the right-hand side of (62) for q small, and integrate. This yields

$$\phi(q) = \begin{cases} O\left(1/q^3\right) & , \quad \operatorname{Im}\mu_r = 0 \\ O\left(1/q^2\right) & , \quad \operatorname{Im}\mu_r \neq 0 \end{cases} \tag{63}$$

For $q \to 0$, the value of $\phi(q)$ goes to ∞ or $-\infty$ very quickly, and this leads to a large number of rotations of a field line around the z axis, as can be seen in Fig. 1(b). For $\operatorname{Im}\varepsilon_r \neq 0$ we find in a similar way

$$\phi(q) = \begin{cases} \dfrac{2}{\alpha}\ln q + O(1) & , \quad z = 0 \\ \dfrac{1}{\alpha}\ln q + O(1) & , \quad z \neq 0 \end{cases} \tag{64}$$

where we have set

$$\alpha = \frac{\operatorname{Im}\varepsilon_r}{\operatorname{Re}\varepsilon_r} \tag{65}$$

For $\operatorname{Im}\varepsilon_r \neq 0$, the approach of $\phi(q)$ to $\pm\infty$ is logarithmic, so much slower than for the case $\operatorname{Im}\varepsilon_r = 0$. Consequently, the windings around the z axis are much thinner than in Fig. 1(b). Due to the damping, it appears as if the field lines wind around the z axis only a few times, as can be seen in Figs. 11 and 12.

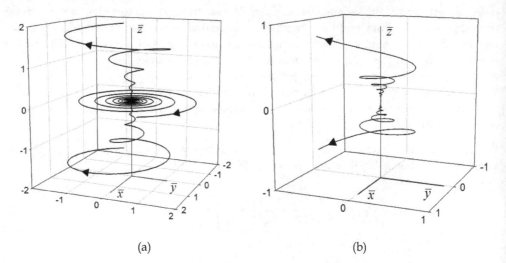

(a)   (b)

Fig. 12. The figure on the left shows two field lines of the Poynting vector for $\varepsilon_r = -1 + 0.1i$ and $\mu_r = 1$. We see that the direction of rotation is reversed, as compared to the direction of rotation in Fig. 1(b). The figure on the right shows two field lines of the Poynting vector for $\varepsilon_r = 1 + 0.07i$ and $\mu_r = 1$. The field lines swirl around on a funnel surface, rather than a cone.

10.2 The funnel vortex

When $\mathrm{Im}\,\varepsilon_r = 0$, field lines lie on a cone, as in Fig. 1(b), and they are very dense near the location of the dipole. We see from Figs. 11 and 12 that when $\mathrm{Im}\,\varepsilon_r \neq 0$ the field lines are not only less dense near the source, but they also do not lie on a cone anymore. We see from the figures that due to the damping the field lines now lie on a funnel-shape surface. It follows from (62) that ϕ increases or decreases monotonically along a field line. We shall now consider ϕ as the independent parameter, and we consider θ to be a function of ϕ. From (61) and (62) we then find

$$\frac{d\theta}{d\phi} = \tfrac{1}{2}\alpha\sin(2\theta) . \tag{66}$$

The solution of this equation is

$$\tan\theta = e^{\alpha(\phi-\phi_i)}\tan\theta_i , \tag{67}$$

with (θ_i, ϕ_i) coordinates of the initial point. We now consider again the behavior of a field line for $q \to 0$, when $\mathrm{Im}\,\varepsilon_r \neq 0$. It follows from (64) that $\phi \to -\infty$ for $\alpha > 0$, and to ∞ for $\alpha < 0$, in the limit $q \to 0$. Therefore, $\alpha\phi \to -\infty$ in both cases, and so $\tan\theta \to 0$ for $q \to 0$. If the initial point is in the region $z > 0$, so $0 < \theta_i < \pi/2$, this implies that $\theta \to 0$ for $q \to 0$. Similarly, if $\pi/2 < \theta_i < \pi$ we have $\theta \to \pi$. For an initial point in the xy plane we have $\theta_i = \pi/2$, and it follows from (58) that then $\theta = \pi/2$ for all points along the field line.

Therefore, the field line lies in the xy plane. It follows from (64) and (67) that $\tan\theta = O(q)$ for $\theta_i \neq \pi / 2$. Consequently, we have for $q \to 0$

$$\theta(q) = \begin{cases} O(q) & , \quad z > 0 \\ \pi / 2 & , \quad z = 0 \\ \pi + O(q) & , \quad z < 0 \end{cases} \tag{68}$$

It follows from (68) that $\theta = 0$, $\theta = \pi / 2$ or $\theta = \pi$ for $q \to 0$. Because the radiated energy is emitted along a field line, we come to the remarkable conclusion that due to the damping all energy is emitted along the z axis or along the xy plane, and this is illustrated in Fig. 12(a). This is in sharp contrast to the situation for $\mathrm{Im}\,\varepsilon_r = 0$, because then field lines lie on any cone around the z axis. In that case, the energy is emitted into the direction θ_o, with θ_o the angle of the cone.

10.3 The displacement

As shown in Fig. 11, a field line approaches asymptotically a straight line ℓ in the far field. The field line runs in the (θ_o, ϕ_o) direction, but the line ℓ does not go through the origin of coordinates. In order to find the line ℓ, we expand the right-hand sides of (61) and (62) in a series in $1 / q$, and integrate the result. We then obtain

$$\theta(q) = \theta_o - Z(\theta_o)\frac{\sin(2\theta_o)}{2q}\mathrm{Im}\,\varepsilon_r + \dots \,, \tag{69}$$

$$\phi(q) = \phi_o - Z(\theta_o)\frac{1}{q}\mathrm{Re}\,\varepsilon_r + \dots \,, \tag{70}$$

where

$$Z(\theta_o) = \frac{1}{|n|^2\,\mathrm{Re}(n / \mu_r)\left(1 - \tfrac{1}{2}\sin^2\theta_o\right)} \tag{71}$$

Along similar steps as in section 5, we now find

$$\ell: \mathbf{q} = t\hat{\mathbf{r}}_o + \mathbf{q}_f \,, \tag{72}$$

with

$$\mathbf{q}_f = -\sin\theta_o Z(\theta_o)[\mathbf{e}_{\phi_o}\,\mathrm{Re}(\varepsilon_r) + \mathbf{e}_{\theta_o}\cos\theta_o\,\mathrm{Im}(\varepsilon_r)] \tag{73}$$

Vector $\mathbf{q}_f$ is perpendicular to $\hat{\mathbf{r}}_o$, so it is a vector in the image plane. It represents the displacement in the far field, and it corresponds to the intersection of ℓ with the image plane, as in Fig. 3. For a dipole in free space, the virtual displacement in the xy plane, $\mathbf{q}_d$, is the same as the displacement $\mathbf{q}_f$ of a field line in the far field. With damping, this is not the case anymore. The intersection of ℓ with the xy plane is now found to be

$$\mathbf{q}_d = -\sin\theta_o Z(\theta_o)[\mathbf{e}_{\phi_o}\,\mathrm{Re}(\varepsilon_r) + \mathbf{e}_{\rho_o}\,\mathrm{Im}(\varepsilon_r)] \quad, \tag{74}$$

with

$$\mathbf{e}_{\rho_o} = \mathbf{e}_x \cos\phi_o + \mathbf{e}_y \sin\phi_o \quad, \tag{75}$$

the radially outward unit vector in the xy plane. The magnitude of the displacement is

$$q_d = \frac{2\sin\theta_o}{|\mu_r|\,\mathrm{Re}(n/\mu_r)(1+\cos^2\theta_o)} \quad, \tag{76}$$

which generalizes (38). In free space, the maximum value of q_d is 2, and this occurs for $\theta_o = \pi/2$. For a dipole embedded in a material, the factor $\mathrm{Re}(n/\mu_r)$ in the denominator can become small, and this would give a large q_d. This could happen, for instance, for $\mu_r = 1$ and ε_r approximately negative. Then n is approximately imaginary, and the displacement is very large.

11. Linear dipole near a mirror

We have studied the field lines of energy flow for an electric dipole with a linear and a rotating dipole moment, both in free space and in an embedding medium. In this section we shall consider the effect of the presence of a boundary. Let the dipole be located on the z axis, a distance H above a mirror. The surface of the mirror is taken as the xy plane. We shall consider a linear dipole, oscillating under an angle γ with the z axis, so

$$\mathbf{u} = \mathbf{e}_y \sin\gamma + \mathbf{e}_z \cos\gamma \tag{77}$$

In the region $z > 0$ the field is identical to the field of the dipole plus the field of its mirror image (which is the reflected field at the interface). The mirror dipole is located a distance H below the xy plane, and it has a dipole moment represented by

$$\mathbf{u}^{im} = -\mathbf{e}_y \sin\gamma + \mathbf{e}_z \cos\gamma \tag{78}$$

The setup is shown in Fig. 13. Let $\mathbf{r}_1$ be the location of a field point, measured from the location of the dipole, and $\mathbf{r}_2$ be the same field point but measured from the image dipole. The electric and magnetic field amplitudes are then given by (4) and (5) with $\mathbf{r}$ replaced by $\mathbf{r}_1$, and for the amplitudes of the image fields we replace $\mathbf{r}$ by $\mathbf{r}_2$ and $\mathbf{u}$ by $\mathbf{u}^{im}$. The Poynting vector can then be constructed, and field lines can be drawn with the method outlined in section 3 (Li & Arnoldus, 2010b). Figure 14 shows the field lines for a perpendicular ($\gamma = 0$) and a parallel ($\gamma = \pi/2$) dipole, both for $h = k_o H = 2\pi$. For these cases, the field line patterns are rotationally symmetric around the z axis. Without the interface, the field lines are straight, as in Fig. 1(a). For the perpendicular dipole, Fig. 14(a), the field lines are more or less straight near the dipole. The field lines that run towards the mirror bend at the mirror, and then run off more or less horizontally. For the parallel dipole, Fig. 14(b), the pattern is much more complex. We see that singularities appear, which are indicated by small white circles. At such a singularity the Poynting vector vanishes. We see from the figure that several field lines end at a singularity, and other field lines bend such as to avoid the singularity. The case of $\gamma = 45°$ is shown in Fig. 15. The field line pattern is not rotationally symmetric around the z axis anymore. At the right of the z axis, the pattern is similar to Fig. 14(a). On the left of the z axis we see numerous singularities, and we now also observe the appearance of three optical vortices in the energy flow pattern. The points labeled a, b and c in the figure are singularities at the centers of the vortices. At the other singularities, except for d, field lines abruptly change direction. It is also interesting to see that there are field lines which start at point a and end at point b. These field lines represent a local energy flow where the energy does not directly originate from the location of the dipole. At point e, some field lines seem to collide, and this leads to a singularity. At singularities f, g and h, field lines split in two directions.

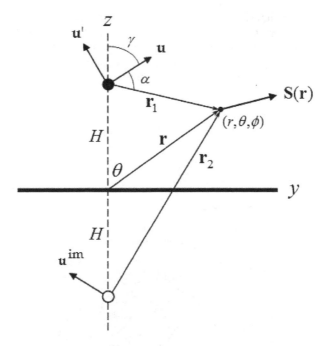

Fig. 13. Setup of the dipole near the mirror.

The vortices and singularities in Figs. (14) and (15) are a result of interference between the radiation that is emitted by the dipole and the radiation that is reflected off the surface. Close to the dipole, the field that is emitted directly by the dipole is much stronger than the reflected field, and it may seem that therefore this field dominates the energy flow pattern.

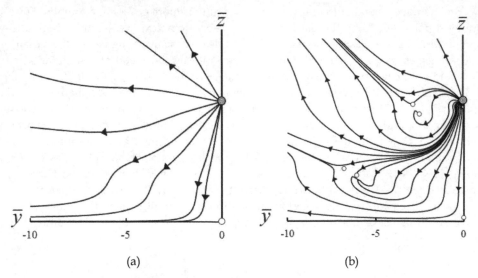

(a)   (b)

Fig. 14. The figure on the left shows the field line pattern for a dipole oscillating perpendicular the the surface. Field lines that are headed towards the surface bend when they come close to the surface. For the figure on the right, the dipole oscillates parallel to the surface, and we see the appearance of singularities. The distance between the dipole and the interface is one wavelength for both figures ($h = 2\pi$).

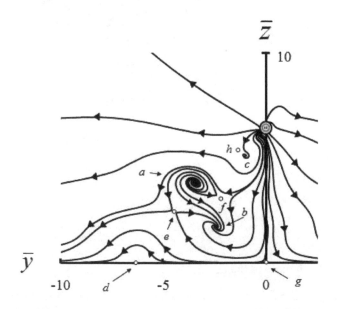

Fig. 15. Field line pattern for a dipole oscillating under 45° with the z axis, and for $h = 2\pi$. Numerous singularities are present, and three vortices appear.

In other words, we may expect that close to the dipole the field lines are straight lines, as in Fig. 1(a), and that away from the dipole interference sets in, leading to complicated flow patterns as in Figs. (14) and (15). We shall now show that this is not the case. In order to study the emission of the radiation, we consider a region close to the dipole. We assume $q_1 \ll 1$ and $q_1 \ll h$. This means that we consider field points that are close to the dipole, compared to a wavelength, and we assume that the distance between the mirror and the dipole is much larger than the distance between the dipole and the field point. The expression for the Poynting vector can be expanded in a series in q_1, and we obtain

$$\sigma(\mathbf{q}) = \frac{\sin\gamma}{q_1} v(h)[(3\cos^2\alpha - 1)\mathbf{u}' - 3\cos\alpha(\hat{\mathbf{q}}_1 \cdot \mathbf{u}')\mathbf{u}] + \hat{\mathbf{q}}_1 \sin^2\alpha + O(1) \tag{79}$$

Here, α is the angle between $\mathbf{u}$ and $\hat{\mathbf{q}}_1$, vector $\mathbf{u}'$ is defined as

$$\mathbf{u}' = -\mathbf{e}_y \cos\gamma + \mathbf{e}_z \sin\gamma , \tag{80}$$

and

$$v(h) = \frac{1}{2h}\left[\frac{\sin(2h)}{2h} - \cos(2h)\right] \tag{81}$$

Without the mirror, this would be $\sigma(\mathbf{q}) = \hat{\mathbf{q}}_1 \sin^2\alpha$, and this is exact at all distances. The corresponding field lines are straight, as in Fig. 1(b). Due to interference, the first term in

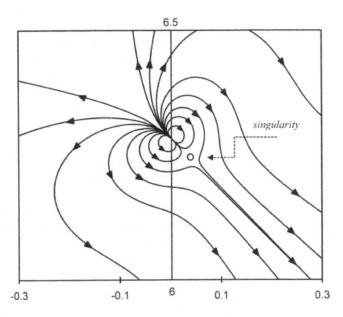

Fig. 16. Field line pattern close to the dipole for $\gamma = 45°$ and $h = 2\pi$.

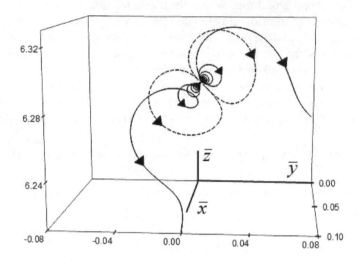

Fig. 17. Field line pattern close to the dipole for $\gamma = 45°$ and $h = 2\pi$, in 3D. It appears that the radiation is emitted as a set of four vortices.

(79) appears, and this term is proportional to $1/q_1$. Close to the dipole, this term dominates. The field line picture close to the dipole is shown in Fig. 16. We see that some field lines are closed loops, and a singularity appears very close to the dipole. When we graph the field lines in 3D, we obtain the result shown in Fig. 17. In front of the yz plane, there are two vortices which come out of the dipole. Two other vortices are behind the yz plane, but these are not shown in the figure. The dashed curves in the figure are closed loops from Fig. 16 in the yz plane. Therefore, radiation is emitted as a set of four vortices, and this is a result of interference between the directly emitted radiation and the reflected field by the mirror. These vortices are present no matter the distance between the dipole and the mirror, but the spatial extent of the vortices diminishes with distance.

12. Conclusions

Energy in an electromagnetic field flows along the field lines of the Poynting vector. We have considered the radiation emitted by an electric dipole. For a linear dipole in free space, the field lines are straight lines coming out of the dipole. When the dipole moment rotates in the xy plane, the field lines are curves that lie on cones around the z axis. Close to the source, the field lines wind around the z axis a large number of times, and at larger distances, in the far field, the field lines approach asymptotically straight lines. Such a line is displaced, as compared to a line that would start at the location of the dipole, and this gives a virtual displacement of the position of the source in the xy plane. Also in the far field, this line is displaced as compared with the radially outward direction, and this gives rise to a shift of the image of the dipole when projected onto an observation plane. When the dipole is embedded in a medium, the damping in the material gives rise to a redistribution of the power flow, as compared to emission in free space. For a linear dipole, the field lines are not straight anymore, and for a circular dipole a field line has a funnel appearance, rather than lying on a cone. Also the spacing of the field lines near the source becomes much less dense due to the damping. It is also shown that when the real part of the permittivity is negative, the field lines reverse in rotation direction, and the flow of energy counterrotates the rotation direction of the dipole. The virtual displacement of the location of the source and the displacement of the image in the far field can be much larger than in free space. When the dipole is located near the surface of a mirror, a host of new phenomena appear due to the interference of the dipole radiation with the reflected radiation by the mirror. In the flow field, singularities and vortices appear in the neighborhood of the dipole and in between the dipole and the mirror. For a linear dipole in free space, the field lines come straight out of the dipole, but when this dipole is located near a mirror, the mechanism of emission is drastically altered. In the oscillation plane of the dipole, all radiation is emitted in a direction perpendicular to the dipole moment, and some field lines form closed loops. Energy flowing along these field lines returns to the dipole. For emission off this plane, the radiation is emitted as a set of four vortices, emanating from the dipole.

13. References

Arnoldus, H. F. & Foley, J. T. (2004), The dipole vortex, *Optics Communications*, Vol. 231, No. 1, (Feb. 2004) (115-128), ISSN: 0030-4018.

Born, M. & Wolf, E. (1980). *Principles of Optics* (6th edition), Pergamon, ISBN: 0-08-026482-6, Oxford, Ch. 3.

Haefner, D., Sukhov, S. & Dogariu, A. (2009), Spin-Hall effect of light in spherical geometry, *Physical Review Letters*, Vol. 102, No. 12, (Mar. 2009) (123903-1-123903-4), ISSN: 0031-9007.

Jackson, J. D. (1998). *Classical Electrodynamics* (3rd edition), Wiley, ISBN: 0-471-30932-X, New York, p. 265.

Li, X. & Arnoldus, H. F. (2010a), Macroscopic far-field observation of the sub-wavelength near-field dipole vortex, *Physics Letters A*, Vol. 374, No. 8, (Feb. 2010) (1063-1067), ISSN: 0375-9601.

Li, X. & Arnoldus, H. F. (2010b), Electric dipole radiation near a mirror, *Physical Review A*, Vol. 81, No. 5, (May 2010) (053844-1-053844-10), ISSN: 1050-2947.

Li, X. & Arnoldus, H. F. (2010c), Reversal of the dipole vortex in a negative index of refraction material, *Physics Letters A*, Vol. 374, No. 43, (Sep. 2010) (4479-4482), ISSN: 0375-9601.

Li, X., Pierce, D. M. & Arnoldus, H. F. (2011a), Redistribution of energy flow in a material due to damping, *Optics Letters*, Vol. 36, No. 3, (Feb. 2011) (349-351), ISSN: 0146-9592.

Li, X., Pierce, D. M. & Arnoldus, H. F. (2011b), Damping of the dipole vortex, *Journal of the Optical Society of America A*, Vol. 28, No. 5, (May 2011) (778-785), ISSN: 1084-7529.

McCall, M. W., Lakhtakia, A. & Weiglhofer, W. S. (2002), The negative index of refraction demystified, *European Journal of Physics*, Vol. 23, No. 3, (May 2002) (353-359), ISSN: 0143-0807.

Shu, J., Li, X. & Arnoldus, H. F. (2008), Energy flow lines for the radiation emitted by a dipole, *Journal of Modern Optics*, Vol. 55, No. 15, (Sep. 2008) (2457-2471), ISSN: 0950-0340.

Shu, J., Li, X. & Arnoldus, H. F. (2009), Nanoscale shift of the intensity distribution of dipole
      radiation, *Journal of the Optical Society of America A*, Vol. 26, No. 2, (Feb. 2009) (395-
      402), ISSN: 1084-7529.

4

Retarded Electromagnetic Interaction and Symmetry Violation of Time Reversal in High Order Stimulated Radiation and Absorption Processes of Lights as Well as Nonlinear Optics – Influence on Fundamental Theory of Laser and Non-Equilibrium Statistical Physics

Mei Xiaochun

Institute of Innovative Physics, Department of Physics, Fuzhou University, Fuzhou, China

1. Introduction

Based on the perturbation method of quantum mechanics and retarded electromagnetic interaction, it is proved that the transition probabilities of light's high order stimulated radiation and absorption are not the same [1]. It indicates that the processes of light's stimulated radiation and absorption as well as nonlinear optics violate time reversal symmetry actually, although the motion equation of quantum mechanics and the interaction Hamiltonian are still invariable. This result can be used to solve the famous irreversibility paradox in the evolution processes of macro-systems which has puzzled physics community for a long time.

Einstein put forward the theory of light's stimulated radiation and absorption in 1917 in order to explain the Planck blackbody radiation formula based on equilibrium theory. According to the Einstein's theory, the parameters of stimulated radiation and absorption are equal to each other with $B_{ml} = B_{lm}$. The same result can also be obtained by means of the calculation of quantum mechanics for the first order process under dipole approximation without considering the retarded interaction (or multiple moment effect) of radiation fields [1]. Because light's stimulated radiation process can be regarded as the time reversal of stimulated absorption process, the result means that light's stimulated radiation and absorption processes have time reversal symmetry.

Nonlinear optics was developed in the 1960's since laser was invented. Also by the dipole approximation without considering the retarded interaction of radiation fields, nonlinear susceptibilities in nonlinear optics are still invariable under time reversal [2]. So the processes of light's radiation and absorption as well as nonlinear optics are considered to be time reversal symmetry at present. In fact, it is a common and wide accepted idea at present that all micro-processes controlled by electromagnetic interaction are symmetrical under time reversal, for the motion equations of quantum mechanics and the Hamiltonian of electromagnetic interaction are unchanged under time reversal.

However, most processes related to laser and nonlinear optics are actually high non-equilibrium ones. As we known that time reversal symmetry will generally be violated in non-equilibrium processes. It is proved in this paper that after the retarded effect of radiation fields is taken into account, the time reversal symmetry will be violated in light's high order stimulated radiation and absorption processes with $B_{ml} \neq B_{lm}$, although the Hamiltonian of electromagnetic interaction is still unchanged under time reversal.

The transition probability of third order process is calculated and the revised formula of nonlinear optics polarizability is deduced in this paper. Many phenomena of time reversal symmetry violation in non-linear optics just as sum frequency, double frequency, different frequencies, double stable states, self-focusing and self-defocusing, echo phenomena, as well as optical self-transparence and self absorptions and so on are analyzed.

The reason to cause symmetry violation is that some filial or partial transition processes of bounding state atoms are forbidden or can't be achieved due to the law of energy conservation. These restrictions can cause the symmetry violation of time reversal of other partial transition processes which can be actualized really. These realizable filial or partial processes which violate time reversal symmetry generally are just the practically observed physical processes. The symmetry violation is also relative to the initial state's asymmetries of bounding atoms before and after time reversal. For the electromagnetic interaction between non-bounding atoms and radiation fields, there is no this kind of symmetry violation of time reversal. For example, in the experiments of particle physics in accelerators, we can not observe the symmetry violation of time reversal.

At last, the influences of symmetry violation of time reversal on the foundation theory of laser and non-equilibrium statistical physics are discussed. The phenomena of producing laser without the reversion of particle population and transition without radiation can be well explained. The result indicates that the irreversibility of evolution processes of macro-systems originates from the irreversibility of micro-processes. The irreversibility paradox can be eliminated thoroughly. By introducing retarded electromagnetic interaction, the forces between classical changed particles are not conservative ones. Based on them, we can establish the revised Liouville equation which is irreversible under time reversal. In this way, we can lay a really rational dynamic foundation for classical non-equilibrium statistical mechanics.

2. The transition probability of the first order process

For simplification, we consider an atom with an electron in its external layer. Electron's mass is μ, charge is q. When there is no external interaction, the Hamiltonian and the wave function of electron are individually

$$\hat{H}_0 = -\frac{\hbar^2}{2m}\nabla^2 + \hat{U}(r) \qquad |\psi_0\rangle = \sum_n e^{-\frac{1}{\hbar}E_n t}|n\rangle \tag{1}$$

After external electromagnetic field is introduced, the interaction Hamiltonian is

$$\hat{H}' = -\frac{q}{c\mu}\hat{A}\cdot\hat{p} + \frac{q^2}{2c^2\mu}\hat{A}^2 + q\phi \tag{2}$$

Because the charge and current densities of radiation field are zero, we can take the gauge condition $\nabla\cdot\hat{A} = 0$ and $\phi = 0$ and write $\hat{H}' = \hat{H}'_1 + \hat{H}'_2$ with

$$\hat{H}'_1 = -\frac{q}{c\mu}\hat{A}\cdot\hat{p} \qquad \hat{H}'_2 = \frac{q^2}{2c^2\mu}\hat{A}^2 \tag{3}$$

Where $\hat{H}'_1$ has the order v/c and $\hat{H}'_2$ has the order v^2/c^2. In the current discussion for light's stimulated radiation and absorption theory, $\hat{H}'_2$ is neglected generally. Because $\hat{H}'_2$ has the same order of magnitude as the second order effects of nonlinear optics, it is remained in the paper. Suppose that electromagnetic wave propagates along $\bar{k}$ direction. Electric field strength is $\bar{E} = \bar{E}_0\sin(\omega t - \bar{k}\cdot\bar{R})$. Here $\bar{R}$ is a direction vector pointing from wave source to observation point. Both $\hat{H}'_1$ and $\hat{H}'_2$ can also be written as

$$\hat{H}'_1 = -\frac{q\bar{E}_0}{2\omega\mu}\left[e^{i(\omega t - \bar{k}\cdot\bar{R})} + e^{-i(\omega t - \bar{k}\cdot\bar{R})}\right]\hat{p}$$

$$\hat{H}'_2 = \frac{q^2\bar{E}_0^2}{2c^2k^2\mu}\left[e^{i(\omega t - \bar{k}\cdot\bar{R})} + e^{-i(\omega t - \bar{k}\cdot\bar{R})}\right]^2 = \frac{q^2\bar{E}_0^2}{2\omega^2\mu}\left[e^{i2(\omega t - \bar{k}\cdot\bar{R})} + e^{-i2(\omega t - \bar{k}\cdot\bar{R})} + 2\right] \tag{4}$$

we can write

$$\hat{H}'_1 = \hat{F}_1 e^{i\omega t} + \hat{F}_1^+ e^{-i\omega t}$$

in which

$$\hat{F}_1 = -\frac{q\bar{E}_0}{2\omega\mu}\cdot e^{-i\bar{k}\cdot\bar{R}}\hat{p} \qquad \hat{F}_1^+ = -\frac{q\bar{E}_0}{2\omega\mu}\cdot e^{i\bar{k}\cdot\bar{R}}\hat{p} \tag{5}$$

Because we always have $\bar{k}\perp\bar{E}_0$ for electromagnetic wave, we have the commutation relation

$$\left[\bar{k}\cdot\bar{R},\ \bar{E}_0\cdot\hat{p}\right] = i\hbar\bar{k}\cdot\bar{E}_0 = 0 \tag{6}$$

So it can be proved that $\bar{E}_0\cdot\hat{p}$ and $\exp(i\bar{k}\cdot\bar{R})$ are also commutative. In this way, (6) can also be written as

$$\hat{F}_1 = -\frac{q\bar{E}_0}{2\omega\mu}\cdot\hat{p}e^{-i\bar{k}\cdot\bar{R}} \qquad \hat{F}_1^+ = -\frac{q\bar{E}_0}{2\omega\mu}\cdot\hat{p}e^{i\bar{k}\cdot\bar{R}} \tag{7}$$

By using perturbation method in quantum mechanics to regard $\hat{H}'$ as perturbation, we write the motion equation and wave function of system as

$$i\hbar\frac{\partial}{\partial t}|\psi\rangle = \left(\hat{H}_0 + \hat{H}'\right)|\psi\rangle \qquad |\psi\rangle = \sum_m a_m(t)e^{-\frac{i}{\hbar}E_m t}|n\rangle \qquad (8)$$

Let

$$a_m(t) = a_m^{(0)}(t) + a_m^{(1)}(t) + a_m^{(2)}(t) + \cdots, \quad E_m - E_n = \hbar\omega_{mn},$$

Substituting them in the motion equation, we can get

$$i\hbar\frac{d}{dt}\left[a_m^{(0)}(t) + a_m^{(1)}(t) + a_m^{(2)}(t) + \cdots\right] = \sum_n \left(\hat{H}'_{1mn} + \hat{H}'_{2mn}\right)\left[a_n^{(0)}(t) + a_n^{(1)}(t) + a_n^{(2)}(t) + \cdots\right]e^{i\omega_{mn}t} \quad (9)$$

The items with same order on the two sides of the equation are taken to be equal to each other. The first four equations are

$$i\hbar\frac{d}{dt}a_m^{(0)}(t) = 0 \quad i\hbar\frac{d}{dt}a_m^{(1)}(t) = \sum_n \hat{H}'_{1mn}a_n^{(0)}(t)e^{i\omega_{mn}t} \qquad (10)$$

$$i\hbar\frac{d}{dt}a_m^{(2)}(t) = \sum_n \hat{H}'_{2mn}a_n^{(0)}(t)e^{i\omega_{mn}t} + \sum_n \hat{H}'_{1mn}a_n^{(1)}(t)e^{i\omega_{mn}t} \qquad (11)$$

$$i\hbar\frac{d}{dt}a_m^{(3)}(t) = \sum_n \hat{H}'_{2mn}a_n^{(1)}(t)e^{i\omega_{mn}t} + \sum_n \hat{H}'_{1mn}a_n^{(2)}(t)e^{i\omega_{mn}t} \qquad (12)$$

Suppose that an electron is in the initial state $|l\rangle$ with energy E_l at time $t=0$, then the electron transits into the final state $|m\rangle$ with energy E_m at time t, we have $a_m^{(0)}(t) = \delta_{ml}$ from the first formula of (10). Put it into the second formula, the probability amplitude of first order process is

$$a_m^{(1)}(t) = \frac{1}{i\hbar}\int_0^t \hat{H}'_{1ml}e^{i\omega_{ml}t}dt = -\frac{\hat{F}_{1ml}\left[e^{i(\omega+\omega_{ml})t}-1\right]}{\hbar(\omega+\omega_{ml})} + \frac{\hat{F}_{1ml}^+\left[e^{-i(\omega-\omega_{ml})t}-1\right]}{\hbar(\omega-\omega_{ml})} \qquad (13)$$

Where

$$\hat{F}_{1ml} = \langle m|\hat{F}_1|l\rangle, \quad \hat{F}_{1ml}^+ = \langle m|\hat{F}_1^+|l\rangle.$$

The formula represents the probability amplitude of an electron transiting from the initial state $|l\rangle$ into the final state $|m\rangle$. In current theory, the so-called rotation wave approximation is used, i.e., only the first item is considered when $\omega = -\omega_{ml}$ and the second one is considered when $\omega = \omega_{ml}$ in (13). But up to now we have not decided which one is the state of high-energy level and which one is the state of low-energy level. In fact, electron can either transit into higher-energy state $|m\rangle$ from low-energy state $|l\rangle$ by absorbing a photon, or transit into low-energy state $|m\rangle$ from high-energy state $|l\rangle$ by emitting a photon. Because photon's energy is always positive, by the law of energy conservation, it is proper for us to think that the condition $\omega = \omega_{ml}$ corresponds to the situation with $E_m > E_l$, in which an electron transits into the high-energy final state $|m\rangle$ from the low-energy initial state $|l\rangle$ by absorbing a photon with energy $\hbar\omega_{ml} = E_m - E_l > 0$. This is just the simulated absorption process with the transition probability in unit time

$$W^{(1)}_{\omega=\omega_{ml}} = \frac{2\pi}{\hbar^2}\left|\hat{F}^+_{1ml}\right|^2 \delta(\omega - \omega_{ml}) \tag{14}$$

Therefore, the condition $\omega = -\omega_{ml}$ corresponds to the situation with $E_m < E_l$, in which an electron transits from the high-energy initial state $|l\rangle$ into the low-energy final state $|m\rangle$ by emitting a photon with energy $-\hbar\omega_{ml} = \hbar\omega_{lm} = E_l - E_m > 0$. This is just the simulated radiation process with the transition probability in unit time

$$W^{(1)}_{\omega=-\omega_{ml}} = \frac{2\pi}{\hbar^2}\left|\hat{F}_{1ml}\right|^2 \delta(\omega + \omega_{lm}) \tag{15}$$

So $W^{(1)}_{\omega=\omega_{ml}}$ and $W^{(1)}_{\omega=-\omega_{ml}}$ represent the different physical processes. It is necessary for us to distinguish the physical meanings of $W^{(1)}_{\omega=\omega_{ml}}$ and $W^{(1)}_{\omega=-\omega_{ml}}$ clearly for the following discussion. As shown in Fig.1, we image a system with three energy levels: medium energy level E_l, high energy $E_m(up)$ and low energy $E_m(down)$. The difference of energy levels between E_l and $E_m(up)$ is the same as that between $E_m(down)$ and E_l. Suppose that the electron is in medium energy level at beginning. Stimulated by radiation field, the electron can either transit up into high-energy level or down into low energy level. In this case, $W^{(1)}_{\omega=\omega_{ml}}$ represents the probability the electron transits up into high-energy level and $W^{(1)}_{\omega=-\omega_{ml}}$ represents the probability the electron transits down into low-energy level.

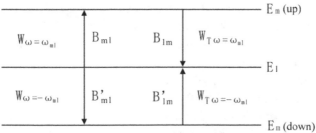

Fig. 1. Electron's transitions among three energy levels.

For visible light with wavelength $\lambda \sim 10^{-7} m$ and common atoms with radius $R \sim 10^{-10} m$, we have $\bar{k} \cdot \bar{R} \sim 10^{-3} \ll 1$. So in the current theory, dipolar approximation $\exp i\bar{k} \cdot \bar{R} \sim 1$ is used. However, it should be noted that the ratio of magnitude between the first order processes and the second order processes in nonlinear optics is just about 10^{-3}. Meanwhile, for the interaction between external fields and electrons in atoms, such as the situations of laser and nonlinear optics, we have $R \approx 0.1 \sim 1m$ so that $\bar{k} \cdot \bar{R} = 10^6 \sim 10^7$ with the macro-order of magnitude. In fact, factor $\bar{k} \cdot \bar{R}$ represents the retarded interaction of electromagnetic field. It can't be neglected in general in the problems of laser and nonlinear optics. It will be seen below that it is just this factor which would play an important role in the symmetry violation of time reversal in light's absorption and radiation processes

Let $\vec{R}_0$ represents the distance vector pointing from radiation source to atomic mass center, $\vec{r}$ represents the distance vector pointing from atomic mass center to electron, we have $\vec{R} = \vec{R}_0 + \vec{r}$. For the interaction process between external electromagnetic field and atom in medium, we have $R_0 = 0.1 \sim 1m$, $\vec{k} \cdot \vec{R}_0 = 10^6 \sim 10^7 >> 1$ and $\vec{k} \cdot \vec{r} << 1$. If radiation fields come from atomic internal, we have $\vec{R}_0 \approx 0$, $\vec{k} \cdot \vec{r} << 1$. In the following discussion, we approximately take:

$$e^{-i\vec{k} \cdot \vec{R}} \approx e^{-i\vec{k} \cdot \vec{R}_0} \left[1 - i\vec{k} \cdot \vec{r} - \left(\vec{k} \cdot \vec{r} \right)^2 / 2 \right] \tag{16}$$

Here $\vec{k} = \omega \vec{\tau} / c$, $\vec{\tau}$ is unit direction vector. By considering relations $\omega_{lm} = -\omega_{ml}$, $\hat{p} = -i\hbar\nabla$ and

$$\langle m|\hat{p}|l \rangle = \mu \langle m|d\hat{r}/dt|l \rangle = \frac{\mu}{i\hbar} \langle m|\left[\hat{r}, \hat{H}_0\right]|l \rangle = i\mu\omega_{ml} \langle m|\hat{r}|l \rangle \tag{17}$$

as well as (7), we can get

$$\hat{F}_{1ml} = \left[-\frac{iq\omega_{ml}}{2\omega} \vec{E}_0 \cdot \langle m|\vec{r}|l \rangle + \frac{q\hbar}{2c\mu} \vec{E}_0 \cdot \langle m|\vec{\tau} \cdot \vec{r}\nabla|l \rangle - \frac{iq\hbar\omega}{4c^2\mu} \vec{E}_0 \cdot \langle m|(\vec{\tau} \cdot \vec{r})^2 \nabla|l \rangle \right] e^{-i\vec{k} \cdot \vec{R}_0} \tag{18}$$

$$\hat{F}_{1ml}^+ = \left[-\frac{iq\omega_{ml}}{2\omega} \vec{E}_0 \cdot \langle m|\vec{r}|l \rangle - \frac{q\hbar}{2c\mu} \vec{E}_0 \cdot \langle m|\vec{\tau} \cdot \vec{r}\nabla|l \rangle - \frac{iq\hbar\omega}{4c^2\mu} \vec{E}_0 \cdot \langle m|(\vec{\tau} \cdot \vec{r})^2 \nabla|l \rangle \right] e^{i\vec{k} \cdot \vec{R}_0} \tag{19}$$

The first item is the result of dipolar moment interaction. The second item is the result of quadrupolar moment interaction and the third item is the result of octupolar moment interaction. The wave functions of stationary states $|m\rangle$ and $|l\rangle$ have fixed parities. The parities of operator $\vec{r}$ and $(\vec{\tau} \cdot \vec{r})^2 \nabla$ are odd and the parity of operator $\vec{\tau} \cdot \vec{r}\nabla$ is even. So by the consideration of symmetry, if matrix element $\langle m|\vec{r}|l \rangle \neq 0$, we would have $\langle m|\vec{\tau} \cdot \vec{r}\nabla|l \rangle = 0$ and $\langle m|(\vec{\tau} \cdot \vec{r})^2 \nabla|l \rangle \neq 0$. Conversely, if $\langle m|\vec{r}|l \rangle = 0$, we would have $\langle m|\vec{\tau} \cdot \vec{r}\nabla|l \rangle \neq 0$ and $\langle m|(\vec{\tau} \cdot \vec{r})^2 \nabla|l \rangle = 0$. Suppose $\langle m|\vec{r}|l \rangle \neq 0$, $\langle m|\vec{\tau} \cdot \vec{r}\nabla|l \rangle = 0$ and $\langle m|(\vec{\tau} \cdot \vec{r})^2 \nabla|l \rangle \neq 0$, we have $\hat{F}_{1ml} \neq \hat{F}_{1ml}^+$ but $\left|\hat{F}_{1ml}\right|^2 = \left|\hat{F}_{1ml}^+\right|^2$. So after retarded interaction is taken into account for the first order processes, we still have $W^{(1)}_{\omega=\omega_{ml}} = W^{(1)}_{\omega=-\omega_{ml}}$, i.e., the transition probabilities of stimulated radiation and stimulated absorption are still the same.

3. The time reversal of the first order process

Let's discuss the time reversal of the first order process below. According to the standard theory of quantum electrodynamics, the time reversal of electromagnetic potential is $\vec{A}(\vec{x},t) \rightarrow -\vec{A}(\vec{x},-t)$. Meanwhile, we have $\hat{p} \rightarrow -\hat{p}$ when $t \rightarrow -t$. The propagation direction of electromagnetic wave should be changed from $\vec{k}$ to $-\vec{k}$ under time reversal (Otherwise retarded wave would become advanced wave so that the law of causality would be violated.). Let subscript T represent time reversal, from (3), (5) and (6), we have $\hat{H}'_{1T}(\bar{x},t) = \hat{H}'_1(\bar{x},t)$ and $\hat{H}'_{2T}(\bar{x},t) = \hat{H}'_2(\bar{x},t)$. The interaction Hamiltonian is unchanged under time reversal. On the other hand, when $t \to -t$, (8) becomes

$$-i\hbar \frac{\partial}{\partial t}|\psi\rangle_T = \left(\hat{H}_{0T} + \hat{H}'_T\right)|\psi\rangle_T \qquad |\psi\rangle_T = \sum_n a_n(-t) e^{\frac{i}{\hbar}E_n t}|n\rangle \qquad (20)$$

Let $a_m(-t) = a_m^{(0)}(-t) + a_m^{(1)}(-t) + a_m^{(2)}(-t) + \cdots$ and put it into the formula above, the motion equation becomes

$$-i\hbar \frac{d}{dt}\left[a_m^{(0)}(-t) + a_m^{(1)}(-t) + a_m^{(2)}(-t) + \cdots\right]$$

$$= \sum_n \left(H'_{1Tmn} + H'_{2Tmn}\right)\left[a_n^{(0)}(-t) + a_n^{(1)}(-t) + a_n^{(2)}(-t) + \cdots\right]e^{-i\omega_{mn}t} \qquad (21)$$

Take index replacements $m \to l$ and $n \to k$ in the formula, then let the items with same order to be equal to each other, we get

$$-i\hbar \frac{d}{dt}a_l^{(0)}(-t) = 0 \qquad -i\hbar \frac{d}{dt}a_l^{(1)}(-t) = \sum_k \hat{H}'_{1Tlk}a_k^{(0)}(-t)e^{-i\omega_{lk}t} \qquad (22)$$

$$-i\hbar \frac{d}{dt}a_l^{(2)}(-t) = \sum_k \hat{H}'_{2Tlk}a_k^{(0)}(-t)e^{-i\omega_{lk}t} + \sum_k \hat{H}'_{1Tlk}a_k^{(1)}(-t)e^{-i\omega_{lk}t} \qquad (23)$$

$$-i\hbar \frac{d}{dt}a_l^{(3)}(-t) = \sum_k \hat{H}'_{2Tlk}a_k^{(1)}(-t)e^{-i\omega_{lk}t} + \sum_k \hat{H}'_{1Tlk}a_k^{(2)}(-t)e^{-i\omega_{lk}t} \qquad (24)$$

On the other hand, under time reversal, the initial state becomes $|m\rangle$ with $a_k^{(0)}(-t) = \delta_{km}$. Put it into the second formula of (22), we get

$$a_l^{(1)}(-t) = -\frac{1}{i\hbar}\int_0^t \hat{H}'_{1Tlm}e^{-i\omega_{lm}t}dt = -\frac{1}{i\hbar}\int_0^t \hat{H}'_{1lm}e^{-i\omega_{lm}t}dt \qquad (25)$$

Because $\hat{H}'_1$ is the Hermitian operator with

$$\hat{H}'_{1lm} = \langle l|\hat{H}'_1|m\rangle = \langle m|\hat{H}'_1|l\rangle^* = \hat{H}'^*_{1ml},$$

we can write

$$\hat{H}'^*_{1ml} = \hat{F}'_{1ml}e^{i\omega t} + \hat{F}'^+_{1ml}e^{-i\omega t}$$

and get

$$\hat{F}'_{1ml} = \left(\hat{F}^+_{1ml}\right)^* = \left[\frac{iq\omega_{ml}}{2\omega}\bar{E}_0 \cdot \langle m|\bar{r}|l\rangle^* - \frac{q\hbar}{2c\mu}\bar{E}_0 \cdot \langle m|\bar{r}\cdot\bar{r}\nabla|l\rangle^* + \frac{iq\hbar\omega}{4c^2\mu}\bar{E}_0 \cdot \langle m|(\bar{r}\cdot\bar{r})^2\nabla|l\rangle^*\right]e^{-i\bar{k}\cdot\bar{R}_0} \quad (26)$$

$$\hat{F}_{1ml}^{\prime+} = \left(\hat{F}_{1ml}\right)^* = \left[\frac{iq\omega_{ml}}{2\omega}\vec{E}_0 \cdot \langle m|\vec{r}|l\rangle^* + \frac{q\hbar}{2c\mu}\vec{E}_0 \cdot \langle m|\vec{\tau}\cdot\vec{r}\nabla|l\rangle^* + \frac{iq\hbar\omega}{4c^2\mu}\vec{E}_0 \cdot \langle m|(\vec{\tau}\cdot\vec{r})^2\nabla|l\rangle^*\right]e^{i\vec{k}\cdot\vec{R}_0} \quad (27)$$

Let $a_{mT}(t)$ represent the time reversal of amplitude $a_m(t)$. Because the original final state becomes $|l\rangle$ and the original initial state becomes $|m\rangle$ under time reversal, we have $a_{mT}^{(1)}(t) = a_l^{(1)}(-t)$. So according to (25), after time reversal, the transition amplitude becomes

$$
\begin{aligned}
a_{mT}^{(1)}(t) &= -\frac{1}{i\hbar}\int_0^t \hat{H}_{1ml}^{\prime*}e^{i\omega_{ml}t}dt = \frac{\hat{F}_{1ml}^\prime\left[e^{i(\omega+\omega_{ml})t}-1\right]}{\hbar(\omega+\omega_{ml})} - \frac{\hat{F}_{1ml}^{\prime+}\left[e^{-i(\omega-\omega_{ml})t}-1\right]}{\hbar(\omega-\omega_{ml})} = \\
&= \frac{\hat{F}_{1ml}^\prime\left[e^{i(\omega-\omega_{lm})t}-1\right]}{\hbar(\omega-\omega_{lm})} - \frac{\hat{F}_{1ml}^{\prime+}\left[e^{-i(\omega+\omega_{lm})t}-1\right]}{\hbar(\omega+\omega_{lm})}
\end{aligned}
\quad (28)
$$

So the condition $\omega = -\omega_{lm} = \omega_{ml}$ corresponds to the situation with $E_m > E_l$, indicating that an electron emits a photon with energy $-\hbar\omega_{lm} = E_m - E_l > 0$ and transits from the initial high-energy state $|m\rangle$ into the final low-energy state $|l\rangle$. This process is the time reversal of stimulated absorption process described by (14). By considering (27), the transition probability in unite time is

$$W_{T\omega=\omega_{ml}}^{(1)} = \frac{2\pi}{\hbar^2}\left|\hat{F}_{1ml}^{\prime+}\right|^2\delta(\omega+\omega_{lm}) = \frac{2\pi}{\hbar^2}\left|\hat{F}_{1ml}^\prime\right|^2\delta(\omega-\omega_{ml}) \quad (29)$$

Comparing with (14) and considering the result $\left|\hat{F}_{1ml}^\prime\right|^2 = \left|\hat{F}_{1ml}^+\right|^2$, we still have $W_{T\omega=\omega_{ml}}^{(1)} = W_{\omega=\omega_{ml}}^{(1)}$. Because we define the time reversal of stimulated absorption process as the stimulated radiation process, the result shows that the transition probability of stimulated absorption is equal to that of stimulated radiation after time reversal for the first order process when retarded interaction is considered. The process is unchanged under time reversal.

Similarly, the condition $\omega = \omega_{lm} = -\omega_{ml}$ corresponds to the situation with $E_m < E_l$, indicating that an electron emits a photon with energy $\hbar\omega_{lm} = E_l - E_m > 0$ and transits from the initial low-energy state $|m\rangle$ into the final high-energy state $|l\rangle$. This process is the time reversal of stimulated radiation process described by (15). By considering (26), the transition probability in unite time is

$$W_{T\omega=-\omega_{ml}}^{(1)} = \frac{2\pi}{\hbar^2}\left|\hat{F}_{1ml}^\prime\right|^2\delta(\omega-\omega_{lm}) = \frac{2\pi}{\hbar^2}\left|\hat{F}_{1ml}^+\right|^2\delta(\omega+\omega_{ml}) \quad (30)$$

After retarded effect is considered, we also have $W_{T\omega=-\omega_{ml}}^{(1)} = W_{\omega=-\omega_{ml}}^{(1)}$ for the first order process. The process is unchanged under time reversal.

The transition relations can be seen clearly in Fig.1. There are two stimulated absorption parameters B_{ml}, $B_{lm}^\prime$ and two stimulated radiation parameters B_{lm}, $B_{ml}^\prime$. The initial and final states of B_{ml} and B_{lm} are opposite. So they are time reversal states. The initial and final states of $B_{ml}^\prime$ and $B_{lm}^\prime$ are also opposite, so they are also time reversal states. Therefore, if B_{ml} is defined as stimulated absorption parameter, B_{lm} should be defined as stimulated radiation parameter for their initial final states are just opposite. Similarly, if B'_{ml} is defined as stimulated absorption parameter, B'_{lm} should be defined as stimulated radiation parameter. Meanwhile, if B_{ml} (or B'_{lm}) is defined as stimulated absorption parameter, B'_{ml} (or B_{lm}) should not be defined as stimulated radiation parameter, for they have same initial and final states and do not describe corresponding stimulated radiation and absorption processes. For the first order process, we have $B_{ml} = B_{lm} = B'_{lm} = B'_{ml}$. But as shown below that in the high order processes, this relation can't hold.

4. The transition probability of the second order process

The second order processes are discussed below. We write $\hat{H}'_2 = \hat{F}_2 e^{2i\omega t} + \hat{F}_2^+ e^{-2i\omega t} + \hat{F}_0$, in which

$$\hat{F}_2 = \frac{q^2 E_0^2}{2\omega^2 \mu} e^{-i2\vec{k}\cdot\vec{R}} \qquad \hat{F}_2^+ = \frac{q^2 E_0^2}{2\omega^2 \mu} e^{i2\vec{k}\cdot\vec{R}} \qquad \hat{F}_0 = \frac{q^2 E_0^2}{\omega^2 \mu} \tag{31}$$

When $m \neq l$ we have $\langle m|l \rangle = 0$. Suppose $\langle m|\vec{\tau}\cdot\vec{r}|l\rangle \neq 0$, we have $\langle m|(\vec{\tau}\cdot\vec{r})^2|l\rangle = 0$. According to (16), we have

$$\hat{F}_{2ml} = -\frac{iq^2 E_0^2}{c\omega\mu} e^{-i2\vec{k}\cdot\vec{R}_0} \langle m|\vec{\tau}\cdot\vec{r}|l\rangle \quad \hat{F}_{2ml}^+ = \frac{iq^2 E_0^2}{c\omega\mu} e^{i2\vec{k}\cdot\vec{R}_0} \langle m|\vec{\tau}\cdot\vec{r}|l\rangle \quad \hat{F}_{0ml} = 0 \tag{32}$$

Similarly, we suppose the initial condition is $a_n^{(0)} = \delta_{nl}$. Substituting (13) into (13) and taking the integral, we can obtain the transition probability amplitude of the second order process

$$a_m^{(2)}(t) = \frac{1}{i\hbar}\int_0^t \hat{H}'_{2ml} e^{i\omega_{ml}t} dt + \frac{1}{i\hbar}\sum_n \int_0^t \hat{H}'_{1mn} a_n^{(1)}(t) e^{i\omega_{mn}t} dt$$

$$= -\frac{\hat{F}_{2ml}\left[e^{i(2\omega+\omega_{ml})t} - 1\right]}{\hbar(2\omega+\omega_{ml})} + \frac{\hat{F}_{2ml}^+\left[e^{-i(2\omega-\omega_{ml})t} - 1\right]}{\hbar(2\omega-\omega_{ml})}$$

$$+ \sum_n \frac{\hat{F}_{1mn}\hat{F}_{1nl}}{\hbar^2(\omega+\omega_{nl})}\left\{\frac{e^{i(2\omega+\omega_{nl}+\omega_{mn})t} - 1}{2\omega+\omega_{nl}+\omega_{mn}} - \frac{e^{i(\omega+\omega_{mn})t} - 1}{\omega+\omega_{mn}}\right\}$$

$$- \sum_n \frac{\hat{F}_{1mn}\hat{F}_{1nl}^+}{\hbar^2(\omega-\omega_{nl})}\left\{\frac{e^{i(\omega_{nl}+\omega_{mn})t} - 1}{\omega_{nl}+\omega_{mn}} - \frac{e^{i(\omega+\omega_{mn})t} - 1}{\omega+\omega_{mn}}\right\}$$

$$+ \sum_n \frac{\hat{F}_{1mn}^+\hat{F}_{1nl}}{\hbar^2(\omega+\omega_{nl})}\left\{\frac{e^{i(\omega_{nl}+\omega_{mn})t} - 1}{\omega_{nl}+\omega_{mn}} + \frac{e^{-i(\omega-\omega_{mn})t} - 1}{\omega-\omega_{mn}}\right\}$$

$$+ \sum_n \frac{\hat{F}_{1mn}^+\hat{F}_{1nl}^+}{\hbar^2(\omega-\omega_{nl})}\left\{\frac{e^{-i(2\omega-\omega_{nl}-\omega_{mn})t} - 1}{2\omega-\omega_{nl}-\omega_{mn}} - \frac{e^{-i(\omega-\omega_{mn})t} - 1}{\omega-\omega_{mn}}\right\} \tag{33}$$

The formula contains the transition processes of single photon's absorption and radiation with $\omega = \pm\omega_{ml}$ as well as the processes of double photon's absorption and radiation with $2\omega = \pm\omega_{ml}$. We only discuss the absorption process of a single photon here. By rotation wave approximation, only the items containing factor $\left(e^{-i(\omega-\omega_{ml})t} - 1\right)/(\omega-\omega_{ml})$ are remained. Let $n=1$ in the fifth and sixth items of the formula, the transition probability amplitude is

$$a_m^{(2)}(t)_{\omega=\omega_{ml}} = \frac{\hat{F}_{1ml}^+ \left(\hat{F}_{1ll} - \hat{F}_{1ll}^+\right)\left[e^{-i(\omega-\omega_{ml})t} - 1\right]}{\hbar^2 \omega (\omega - \omega_{ml})} = \frac{\hat{F}_{1ml}^+ \left(\hat{F}_{1ll} - \hat{F}_{1ll}^+\right)\left[e^{-i(\omega-\omega_{ml})t} - 1\right]}{\hbar^2 \omega_{ml} (\omega - \omega_{ml})} \tag{34}$$

Therefore, after the second order process is considered, the total transition probability amplitude is

$$a_m(t)_{\omega=\omega_{ml}} = a_m^{(1)}(t)_{\omega=\omega_{ml}} + a_m^{(2)}(t)_{\omega=\omega_{ml}} = \frac{\hat{F}_{1ml}^+ \left[e^{-i(\omega-\omega_{ml})t} - 1\right]}{\hbar(\omega - \omega_{ml})}\left(1 + \frac{\hat{F}_{1ll} - \hat{F}_{1ll}^+}{\hbar\omega_{ml}}\right) \tag{35}$$

Because we have $\omega_{ll} = 0$ and $\langle l|\vec{r}|l\rangle = 0$, when $m = l$, the first item in (18) and (19) are equal to zero, but the second items are not equal to zero in general. By the consideration of symmetry, the third item are also zero, or can be neglected by comparing with the second item. So it is enough for us only to consider quadrupole moment interaction in this case. We have

$$\hat{F}_{1ll} - \hat{F}_{1ll}^+ = \frac{q\hbar}{c\mu}\vec{E}_0 \cdot \langle l|\vec{\tau}\cdot\vec{r}\nabla|l\rangle \cos\vec{k}\cdot\vec{R}_0 = B_{1l} + iB_{2l} \tag{36}$$

Let

$$A_l^2 = B_{1l}^2 + 2\hbar\omega_{ml}B_{1l} + B_{2l}^2 \tag{37}$$

When $\omega = \omega_{ml}$ we get the transition probability of the second order stimulated absorption process

$$W_{\omega=\omega_{ml}}^{(2)} = \frac{2\pi}{\hbar^2}\left|\hat{F}_{1ml}\right|^2 \left\{1 + \frac{A_l^2}{\hbar^2\omega_{ml}^2}\right\}\delta(\omega - \omega_{ml}) \tag{38}$$

The magnitude order of the revised value of the second order process is estimated below. The wave function of bounding state's atoms can be developed into series with form $|l\rangle \sim \sum b_n(\theta,\phi)r^n$ in general. If $\langle l|\vec{\tau}\cdot\vec{r}\nabla|l\rangle \neq 0$, we have $\vec{\tau}\cdot\vec{r}\nabla|l\rangle \sim r\partial/(\partial r)|l\rangle \sim |l\rangle$ approximately, or $\langle l|\vec{\tau}\cdot\vec{r}\nabla|l\rangle \sim \langle l|l\rangle \sim 1$. By taking $\omega_{ml} = 10^{16}$, we have

$$\frac{A_l^2}{\hbar\omega_{ml}^2} \sim 4q^2 E_0^2 \left|\langle l|\vec{\tau}\cdot\vec{r}\nabla|l\rangle\right|^2 / (c\mu\omega_{ml})^2 \sim 1.4\times 10^{-26} E_0^2 \tag{39}$$

In weak electromagnetic fields with $E_0 \ll 10^{13} V / m$, the revised values of the second order processes can be neglected. When the fields are strong enough with $E_0 \approx 10^{12} \sim 10^{13} V / m$, the revised value is big enough to be observed. The revised factor A_l of the second order processes is only related to initial state, having nothing to do with final states. On the other hand, if the retarded effect of radiation fields is neglected with $\vec{k} \cdot \vec{r} \sim \vec{\tau} \cdot \vec{r} = 0$, we have $\langle l | \vec{\tau} \cdot \vec{r} \nabla | l \rangle = 0$, the revised value of the second order process vanishes.

5. The time reversal of the second order process

The time reversal of the second order process is discussed now. Under time reversal, the initial state becomes $a_k^{(0)}(-t) = \delta_{kn}$. By means of relations $\hat{H}'_{1Tlm} = \hat{H}'^{*}_{1ml}$ and $\hat{H}'_{2Tlm} = \hat{H}'^{*}_{2ml}$, we get the time reversal of transition amplitude for the second order processes according to(11)

$$a_l^{(2)}(-t) = -\frac{1}{i\hbar} \int_0^t \hat{H}'_{2Tlm} e^{-i\omega_{lm} t} dt - \frac{1}{i\hbar} \sum_k \int_0^t \hat{H}'_{1Tlk} a_k^{(1)}(-t) e^{-i\omega_{lk} t} dt$$

$$= -\frac{1}{i\hbar} \int_0^t \hat{H}'^{*}_{2ml} e^{i\omega_{ml} t} dt - \frac{1}{i\hbar} \sum_k \int_0^t \hat{H}'^{*}_{1kl} a_k^{(1)}(-t) e^{i\omega_{kl} t} dt \tag{40}$$

Let

$$\hat{H}'^{*}_{2ml} = \hat{F}'_{2ml} e^{i2\omega t} + \hat{F}'^{+}_{2ml} e^{-i2\omega t} + \hat{F}'_{0ml},$$

when $m \neq l$, we have $\langle m | l \rangle = 0$ and

$$\hat{F}'_{2ml} = -\frac{iq^2 E_0^2}{c\omega\mu} e^{-i2\vec{k}\cdot\vec{R}_0} \langle m | \vec{\tau} \cdot \vec{r} | l \rangle^* \quad \hat{F}'^{+}_{2ml} = \frac{iq^2 E_0^2}{c\omega\mu} e^{i2\vec{k}\cdot\vec{R}_0} \langle m | \vec{\tau} \cdot \vec{r} | l \rangle^* \quad \hat{F}'_{0ml} = 0 \tag{41}$$

So the time reversal of transition probability amplitude of the second process is

$$a_{Tm}^{(2)}(t) = a_l^{(2)}(-t) = \frac{\hat{F}'_{2ml}\left[e^{i(2\omega+\omega_{ml})t} - 1\right]}{\hbar(2\omega+\omega_{ml})} - \frac{\hat{F}'^{+}_{2ml}\left[e^{-i(2\omega-\omega_{ml})t} - 1\right]}{\hbar(2\omega-\omega_{ml})}$$

$$+ \sum_k \frac{\hat{F}'^{+}_{1mk}\hat{F}'^{+}_{1kl}}{\hbar^2(\omega-\omega_{mk})} \left\{ \frac{e^{-i(2\omega-\omega_{mk}-\omega_{kl})t} - 1}{2\omega-\omega_{mk}-\omega_{kl}} - \frac{e^{-i(\omega-\omega_{kl})t} - 1}{\omega-\omega_{kl}} \right\}$$

$$- \sum_k \frac{\hat{F}'^{+}_{1mk}\hat{F}'_{1kl}}{\hbar^2(\omega-\omega_{mk})} \left\{ \frac{e^{i(\omega_{mk}+\omega_{kl})t} - 1}{\omega_{mk}+\omega_{kl}} - \frac{e^{i(\omega+\omega_{kl})t} - 1}{\omega+\omega_{kl}} \right\}$$

$$+ \sum_k \frac{\hat{F}'_{1mk}\hat{F}'^{+}_{1kl}}{\hbar^2(\omega+\omega_{mk})} \left\{ \frac{e^{i(\omega_{mk}+\omega_{kl})t} - 1}{\omega_{mk}+\omega_{kl}} + \frac{e^{-i(\omega-\omega_{kl})t} - 1}{\omega-\omega_{kl}} \right\}$$

$$+ \sum_k \frac{\hat{F}'_{1mk}\hat{F}'_{1kl}}{\hbar^2(\omega+\omega_{mk})} \left\{ \frac{e^{i(2\omega+\omega_{mk}+\omega_{kl})t} - 1}{2\omega+\omega_{mk}+\omega_{kl}} - \frac{e^{i(\omega+\omega_{kl})t} - 1}{\omega+\omega_{kl}} \right\} \tag{42}$$

When $\omega = \omega_{ml}$, we take $k = m$ in the third and fifth items of the formula. Also by rotation wave approximation, the time reversal of probability amplitude is

$$a_{mT}^{(2)}(t)_{\omega=\omega_{ml}} = \frac{\hat{F}_{1ml}'^{+}\left(F_{1mm}' - \hat{F}_{1mm}'^{+}\right)\left[e^{-i(\omega-\omega_{ml})t} - 1\right]}{\hbar^2\omega(\omega - \omega_{ml})} = \frac{\hat{F}_{1ml}'^{+}\left(F_{1mm}' - \hat{F}_{1mm}'^{+}\right)\left[e^{-i(\omega-\omega_{ml})t} - 1\right]}{\hbar^2\omega_{ml}(\omega - \omega_{ml})}$$ (43)

The time reversal of the total stimulated absolution process is

$$a_{Tm}(t)_{\omega=\omega_{ml}} = a_{Tm}^{(1)}(t)_{\omega=\omega_{ml}} + a_{Tm}^{(2)}(t)_{\omega=\omega_{ml}} = \frac{\hat{F}_{1ml}'^{+}\left[e^{-i(\omega-\omega_{ml})t} - 1\right]}{\hbar(\omega - \omega_{ml})}\left(1 + \frac{\hat{F}_{1mm}' - \hat{F}_{1mm}'^{+}}{\hbar\omega_{ml}}\right)$$ (44)

Comparing with (34), because of $\hat{F}_{mm}' \neq \hat{F}_{ll}$, we have

$$a_{mT}(t)_{\omega=\omega_{ml}} \neq a_m(t)_{\omega=\omega_{ml}},$$

the transition probability amplitude can not keep unchanged. Similarly, by considering $\omega_{mm} = 0$ and from (26) and (27), we have

$$\hat{F}_{1mm}' - \hat{F}_{1mm}'^{+} = -\frac{q\hbar}{c\mu}\bar{E}_0 \cdot \langle m|\bar{\tau} \cdot \bar{r}\nabla|m\rangle^* \cos\bar{k} \cdot \bar{R}_0 = -B_{1m} + iB_{2m}$$ (45)

Let

$$A_m'^2 = B_{1m}^2 - 2\hbar\omega_{ml}B_{1m} + B_{2m}^2$$ (46)

When $\omega = \omega_{ml}$, the time reversal of stimulated absorption probability of second process is

$$W_{T\omega=\omega_{ml}}^{(2)} = \frac{2\pi}{\hbar^2}\left|\hat{F}_{1ml}'^{+}\right|^2\left\{1 + \frac{A_m'^2}{\hbar^2\omega_{ml}^2}\right\}\delta(\omega - \omega_{ml}) = \frac{2\pi}{\hbar^2}\left|\hat{F}_{1ml}'\right|^2\left\{1 + \frac{A_m'^2}{\hbar^2\omega_{ml}^2}\right\}\delta(\omega - \omega_{ml})$$ (47)

The revised factor A_m is also only relative to initial state. Because $A_m' \neq A_l$, we have

$$W_{\omega=\omega_{ml}}^{(2)}\left(\hat{F}_{1ml}, A_l\right) \neq W_{T\omega=\omega_{ml}}^{(2)}\left(\hat{F}_{1ml}, A_m'\right)$$ (48)

The second process of stimulated absolution violates time reversal symmetry. The parameter of symmetry violation of the second order process can be is defined as

$$\beta = \frac{W_{T\omega=\omega_{ml}}^{(2)} - W_{\omega=\omega_{ml}}^{(2)}}{W_{\omega=\omega_{ml}}^{(2)}} \sim \frac{\left(A_m^2 - A_l^2\right)}{\hbar^2\omega_{ml}^2} \sim 10^{-26}E_0^2$$ (49)

When the radiation fields are strong enough with $E_0 \approx 10^{12} \sim 10^{13}V/m$, the time reversal symmetry violation of the second order process would be great.

Meanwhile, by means of (33) and (42), for the second order process with $\omega = -\omega_{ml}$, the transition amplitude and probability can be obtained with

$$a_m^{(2)}(t)_{\omega=-\omega_{ml}} = \frac{\hat{F}_{1ml}\left(\hat{F}_{1ll} - \hat{F}_{1ll}^+\right)\left[e^{i(\omega+\omega_{ml})t} - 1\right]}{\hbar^2 \omega_{ml}(\omega + \omega_{ml})} \tag{50}$$

$$W_{\omega=-\omega_{ml}}^{(2)} = \frac{2\pi}{\hbar^2}\left|\hat{F}_{1ml}\right|^2 \left\{1 + \frac{A_l^2}{\hbar^2 \omega_{ml}^2}\right\} \delta(\omega + \omega_{ml}) \tag{51}$$

Their time reversals are

$$a_m^{(2)}(t)_{T\omega=-\omega_{ml}} = \frac{\hat{F}_{1ml}'\left(\hat{F}_{1mm}' - \hat{F}_{1mm}'^+\right)\left[e^{i(\omega+\omega_{ml})t} - 1\right]}{\hbar^2 \omega_{ml}(\omega + \omega_{ml})} \tag{52}$$

$$W_{T\omega=-\omega_{ml}}^{(2)} = \frac{2\pi}{\hbar^2}\left|\hat{F}_{1ml}\right|^2 \left\{1 + \frac{A_m''^2}{\hbar^2 \omega_{ml}^2}\right\} \delta(\omega + \omega_{ml}) \tag{53}$$

Also, the process violates time reversal symmetry. It is easy to prove that for the second order processes of double photon absorptions with $2\omega = \pm\omega_{ml}$, the transition probabilities are unchanged under time reversal. The symmetry violation appears in the third order processes.

6. Accumulate solution of double energy level system and its time reversal

What has been discussed above is that the radiation fields are polarized and monochromatic light. It is easy to prove that when the radiation fields are non-polarized and non-monochromatic light, time reversal symmetry is still violated after the retarded effect of radiation field is taken into account in the high order processes. But we do not discuss this problem more here. The approximation method of perturbation is used in the discussion above. In order to prove that symmetry violation of time reversal is not introduced by the approximate method, we discuss double energy level system below. The wave function of double energy lever system can be written as

$$|\psi\rangle = a(t)e^{-\frac{i}{\hbar}E_1 t}|1\rangle + b(t)e^{-\frac{i}{\hbar}E_2 t}|2\rangle \tag{54}$$

Thus the motion equations of quantum mechanics are

$$i\hbar\dot{a}(t) = \hat{H}_{11}'a(t) + \hat{H}_{12}'e^{-i\omega_{21}t}b(t) \qquad i\hbar\dot{b}(t) = \hat{H}_{21}'e^{i\omega_{21}t}a(t) + \hat{H}_{22}'b(t) \tag{55}$$

For simplicity, we only consider the first item of the Hamiltonian (2) to take $\hat{H}_1' \neq 0$, $\hat{H}_2' = 0$ and $\hat{H}' = \hat{H}_1'$. By taking dipolar approximation with $\vec{k} \cdot \vec{R} = 0$, we have $\langle 1|\hat{H}'|1\rangle = \langle 2|\hat{H}'|2\rangle = 0$. By the rotation wave approximation again, the motion equations becomes

$$\ddot{a}(t) - i(\omega - \omega_{21})\dot{a}(t) + A^2 a(t) = 0 \qquad \ddot{b}(t) + i(\omega - \omega_{21})\dot{b}(t) + A^2 b(t) = 0 \tag{56}$$

Here $A = \left|\hat{F}_{21}\right|^2 / \hbar^2$. These two equations have accurate solutions, i.e., the so-called Rabi solutions. Suppose that atom is in the state $|1\rangle$ at beginning with $a(t=0) = 1$ and $b(t=0) = 0$, we can get [1]

$$\left|b(t)\right|^2 = \frac{4A^2 \sin^2 \sqrt{(\omega - \omega_{21})^2 + 4A^2} \, t / 2}{(\omega - \omega_{21})^2 + 4A^2} \tag{57}$$

If the atom is in the state $|2\rangle$ at beginning with $b(t=0) = 1$ and $a(t=0) = 0$, we have

$$\left|a(t)\right|^2 = \frac{4A^2 \sin^2 \sqrt{(\omega - \omega_{21})^2 + 4A^2} \, t / 2}{(\omega - \omega_{21})^2 + 4A^2} \tag{58}$$

So for the Rabi process, the probability that atom transits from $|1\rangle$ into $|2\rangle$ is the same as that atom transits from $|2\rangle$ into $|1\rangle$. The processes are symmetrical under time reversal. In fact, let $t \to -t$ in (56), we get

$$\ddot{a}(-t) + i(\omega - \omega_{21})\dot{a}(-t) + V^2 a(-t) = 0 \quad \ddot{b}(-t) - i(\omega - \omega_{21})\dot{b}(-t) + V^2 b(-t) = 0 \tag{59}$$

Comparing these two formulas with (56), we know that as long as let $a(-t) = b(t)$ and $b(-t) = a(t)$, the motion equations are the same under time reversal.

However, if retarded effect is considered with $\bar{k} \cdot \bar{R} \neq 0$, we have $\langle 1|\hat{H}'|1\rangle \neq 0$ and $\langle 2|\hat{H}'|2\rangle \neq 0$. By taking $\hat{H}_1' \neq 0$, $\hat{H}_2' = 0$ similarly, we have

$$\ddot{a}(t) - i\left\{ \frac{1}{\hat{H}_{12}'}\left[(\omega - \omega_{21})\hat{F}_{12}e^{i\omega t} - (\omega + \omega_{21})\hat{F}_{12}e^{-i\omega t}\right] - \frac{1}{\hbar}\left(\hat{H}_{11}' + \hat{H}_{22}'\right) \right\}\dot{a}(t)$$

$$- \left\{ \frac{\hat{H}_{11}'}{\hbar \hat{H}_{12}'}\left[(\omega - \omega_{21})\hat{F}_{12}e^{i\omega t} - \frac{1}{\hbar}(\omega + \omega_{21})\hat{F}_{12}e^{-i\omega t}\right] \right.$$

$$\left. + \frac{\omega}{\hbar}\left(\hat{F}_{12}e^{i\omega t} - \hat{F}_{12}^+ e^{-i\omega t}\right) + \frac{1}{\hbar^2}\left(\hat{H}_{11}'\hat{H}_{22}' - \hat{H}_{12}'\hat{H}_{21}'\right) \right\}a(t) = 0 \tag{60}$$

$$\ddot{b}(t) + i\left\{ \frac{1}{\hat{H}_{12}'}\left[(\omega - \omega_{21})\hat{F}_{12}^+ e^{-i\omega t} - (\omega + \omega_{21})\hat{F}_{12}e^{i\omega t}\right] - \left(\hat{H}_{11}' + \hat{H}_{22}'\right) \right\}\dot{b}(t)$$

$$- \left\{ \frac{\hat{H}_{22}'}{\hbar \hat{H}_{12}'}\left[(\omega - \omega_{21})\hat{F}_{12}^+ e^{-i\omega t} - \frac{1}{\hbar}(\omega + \omega_{21})\hat{F}_{12}e^{i\omega t}\right] \right.$$

$$\left. + \frac{\omega}{\hbar}\left(\hat{F}_{12}e^{i\omega t} - \hat{F}_{12}^+ e^{-i\omega t}\right) + \frac{1}{\hbar^2}\left(\hat{H}_{11}'\hat{H}_{22}' - \hat{H}_{12}'\hat{H}_{21}'\right) \right\}b(t) = 0 \tag{61}$$

Retarded Electromagnetic Interaction and Symmetry Violation of Time Reversal in High Order Stimulated
Radiation and Absorption Processes of Lights as Well as Nonlinear Optics...

87

The equations have no accurate solutions in this case. Under time reversal, we have $t \to -t$, $\hat{H}_{T12} = \hat{H}_{21}^{*}$, $\hat{F}_{T12} = \hat{F}_{21}'$, $\hat{F}_{T12}^{+} = \hat{F}_{21}'^{+}$, the formulas above become

$$
\ddot{a}(-t) + i \left\{ \frac{1}{\hat{H}_{21}'^{*}} \left[\left(\omega - \omega_{21} \right) \hat{F}_{21}' e^{-i\omega t} - \left(\omega + \omega_{21} \right) \hat{F}_{21}' e^{i\omega t} \right] - \frac{1}{\hbar} \left(\hat{H}_{11}'^{*} + \hat{H}_{22}'^{*} \right) \right\} \dot{a}(-t)
$$

$$
- \left\{ \frac{\hat{H}_{11}'^{*}}{\hbar \hat{H}_{21}'^{*}} \left[\left(\omega - \omega_{21} \right) \hat{F}_{21}' e^{-i\omega t} - \frac{1}{\hbar} \left(\omega + \omega_{21} \right) \hat{F}_{21}^{*} e^{i\omega t} \right] \right.
$$

$$
\left. + \frac{\omega}{\hbar} \left(\hat{F}_{21}' e^{-i\omega t} - \hat{F}_{21}'^{+} e^{i\omega t} \right) + \frac{1}{\hbar^2} \left(\hat{H}_{11}'^{*} \hat{H}_{22}'^{*} - \hat{H}_{21}'^{*} \hat{H}_{21}'^{*} \right) \right\} a(-t) = 0 \qquad (62)
$$

$$
\ddot{b}(-t) - i \left\{ \frac{1}{\hat{H}_{21}'^{*}} \left[\left(\omega - \omega_{21} \right) \hat{F}_{21}'^{+} e^{i\omega t} - \left(\omega + \omega_{21} \right) \hat{F}_{21}' e^{-i\omega t} \right] - \left(\hat{H}_{11}'^{*} + \hat{H}_{22}'^{*} \right) \right\} \dot{b}(-t)
$$

$$
- \left\{ \frac{\hat{H}_{22}'^{*}}{\hbar \hat{H}_{21}'^{*}} \left[\left(\omega - \omega_{21} \right) \hat{F}_{21}'^{+} e^{i\omega t} - \frac{1}{\hbar} \left(\omega + \omega_{21} \right) \hat{F}_{21}' e^{-i\omega t} \right] \right.
$$

$$
\left. + \frac{\omega}{\hbar} \left(\hat{F}_{21}' e^{-i\omega t} - \hat{F}_{21}'^{+} e^{i\omega t} \right) + \frac{1}{\hbar^2} \left(\hat{H}_{11}'^{*} \hat{H}_{22}'^{*} - \hat{H}_{21}'^{*} \hat{H}_{12}'^{*} \right) \right\} b(-t) = 0 \qquad (63)
$$

Because of $\hat{H}_{11}' \neq \hat{H}_{22}'$, $\hat{F}_{21}' \neq \hat{F}_{21}'^{+}$ and $\hat{F}_{21}' \neq \hat{F}_{21}'^{*}$, even by taking $a(-t) \to b(t)$, $b(-t) \to a(t)$, the motion equations can't yet keep unchanged under time reversal. So after retarded effect of radiation field is considered, the double energy level system can't keep unchanged under time reversal. It means that symmetry violation of time reversal is an inherent character of systems, not originates from the approximate method of perturbation.

7. The time reversal of the third order process of double photons

According to (33), by considering rotation wave approximation, when $2\omega = \omega_{ml}$ we have the transition probability amplitude of the second order process of double photon stimulated absorption

$$
a_{m}^{(2)}(t)_{2\omega=\omega_{ml}} = \frac{\left[e^{-i(2\omega-\omega_{ml})t} - 1 \right]}{\hbar \left(2\omega - \omega_{ml} \right)} \left[\hat{F}_{2ml}^{+} + \frac{2\hat{F}_{1ml}^{+} \left(\hat{F}_{1ll}^{+} - \hat{F}_{1mm}^{+} \right)}{\hbar \omega_{ml}} \right] \qquad (64)
$$

So the total transition probability of stimulated absorption of double photons in the first and second processes in unit time is

$$
W_{2\omega=\omega_{ml}}^{(2)} = \frac{2\pi}{\hbar^2} \left\{ \left| \hat{F}_{2ml}^{+} \right|^2 + \frac{4}{\hbar^2 \omega_{ml}^2} \left| \hat{F}_{1ml}^{+} \right|^2 \left| \hat{F}_{1ll}^{+} - \hat{F}_{1mm}^{+} \right|^2 \right.
$$

$$+\frac{4}{\hbar\omega_{ml}}\text{Re}\left[\left(\hat{F}_{2ml}^{+}\right)^{*}\hat{F}_{1ml}^{+}\left(\hat{F}_{1ll}^{+}-\hat{F}_{1mm}^{+}\right)\right]\right\}\delta\left(2\omega-\omega_{ml}\right)\tag{65}$$

Here "Re" represents the real part of the function. Similarly, we can obtain the time reversal of transition probability of stimulated absorption of double photons in the second process

$$a_{Tm}^{(2)}\left(t\right)_{2\omega=\omega_{ml}}=-\frac{\left[e^{-i\left(2\omega-\omega_{ml}\right)t}-1\right]}{\hbar\left(2\omega-\omega_{ml}\right)}\left[\hat{F}_{2ml}'^{+}+\frac{2\hat{F}_{1ml}'^{+}\left(\hat{F}_{1ll}'^{+}-\hat{F}_{1mm}'^{+}\right)}{\hbar\omega_{ml}}\right]\tag{66}$$

We have relations

$$\left|\hat{F}_{2ml}'^{+}\right|^{2}=\left|\hat{F}_{2ml}^{+}\right|^{2}\quad\left|\hat{F}_{1ml}'^{+}\right|^{2}=\left|\hat{F}_{1ml}^{+}\right|^{2}\quad\hat{F}_{1ml}'=\left(\hat{F}_{1ml}^{+}\right)^{*}$$

$$\hat{F}_{1ml}'^{+}=\hat{F}_{1ml}^{*}\quad\hat{F}_{2ml}'=\left(\hat{F}_{2ml}^{+}\right)^{*}\quad\hat{F}_{2ml}'^{+}=\hat{F}_{2ml}^{*}\tag{67}$$

The time reversal of total transition probability of stimulated absorption of double photons in the first and second processes is

$$W_{T2\omega=\omega_{ml}}^{(2)}=\frac{2\pi}{\hbar^{2}}\left\{\left|\hat{F}_{2ml}\right|^{2}+\frac{4}{\hbar^{2}\omega_{ml}^{2}}\left|\hat{F}_{1ml}\right|^{2}\left|\hat{F}_{1ll}-\hat{F}_{1mm}\right|^{2}\right.$$

$$\left.+\frac{4}{\hbar\omega_{ml}}\text{Re}\left[\hat{F}_{2ml}^{*}\hat{F}_{1ml}\left(\hat{F}_{1ll}-\hat{F}_{1mm}\right)\right]\right\}\delta\left(2\omega-\omega_{ml}\right)\tag{68}$$

According to the formulas (18), (19) and (32), we have

$$\left(\hat{F}_{2ml}^{+}\right)^{*}\hat{F}_{1ml}^{+}\left(\hat{F}_{1ll}^{+}-\hat{F}_{1mm}^{+}\right)=\hat{F}_{2ml}^{*}\hat{F}_{1ml}\left(\hat{F}_{1ll}-\hat{F}_{1mm}\right)\tag{69}$$

as well as $\left|\hat{F}_{2ml}^{+}\right|^{2}=\left|\hat{F}_{2ml}^{+}\right|^{2}$, $\left|\hat{F}_{1ml}^{+}\right|^{2}=\left|\hat{F}_{1ml}^{+}\right|^{2}$, $\left|\hat{F}_{1ll}^{+}-\hat{F}_{1mm}^{+}\right|^{2}=\left|\hat{F}_{1ll}-\hat{F}_{1mm}\right|^{2}$, we have $W_{T2\omega=\omega_{ml}}^{(2)}=W_{2\omega=\omega_{ml}}^{(2)}$ .So there is no symmetry violation of time reversal in the first and second order processes of double photons. We should consider the third order process. According to (12), the transition probability amplitude of the third order process is

$$a_{m}^{(3)}\left(t\right)=\frac{1}{i\hbar}\sum_{n}\int_{0}^{t}\hat{H}_{2mn}'a_{n}^{(1)}\left(t\right)e^{i\omega_{mn}t}dt+\frac{1}{i\hbar}\sum_{n}\int_{0}^{t}\hat{H}_{1mn}'a_{n}^{(2)}\left(t\right)e^{i\omega_{mn}t}dt\tag{70}$$

By taking the integral of the formula, we can obtain the probability amplitude of the third order processes. The result is shown in appendix. For the double photon absorption processes with $2\omega=\omega_{ml}$, by rotation wave approximation, the probability amplitude and transition probability in unit time are individually

$$a_m^{(3)}(t)_{2\omega=\omega_{ml}} = \frac{\left[e^{-i(2\omega-\omega_{ml})t}-1\right]}{\hbar(2\omega-\omega_{ml})} \frac{4\hat{F}_{1ml}^+\left(\hat{F}_{1ll}^+-\hat{F}_{1mm}^+\right)\left(\hat{F}_{1ll}^+-\hat{F}_{1ll}^+\right)}{\hbar^2\omega_{ml}^2} \tag{71}$$

$$W_{2\omega=\omega_{ml}}^{(3)} = \frac{2\pi}{\hbar^2}\left\{\left|\hat{F}_{2ml}^+\right|^2 + \frac{4\left(1+4A_l^2\right)}{\hbar^2\omega_{ml}^2}\left|\hat{F}_{1ml}^+\right|^2\left|\hat{F}_{1ll}^+-\hat{F}_{1mm}^+\right|^2\right.$$

$$\left. +\frac{4}{\hbar\omega_{ml}}\mathrm{Re}\left[\left(1+i2A_l\right)\left(\hat{F}_{2ml}^+\right)^*\hat{F}_{1ml}^+\left(\hat{F}_{1ll}^+-\hat{F}_{1mm}^+\right)\right]\right\}\delta(2\omega-\omega_{ml}) \tag{72}$$

Here $iA_l = \hat{F}_{1ll}^+ - \hat{F}_{1ll}^+$. On the other hand, by taking $k \to j$ in (24), we can get the time reversal of (70)

$$a_{Tm}^{(3)}(t) = a_l^{(3)}(-t) = -\frac{1}{i\hbar}\sum_j\int_0^t\hat{H}'_{2Tlj}a_j^{(1)}(-t)e^{i\omega_{jl}t}dt - \frac{1}{i\hbar}\sum_j\int_0^t\hat{H}'_{1Tlj}a_j^{(2)}(-t)e^{i\omega_{jl}t} \tag{73}$$

By considering relations $\hat{H}'_{2Tlj} = \hat{H}''^*_{2jl}$, $\hat{H}'_{1Tlj} = \hat{H}''^*_{1jl}$, $\hat{F}'_{1ml} = \left(\hat{F}^+_{1ml}\right)^*$, $\hat{F}'^+_{1ml} = \hat{F}^*_{1ml}$, $\hat{F}'_{2ml} = \left(\hat{F}^+_{2ml}\right)^*$ and $\hat{F}'^+_{2ml} = \hat{F}^*_{2ml}$ as well as by the same method to do integral and take rotation wave approximation,, we get the time reversals of probability amplitude and transition probability in the third order process of double photon absorption individually

$$a_{Tm}^{(3)}(t)_{2\omega=\omega_{ml}} = \frac{\left[e^{-i(2\omega-\omega_{ml})t}-1\right]}{\hbar(2\omega-\omega_{ml})} \frac{4\hat{F}'^+_{1ml}\left(\hat{F}'^+_{1ll}-\hat{F}'^+_{1mm}\right)\left(\hat{F}'_{1mm}-\hat{F}'^+_{1mm}\right)}{\hbar^2\omega_{ml}^2}$$

$$= -\frac{\left[e^{-i(2\omega-\omega_{ml})t}-1\right]}{\hbar(2\omega-\omega_{ml})} \frac{4\hat{F}^*_{1ml}\left(\hat{F}^*_{1ll}-\hat{F}^*_{1mm}\right)\left(\hat{F}_{1mm}-\hat{F}^+_{1mm}\right)^*}{\hbar^2\omega_{ml}^2} \tag{74}$$

$$W_{T2\omega=\omega_{ml}}^{(3)} = \frac{2\pi}{\hbar^2}\left\{\left|\hat{F}_{2ml}\right|^2 + \frac{4\left(1+4A_m^2\right)}{\hbar^2\omega_{ml}^2}\left|\hat{F}_{1ml}\right|^2\left|\hat{F}_{1ll}-\hat{F}_{1mm}\right|^2\right.$$

$$\left. +\frac{4}{\hbar\omega_{ml}}\mathrm{Re}\left[\left(1+i2A_m\right)\hat{F}_{2ml}\hat{F}^*_{1ml}\left(\hat{F}^*_{1ll}-\hat{F}^*_{1mm}\right)\right]\right\}\delta(2\omega-\omega_{ml}) \tag{75}$$

Here $iA_m = \hat{F}_{1mm} - \hat{F}^+_{1mm}$. Comparing with (72), we know that the difference $A_l \neq A_m$ leads to time reversal symmetry violation. But if the high order multiple moment effects are omitted, symmetry violations of time reversals in the third order processes would not exist.

8. Sum frequency process and its time reversal

The process of sum frequency process in non-linear optics is that an electron translates into higher energy lever $|m\rangle$ from low energy level $|l\rangle$ by absorbing two photons with frequencies ω_1 and ω_2 individually, then emits out a photon with frequency $\omega_3 = \omega_1 + \omega_2$ and translates from higher energy level $|m\rangle$ into low energy level $|l\rangle$ again. Suppose that incident light is parallel one containing frequencies ω_1, ω_2 and $\omega_3 = \omega_1 + \omega_2$, and the strength of electrical field is $\vec{E}_0$, the interaction Hamiltonians between electron and radiation field are

$$\hat{H}'_1 = \sum_{\lambda=1}^{3} \left(\hat{F}_{1\lambda} e^{i\omega_\lambda t} + \hat{F}^+_{1\lambda} e^{-i\omega_\lambda t} \right) \quad \hat{H}'_2 = \sum_{\lambda=1}^{3} \left(\hat{F}_{2\lambda} e^{i\omega_\lambda t} + \hat{F}^+_{1\lambda} e^{-i\omega_\lambda t} + \hat{F}_{0\lambda} \right) \tag{76}$$

Here

$$\hat{F}_{1\lambda} = -\frac{q\vec{E}_0}{2\omega_\lambda \mu} \cdot e^{-i\vec{k}_\lambda \cdot \vec{R}} \hat{p} \quad \hat{F}^+_{1\lambda} = -\frac{q\vec{E}_0}{2\omega_\lambda \mu} \cdot \hat{p}^+ e^{i\vec{k}_\lambda \cdot \vec{R}} \tag{77}$$

$$\hat{F}_{2\lambda} = \frac{q^2 E_0^2}{2\omega_\lambda^2 \mu} e^{-i2\vec{k}_\lambda \cdot \vec{R}} \quad \hat{F}^+_{2\lambda} = \frac{q^2 E_0^2}{2\omega_\lambda^2 \mu} e^{i2\vec{k}_\lambda \cdot \vec{R}} \quad \hat{F}_{0\lambda} = \frac{q^2 E_0^2}{\omega_\lambda^2 \mu} \tag{78}$$

When we calculate probability amplitude, the result corresponds to let $\omega \to \omega_\lambda$ and take sum over index λ in (33) and the formula in the appendix of this paper. For the process an electron absorbs two photons with frequencies ω_1 and ω_2, transits from low energy level $|l\rangle$ into high energy level $|m\rangle$, transition probability corresponds to let $2\omega = \omega_1 + \omega_2 = \omega_{ml}$ in (33) of double photon absorption process. When the electron transits back from high energy level $|m\rangle$ into low energy level $|l\rangle$ by emitting a photon with frequency ω_3, after the retarded effect and high order processes are considered, according to (47), the transition probability is

$$W^{(2)}_{\omega_3 = \omega_{ml}} = \frac{2\pi}{\hbar^2} \left| \hat{F}_{1ml} \right|^2 \left\{ 1 + \frac{A_m'^2}{\hbar^2 \omega_{ml}^2} \right\} \delta(\omega_3 - \omega_{ml}) \tag{79}$$

Therefore, for the sum frequency process that an electron translates into state $|m\rangle$ from state $|l\rangle$ by absorbing two photons with frequencies ω_1 and ω_2, then translates back into original state $|l\rangle$ from state $|m\rangle$ by emitting out a photon with frequencies $\omega_3 = \omega_1 + \omega_2$, the total transition probability is

$$W^{(2)}_{\omega_1 + \omega_2 = \omega_{ml}} + W^{(2)}_{\omega_3 = \omega_{ml}} = \frac{2\pi}{\hbar^2} \left\{ \left| \hat{F}^+_{2ml} \right|^2 + \frac{4\left(1 + 4A_l^2\right)}{\hbar^2 \omega_{ml}^2} \left| \hat{F}^+_{1ml} \right|^2 \left| \hat{F}^+_{1ll} - \hat{F}^+_{1mm} \right|^2 \right.$$

$$\left. + \frac{4}{\hbar \omega_{ml}} \mathrm{Re}\left[\left(1 + i2A_l\right)\left(\hat{F}^+_{2ml}\right)^* \hat{F}^+_{1ml}\left(\hat{F}^+_{1ll} - \hat{F}^+_{1mm}\right) \right] \right\} \delta(2\omega - \omega_{ml})$$

$$+\frac{2\pi}{\hbar^2}\left|\hat{F}_{1ml}\right|^2\left\{1+\frac{A_m'^2}{\hbar^2\omega_{ml}^2}\right\}\delta(\omega_3-\omega_{ml})\qquad(80)$$

Meanwhile, according to the calculation in (38), after the retarded effects and high order processes are considered, the transition probability that an electron transits from state $|l\rangle$ into state $|m\rangle$ by absorbing a photon with frequency ω_3 is

$$W_{T\omega_3=\omega_{ml}}^{(2)}=\frac{2\pi}{\hbar^2}\left|\hat{F}_{1ml}\right|^2\left\{1+\frac{A_l^2}{\hbar^2\omega_{ml}^2}\right\}\delta(\omega_3-\omega_{ml})\qquad(81)$$

So for the time reversal of sum frequency process that an electron absorbs a photon with frequency ω_3 and transits from state $|l\rangle$ into state $|m\rangle$, then emits two photons with frequencies ω_1 and ω_2, and transits back into original state $|l\rangle$, the total transition probability is

$$W_{T\omega_1+\omega_2=\omega_{ml}}^{(2)}+W_{T\omega_3=\omega_{ml}}^{(2)}=\frac{2\pi}{\hbar^2}\left\{\left|\hat{F}_{2ml}\right|^2+\frac{4\left(1+4A_m'^2\right)}{\hbar^2\omega_{ml}^2}\left|\hat{F}_{1ml}\right|^2\left|\hat{F}_{1ll}-\hat{F}_{1mm}\right|^2\right.$$

$$\left.+\frac{4}{\hbar\omega_{ml}}\mathrm{Re}\left[\left(1+i2A_m'\right)\hat{F}_{2ml}\hat{F}_{1ml}^*\left(\hat{F}_{1ll}^*-\hat{F}_{1mm}^*\right)\right]\right\}\delta(2\omega-\omega_{ml})$$

$$+\frac{2\pi}{\hbar^2}\left|\hat{F}_{1ml}^+\right|^2\left\{1+\frac{A_l^2}{\hbar^2\omega_{ml}^2}\right\}\delta(\omega_3-\omega_{ml})\qquad(82)$$

Similarly, because of $A_l\neq A_m'$, sum frequency process violates time reversal symmetry.

By the same method, we can prove that the other processes of non-linear optics just as double frequency, difference frequency, parametric amplification, Stimulated Raman scattering, Stimulated Brillouin scattering and so on are also asymmetric under time reversal. The reason is the same that the light's high order stimulated radiation and absorption processes are asymmetric under time reversal after retarded effect of radiation fields are taken into account.

9. Non-linear polarizations and symmetry violation of time reversal

What is discussed above is based on quantum mechanics. But in non-linear optics, we often calculate practical problems based on classical equations of electromagnetic fields. So we need to discuss the revised non-linear polarizations when the retarded effect of radiation field is taken into account. According to the current theory of nonlinear optics, polarizations are unchanged under time reversal. This does not coincident with practical situations. The reason is that current theory only considers the dipolar approximation without considering the high order processes and the retarded effects of radiation fields. We now discuss the revision of non-linear polarizations after the retarded effect of radiation fields and the high order perturbation processes are taken into account.

Let B_{ml} represent the stimulated absorption probability of an electron (unit radiation density and unit time) transiting from initial low-energy state $|l\rangle$ to final high-energy state $|m\rangle$, B_{lm} represent the probability of stimulated radiation of an electron (unit radiation density and unit time) transiting from the initial high-energy state $|m\rangle$ into the final low-energy state to $|l\rangle$, we have

$$B_{ml} = \frac{4\pi^2}{3\hbar^2}\left|\bar{D}_{ml}\right|^2 (1+\lambda_{ml}) \qquad B_{lm} = (B_{ml})_T = \frac{4\pi^2}{3\hbar^2}\left|\bar{D}_{ml}\right|^2 (1+\lambda'_{ml}) \tag{83}$$

Here $\bar{D}_{ml}$ is the dipolar moment of an electron. We have $\lambda_{ml} \neq \lambda'_{ml}$ and $B_{ml} \neq B_{lm}$ in general, i.e., the parameters of light's stimulated radiation and absolution are not the same. It also means that the nonlinear optical processes would violate time reversal symmetry in general. Let $\bar{D}'_{ml} = \sqrt{1+\lambda_{ml}}\,\bar{D}_{ml}$ representing the revised dipolar moment after retarded effect is considered, $\bar{D}'_{lm} = \sqrt{1+\lambda_{lm}}\,\bar{D}_{lm}$ representing the time reversal of $\bar{D}'_{ml}$, we have $\lambda_{ml} \neq \lambda_{lm}$ and $\bar{D}'_{ml} \neq \bar{D}'_{lm}$ in general. Therefore, as long as we let $\bar{D}_{ml} \rightarrow \bar{D}'_{ml}$ in the current formula $\chi^{(n)}_{ij...k}$ of non-linear polarizations, we obtain the revised formula after the retarded effect of radiation fields and the high order perturbation processes are taken into account. Correspondingly, we let $\bar{D}_{lm} \rightarrow \bar{D}'_{lm}$ and obtain the time reversal formula $\chi^{(n)}_{Tij...k}$. It is obvious that non-linear polarizations can not keep unchanged under time reversal. For example, for the non-linear polarizations of the second order processes, we have

$$\chi^{(2)}_{ijk} = (-\omega_1 - \omega_2, \omega_1, \omega_2) = \frac{N}{4\varepsilon_0\hbar^2}\sum_{nn'}\sum_p \frac{(D'_i)_{\beta n'}(D'_j)_{n'n}(D'_k)_{n'\beta}}{(\omega_{n'\beta}+\omega_2)(\omega_{n\beta}+\omega_1+\omega_2)} \tag{84}$$

Its time reversal is

$$\chi^{(2)}_{Tijk} = (-\omega_1 - \omega_2, \omega_1, \omega_2) = \frac{N}{4\varepsilon_0\hbar^2}\sum_{nn'}\sum_p \frac{(D'_i)_{n'\beta}(D'_j)_{n\,n'}(D'_k)_{\beta\,n'}}{(\omega_{n'\beta}+\omega_2)(\omega_{n\beta}+\omega_1+\omega_2)} \tag{85}$$

In general, we have $\bar{D}'_{ml} \neq \bar{D}'_{lm}$ and $\chi^{(2)}_{ijk} \neq \chi^{(2)}_{Tijk}$, so we have $\chi^{(n)}_{ij...k} \neq \chi^{(n)}_{Tij...k}$. Therefore, the polarization formula of electrical medium and its time reversal are

$$\bar{P} = \varepsilon_0\left(\chi^{(1)}_i E_i + \chi^{(2)}_{ij}E_iE_j + \chi^{(3)}_{ijk}E_iE_jE_k \cdots\right) \tag{86}$$

$$\bar{P}_T = \varepsilon_0\left(\chi^{(1)}_{Ti}E_i + \chi^{(2)}_{Tij}E_iE_j + \chi^{(3)}_{Tijk}E_iE_jE_k \cdots\right) \tag{87}$$

We have $P_T \neq P$ in general. So the motion equation of classical electrical field and its time reversal are also asymmetrical in general with forms

$$\nabla^2\bar{E} - \mu_0\varepsilon_0\frac{\partial^2}{\partial t^2}\bar{E} = \mu_0\frac{\partial^2}{\partial t^2}\bar{P} \qquad \nabla^2\bar{E} - \mu_0\varepsilon_0\frac{\partial^2}{\partial t^2}\bar{E} = \mu_0\frac{\partial^2}{\partial t^2}\bar{P}_T \tag{88}$$

Because (86) ~ (88) are the basic equations of nonlinear optics, we can see that the general nonlinear optical processes violate time reversal symmetry.

In fact, by analyzing nonlinear optics phenomena without complex calculations, we can know the irreversibility of nonlinear optical processes. Although the irreversibility concept is not completely the same as that of the asymmetry of time reversal, they coincide in essence. So let's analyze some practical examples to exposure the irreversibility of nonlinear optics processes below.

10. Irreversibility of nonlinear optics processes

As we know that the processes of linear optics such as light's propagations, reflection, refraction, polarization and so on in uniform mediums are reversible. For example, light's focus through a common convex mirror shown in Fig.6. When a beam of parallel light is projected into a convex mirror, it is focused at point O. If we put a same convex mirror at point B and O is also the focus of convex mirror B, light emitted from point O will become a beam of parallel light again when it passes the convex mirror B. The process that light moving from $O \rightarrow B$ can be regarded as the time reversal process of light moving from $A \rightarrow O$. It is obvious that the process is reversible. The second example is that a beam of white sunlight can be decomposed to a spectrum with different colors by a prism. When these lights with difference colors are reflected back into prism along same paths, white sunlight will be formed again. The third example is that a beam of light can become two different polarization lights with different propagation directions when the light is projected into a double refraction crystal. If these two polarization lights are reflected back into the crystal along same path again, the original light is formed. All of these processes are reversible. But in the processes of non-linear optics, reversibility does not exist. Some examples are shown below.

10.1 Light's multiple frequency, difference frequency and parameter amplification

As shown in Fig.2, a beam of laser with frequency ω is projected into a proper medium and proper phase matching technology is used. The light with double frequency 2ω is found in out going light besides original light with frequency ω. If the lights with frequencies ω and 2ω are reflected back into the same medium, as shown in Fig.2, they can't be completely synthesized into the original light with a single frequency ω. Some light with frequency ω will become light with multiple frequency again by multiple frequency process and some light with frequency 2ω will become the light with frequency ω by difference frequency process. Meanwhile, some light with frequencies ω and 2ω will penetrate medium without being changed as shown in Fig. 3. So the original input light can' be recovered and the reversibility of process is broken. The situations are the same for sum frequency, difference frequency and parameter amplification processes and so on.

10.2 Bistability of optics [3]

As shown in Fig. 4 and 5, the processes of optical bistability are similar to the polarization and magnetization processes of ferroelectrics and ferromagnetic. In the processes the hysteretic loops are formed between incident and outgoing electrical field strengths. In the polarization   and   magnetization   processes   of   ferroelectrics   and   ferromagnetic, electromagnetic fields changing along positive directions can be regarded as the time reversal of fields changing along negative directions. There exists electric and magnetic hysteresis. The hysteretic loops are similar to heat engine cycling loops. After a cycling, heat dissipation is produced and the reversibility of process is violated.

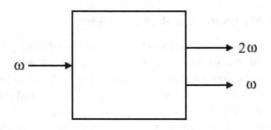

Fig. 2. Process of light's multiple frequency

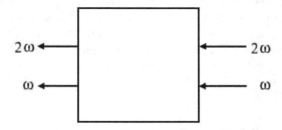

Fig. 3. Time Reversal process of Light's multiple frequency

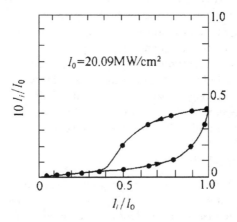

Fig. 4. Optical bistability of nitrobenzene

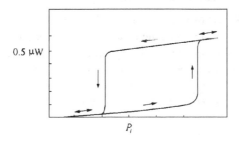

0.5 μW

P_i

Fig. 5. Optical bistability of mixing type

10.3 Self-focusing and self-defocusing processes of light [4]

Medium's refractive index will change nonlinearly when a beam of laser with uneven distribution on its cross section, for example the Gauss distribution, is projected into a proper medium. The result is that medium seems becoming a convex or concave mirror so that parallel light is focused or defocused. This is just the processes of self-focusing and self-defocusing of lights. The stationary self-focusing process is shown in Fig.7. Parallel light is focused at point O. Then it becomes a thin beam of light projecting out medium. We compare it with common focusing process shown in Fig. 6. If the self-focusing process is reversible, the light focused at pint O would become parallel light again when it projecting out the medium as shown in dotted lines in Fig.7. But it dose not do actually. So the self-focusing process is irreversible. And so do for the self-defocusing process of light.

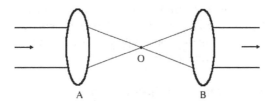

A   B

Fig. 6. Focusing process of light.

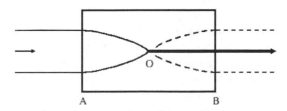

A   B

Fig. 7. Self-focusing process of light.

10.4 Double and multi-photon absorption [5]

In double absorption process of photons, an electron in low-energy level will absorb two photons with frequencies ω_1 and ω_2, then transits to high-energy level. But if the electron at high-energy level transits back to low-energy level, it either gives out only a photon at frequency $\omega_3 = \omega_1 + \omega_2$, or two photons at frequencies $\omega_1' \neq \omega_1$, $\omega_2' \neq \omega_2$ in general. It will not produce two photons with original frequencies ω_1 and ω_2. Double photon absorption process is irreversible. And so is for multi-photon absorption.

10.5 Photon echo phenomena [6]

Under certain temperature and magnetic field condition, a beam of laser can be split into two lights with a time difference by using a time regulator of optics. Then two lights are emitted into a proper crystal. Thus three light signals can be observed when they pass through the crystal. The last signal is photon echo. This is a kind of instant coherent phenomena of light. If these three lights signals are imported into same medium again, they can't return into origin two lights. Either three signals are observed (no new echo is produced) or more signals are observed (new signals are produced). In fact, besides photon echo, there are electron spin echo, ferromagnetic echo and plasma echo and so on. All of them are irreversible and violate time reversal symmetry.

10.6 Light's spontaneous radiation processes

As we known that there exist two kinds of different processes for light's radiations, i.e., spontaneous radiation and stimulated radiation. However, there exists only one kind of absorption process, i.e., stimulated radiation without spontaneous absorption in nature. An electron can only transform from high energy level into low energy level by emitting a photon spontaneously, but it can not transform from low energy level into high energy level by absorbing a photon spontaneously. So the processes of light's absorptions themselves are obviously asymmetrical under time reversal.

11. Influence on the fundamental theory of laser

The influence of higher order revision on the fundamental theory of laser is discussed below. Let us first discuss the double energy level system. Let N_1 represent the number of electrons on lower energy level and N_2 represent the number of electrons on higher energy level. With higher order revision, we have $B_{12} \neq B_{21}$. Under the circumstance of having no electron population revision $N_2 < N_1$, as long as B_{21} is larger enough than B_{12}, we still have $B_{21}N_2 > B_{12}N_1$. This means that without the reversion of electron population, laser can still be produced. At present, many experiments have verified this result [7]. In fact, the number of electrons on an energy level can not be determined directly at present. What can be determined by experiments is the number of photons emitted by atoms. And the numbers of photons are calculated by utilizing $\rho B_{21}N_2$, $\rho B_{12}N_1$ and $A_{21}N_2$. By considering the high order retarded effect of radiation fields, the condition of stimulated amplification to produce laser should be changed from $N_2 > N_1$ to $B_{21}N_2 > B_{12}N_1$.

Secondly, according to the current theory, we must have at least three energy levels to produce laser. For the systems with two energy levels, there is a so-called fine balance with $B_{12}N_1\rho(\nu) = A_{21}N_2 + B_{21}N_2\rho(\nu)$. If $B_{12} = B_{21}$ and $A_{21} = k_{21}B_{21}$, we have

$$\frac{N_2}{N_1} = \frac{\rho}{(\rho + \kappa_{21})} < 1 \tag{89}$$

That is $N_2 < N_1$. In this case, we do not have population reversion, therefore no laser is produced. According this paper, suppose we still have $A_{21} = k_{21}B_{21}$, when the balance is reached, we still have

$$\frac{N_2}{N_1} = \frac{\rho B_{12}}{(\rho + \kappa_{21})B_{21}} \tag{90}$$

Because of $B_{12} \neq B_{21}$, as long as relation $\rho B_{12} > (\rho + \kappa_{21})B_{21}$ is satisfied, we still have $N_2 > N_1$ so that population reversion can still be caused. But in this case, we have $B_{21}N_2 < B_{12}N_1$. That is to say, for the steady system of double energy levels, even under the condition of population reversion, laser can still not being produced. For the non-steady system of double energy levels, we have two cases

$$\frac{dN_2}{dt} = B_{12}N_1\rho - A_{21}N_2 - B_{21}N_2\rho > 0 \tag{91}$$

$$\frac{dN_2}{dt} = B_{12}N_1\rho - A_{21}N_2 - B_{21}N_2\rho < 0 \tag{92}$$

When $dN_2 / dt > 0$, we have $B_{12}N_1\rho > A_{21}N_2 + B_{21}N_2\rho$, so $B_{12}N_1\rho > B_{21}N_2\rho$, no laser is produced. When $dN_2 / dt < 0$, we have $B_{12}N_1\rho < A_{21}N_2 + B_{21}N_2\rho$. In this case, if $B_{12}N_1\rho < B_{21}N_2\rho$, laser can be produced. If $B_{12}N_1\rho > B_{21}N_2\rho$, laser can be produced.

Next, we discuss the influence on the system of three energy levels. The standard stimulated radiation and absorption process in the system of three energy levels is shown in Fig.8. In the current theory, however, the processes to produce laser is actually simplified as shown in Fig 9. By analyzing the difference between them, we know the significance of this paper's revision. According to Fig.8, when particles which are located on ground state E_1 at the beginning are pumped into E_3 energy level, they can transit into E_2 energy level through both radiation transition and non-radiation transition. The population reversion can be achieved between E_1 and E_2, so that the laser with frequency ω_{21} can be produced. Comparing with Fig.8, the process shown in Fig.9 omits the spontaneous radiation and stimulated radiation transitions, as well as particle's transition from E_2 level into E_3 level. According to the Einstein's theory, we have $B_{13} = B_{31}$. The possibility is the same for a particle transiting from ground state into E_3 energy level and transiting from E_3 energy level back into ground state. Suppose that the number of particles transiting in unit time from ground state into E_3 energy level is $\rho(\nu_{13})B_{13}N_1$. There will be $\rho B_{31}N_3$ particles transiting back to ground state from E_3 energy level by stimulated radiation, and

$A_{31}N_3 = \kappa_{31}B_{31}N_3$ particles transiting back to ground state from E_3 energy level by spontaneous radiation. Therefore, most particles which have transited to E_3 energy level will come back to original ground state by emitting photons with frequency ω_{31}, so that population reversion between E_1 and E_2 energy levels will be affected greatly. Meanwhile, because of $B_{23} = B_{32}$, some particles on E_2 energy level coming from E_3 energy level will transit back into E_3 energy level by stimulated absorption, so that population reversion between E_1 and E_2 will also be reduced. These results indicate that the Einstein's theory is only suitable for equilibrium processes, rather than the non-equilibrium process of laser production.

The current theory of laser uses a fussy method to avoid theses problems. The probability a particle transits back to ground state from E_3 energy level is not considered directly. In stead, we use a pumping speed R replaces $\rho_{13}B_{31}N_1 - (\rho_{13} + \kappa_{31})B_{31}N_3$. On the other hand, the non-radiation transition is used to replace $(\rho_{23} + \kappa_{32})B_{32}N_3 - \rho_{23}B_{32}N_2$. In this way, the complexity of process is simplified.

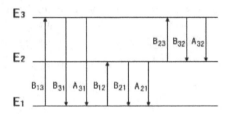

Fig. 8. Transition among three energy levels.

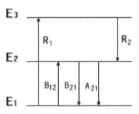

Fig. 9. Simplification of Fig.8.

According to the revision in this paper, we have $B_{ml} \neq B_{lm}$. We can provide a simpler and rational picture for the production of laser in the system of three energy levels. In this case, we can have $B_{31} \ll B_{13}$ and $(\rho_{13} + \kappa_{31})B_{31}N_3 \ll \rho_{13}B_{31}N_1$, so that only a few particles can transit back to ground state from E_3 energy level by stimulated radiation and spontaneous radiation after they transit from ground state into E_3 energy level. Most of particles on E_3 energy level will transit into E_2 energy level. On the other hand, because of $B_{23} \ll B_{32}$, we have $\rho_{23}B_{23}N_2 \ll (\rho_{23} + \kappa_{32})B_{32}N_3$. Most of particles can not transit back into E_3 energy level after they transit to E_2 energy from E_3 energy level. Meanwhile, because of $B_{12} \ll B_{21}$, it is difficult for particles on ground state to transit into E_2 energy level from ground state by stimulated absorption, but is easy to transit to ground state from E_2 energy level by stimulated radiation. Therefore, a high effective and ideal laser system of three energy levels should satisfy the conditions $B_{21} \ll B_{12}$, $B_{31} \ll B_{13}$ and $B_{23} \ll B_{32}$. It is obvious that as long as $B_{ml} \neq B_{lm}$, we can simply and rationally explain the production of laser of the system of three energy levels.

In this way, we can also well explain the phenomenon of optical self-transparence and self absorptions [8]. Experiments show that in strong electric fields, some medium can have the saturated absorption of light, so that the medium will become transparent for light. The current explanation of saturated absorption is that the number N_1 of particles located on low energy level becomes smaller and the absorption of light is proportional to the number of particles located on low energy level, therefore the stimulated absorption becomes smaller. Meanwhile, the transmission light increases due to the stimulated radiation of particles located on high energy level, i.e., the self-transparence phenomena of saturated absorption appears. The problem of this explanation is that if the number N_1 of particles located on low energy level decreases and the number N_2 of particles located on high energy level increase, the spontaneous radiation will also increase. When stationary states are reached, we always have $A_{21}N_2$ photons emitted in the form of spontaneous radiation in unit time. Because spontaneous radiation is in all directions of space, it is difficult for medium to achieve real transparence.

According to the revised theory of this paper, the revised factor is $\alpha_{ml} \sim E_0^2$. If $\alpha_{ml} < 0$, in strong field, we may have $\alpha_{ml} \sim -1$ for some mediums, so that the stimulated absorption parameter B_{ml} is very small even becomes $B_{ml} \sim 0$. In this case, even though a great number of particles are still located on low energy level, the saturated absorption of light is still possible so that the medium become transparence. However, according to current theory, we have $B_{ml} \sim E_0^2$. When E_0 increases, the stimulated absorption parameter will increase so that it is impossible for us to have $B_{ml} \sim 0$. Conversely, if $\alpha_{ml} > 0$ with $\alpha_{ml} \sim E_0^2$, light's absorption for some mediums will increase greatly in strong field. This is just the phenomena of self absorption. In the current non-linear optics, we explain the phenomena of self absorption with the absorptions of double photons or multi-photons, as well as stimulated scattering. Based on this paper, besides the absorptions of double photons or multi-photons, the process of single photon can also cause trans-normal absorption. It is obvious that the revised theory can explain theses phenomena more rationally.

12. Discussion on the reasons of symmetry violation of time reversal

We need to discuss the reason of the symmetry violation of time reversal In the paper, semi-classical method is used, i.e., quantum mechanics is used to describe charged particles and classical electromagnetic theory is used to describe radiation fields. The limitation of this method is that spontaneous radiation can not be deduced automatically from the theory. The spontaneous radiation formula has to be obtained indirectly by means of the Einstein's theory of light's radiation and absorption. Strictly, we should discuss the problems using complete quantum theory, from which we can deduce spontaneous radiation probability automatically.

However, as we known, except the spontaneous radiation, the results are exactly the same by using both the semi-classical method and the complete quantum method to calculate light's stimulated radiation and absorption probabilities. It also means that if we use complete quantum mechanics to discuss light's stimulated radiation and absorption, time reversal symmetry will also be violated after the retarded effects of radiation fields are taken into account. It is just the spontaneous radiation which indicates the asymmetry of time reversal in the processes of interaction between light and charged particles, for there exits only light's spontaneous radiation without light's spontaneous absolution in nature. This result is completely asymmetrical.

In fact, in complete quantum mechanics, we use photon's creation and annihilation operator $\hat{a}^+$ and $\hat{a}$ to replace the factor $-q\bar{E}_0 / 2\omega\mu$ in semi-classical theory. This kind of correspondence does not change the results of time reversal symmetry violation in calculation processes. The problem is that if photon's creation or annihilation operators are used, some complexity and problem will be caused in high order processes so that it may be too difficult to calculate. So in the problems of light's stimulated radiation and absorption and nonlinear optics, we use actually semi-classical or even complete classical theory and methods and always obtain satisfied results at present.

Because the interaction Hamiltonian and the motion equation of quantum mechanics keep the same under time reversal, what causes the symmetry violation of time reversal? The method of rotation wave approximation is used in the paper. Does this approximation method cause the symmetry violation of time reversal? Let's discuss this issue in the following.

Suppose that micro-states are described by $|\psi\rangle$ and $|\phi\rangle$. Their time reversal are $|\psi_T\rangle = T|\psi\rangle$ and $|\phi_T\rangle = T|\phi\rangle$. Suppose that the interaction Hamiltonian remains the same under time reversal, according to quantum mechanics, we have the so-called detail balance formula

$$\langle\psi|\hat{H}|\phi\rangle = \langle\phi_T|\hat{H}|\psi_T\rangle \tag{93}$$

It indicates that the probability amplitude keeps the same under time reversal in the quantum transition process. For the problem of light's stimulated radiation and absolution, if the radiation field is only one with a single frequency, the interaction Hamiltonian is

$$\hat{H} = \hat{F}_0 + \hat{F}_1 e^{i\omega t} + \hat{F}_1^+ e^{-i\omega t} + \hat{F}_2 e^{i2\omega t} + \hat{F}_1^+ e^{-i2\omega t} \tag{94}$$

Meanwhile, if only a single particle state is considered, we have

$$|\psi\rangle = |\phi\rangle = \sum_m a_m(t) e^{-\frac{i}{\hbar}E_m t}|m\rangle \quad |\psi_T\rangle = |\phi_T\rangle = \sum_m a_m(-t) e^{\frac{i}{\hbar}E_m t}|m\rangle \tag{95}$$

Substitute (94) and (95) into (93), we obtain

$$\sum_{m,l} e^{\frac{i}{\hbar}(E_m - E_l)t} a_m(t) a_l(t) \langle m|\hat{F}_0 + \hat{F}_1 e^{i\omega t} + \hat{F}_1^+ e^{-i\omega t} + \hat{F}_2 e^{i2\omega t} + \hat{F}_1^+ e^{-i2\omega t}|l\rangle$$

$$= \sum_{m,l} e^{-\frac{i}{\hbar}(E_m - E_l)t} a_m(-t) a_l(-t) \langle m|\hat{F}_0 + \hat{F}_1 e^{i\omega t} + \hat{F}_1^+ e^{-i\omega t} + \hat{F}_2 e^{i2\omega t} + \hat{F}_1^+ e^{-i2\omega t}|l\rangle \tag{96}$$

The formula is the sum of multinomial. It means that the total probability amplitude is unchanged under time reversal. However, by the constraint of energy conservation law, in the formula above, only a few items which satisfy the condition $E_m - E_l = \pm n\hbar\omega$ can be realized really. Those items which do not satisfy the condition are forbidden actually. Keeping the items which satisfy the condition of energy conservation and giving up the items which do not, the procedure is just the so-called rotation wave approximation. It is obvious that the two sides of equation (96) will not equal to each other after going through the procedure, i.e., the symmetry of time reversal will be violated.

The paper calculates the transition and time reversal problems of partial items corresponding to the operators $\hat{F}_1 e^{-i\omega t}$ and $\hat{F}_1^+ e^{i\omega t}$ under the situation $n = 1$. Because (68) can not be accurately calculated, we use approximation method and let $a_m(t) = a_m^{(0)}(t) + a_m^{(1)}(t) + a_m^{(2)}(t) + \cdots$. For the first order approximation, we have $a_m(t) = a_m^{(0)}(t) + a_m^{(1)}(t)$. Suppose that an atom transits from state $l\rangle$ into state $m\rangle$, we get transition probability (14) and its time reversal (29). It indicates that the first process is unchanged under time reversal after retarded effect of radiation field is considered. For the second processes, let $a_m(t) = a_m^{(0)}(t) + a_m^{(1)}(t) + a_m^{(2)}(t)$ and assume in the same way that an atom transits from state $l\rangle$ into state $m\rangle$, we get (38) and its time reversal (47). The result violates the symmetry of time reversal and the symmetry violation is relative to the asymmetry of initial states of bounding state atoms before and after time reversal. The uniform values of the Hamiltonian operator for the initial states of an atom before and after time reversal are not equal to each other. Therefore, one reason that causes the symmetry violation of time reversal is that the condition of energy conservation forbids some transition processes between bounding state atoms, so that realizable processes violates time reversal symmetry with

$$W_{\omega=\pm\omega_{ml}}\left(\hat{F}_{1ml}\right) + W_{2\omega=\pm\omega_{ml}}\left(\hat{F}_{1ml}, A_l\right) + W_{3\omega=\pm\omega_{ml}}\left(\hat{F}_{1ml}, A_l\right) + \cdots$$

$$\neq W_{T\omega=\pm\omega_{ml}}\left(\hat{F}_{1ml}\right) + W_{T2\omega=\pm\omega_{ml}}\left(\hat{F}_{1ml}, A'_m\right) + W_{T3\omega=\pm\omega_{ml}}\left(\hat{F}_{1ml}, A'_m\right) + \cdots \qquad (97)$$

Meanwhile, for concrete atoms, the other restriction conditions just like the wave function's symmetries should be also considered. So only a few and specific transitions can be achieved actually. Most processes in (96) can not be completed. These realizable processes are just what we can observe and measure. They are irreversible in general. Therefore, the symmetry violation of time reversal in the filial or partial processes of light's stimulated radiation and absolution do not contradict with the fine balance formula (93) actually.

On the other hand, the symmetry violation of time reversal is also related to the asymmetry of initial states of bounding state's atoms before and after time reversal. For the interaction between radiation fields and the non-bounding state's atoms with continuous energy levels, there exists no symmetry violation of time reversal. In this case, there is no the asymmetry problem of the initial states before and after time reversal. This is why we can not find symmetry violation of time reversal in the particle collision experiments for changed particles created by accelerators are non-bounding ones.

Meanwhile, there is a difference of negative sign between A_l shown in (37) and A'_m shown in (46). This difference is caused by the interference of amplitudes between the first order and the second order processes before and after time reversal. But if the retarded effect of radiation field is neglected, the processes of light's stimulated radiation and absolution will be symmetrical under time reversal. So the reasons of symmetry violation of time reversal are caused by multi-factors and are quite complex.

13. Influence on non-equilibrium statistical mechanics

As well-known that although classical equilibrium state statistical physics has been a very mature one, the foundation of non-equilibrium state statistical physics has not be established up to now day. The key is that the evolution processes of macro-systems controlled by the second law of thermodynamics are irreversible under time reversal, but the processes of micro-physics are considered reversible. Because macro-systems are composed of micro-particles, there exists a sharp contradiction here. This is so-called reversibility paradox which has puzzled physics community for a long time [9]. Though many theories have been proposed trying to resolve this problem, for example, the theories of coarseness and mixing current and so on [10], none is satisfied.

The significance of this paper is to provide us a method to solve this problem. We known that macro-systems are composed of atoms and molecules, and atoms and molecules are composed of charged particles. By the photon's radiations and absorptions, charged particles of bounding states and radiation fields interact. According to the discussion in the paper, after the retarded effects of radiation fields are considered, the time reversal symmetry of light's stimulated radiation and absolution is violated, even though the interaction Hamiltonian is unchanged under time reversal. Only when the system reaches macro-equilibrium states, or the probabilities of micro-particles radiating and absorbing photons are the same from the point of view of statistical average, the processes are reversible under time reversal. Therefore, it can be said that irreversibility of macro-processes originates from the irreversibility of micro-processes actually.

By introducing retarded electromagnetic interaction, the forces between charged particles will become non-conservative ones. Based on it, we can establish the revised Liouville equation which is irreversible under time reversal. In this way, we can lay a really rational dynamic foundation for classical non-equilibrium statistical mechanics. The united description can be reached for classical equilibrium and non-equilibrium statistical mechanics. The detail will be provided later.

14. Acknowledgment

The author gratefully acknowledges the valuable discussions of Professors Qiu Yishen in Physical and Optical Technology College, Fujian Normal University and Zheng Shibiao in Physical Department, Fuzhou University.

15. Appendix

15.1 The transition probability amplitude of the third order process for light's stimulated radiation and absorption

$$
a_m^{(3)}(t) = \frac{1}{\hbar^2} \sum_n \left\{ \frac{\hat{F}_{2mn}\hat{F}_{1nl}}{\omega + \omega_{nl}} \left[\frac{e^{i(3\omega+\omega_{nl}+\omega_{mn})t}-1}{3\omega+\omega_{nl}+\omega_{mn}} - \frac{e^{i(2\omega+\omega_{mn})t}-1}{2\omega+\omega_{mn}} \right] \right.
$$

$$
- \frac{\hat{F}_{2mn}\hat{F}_{1nl}^+}{\omega - \omega_{nl}} \left[\frac{e^{i(\omega+\omega_{nl}+\omega_{mn})t}-1}{\omega+\omega_{nl}+\omega_{mn}} - \frac{e^{i(2\omega+\omega_{mn})t}-1}{2\omega+\omega_{mn}} \right] - \frac{\hat{F}_{2mn}^+\hat{F}_{1nl}}{\omega - \omega_{nl}} \left[\frac{e^{-i(\omega-\omega_{nl}-\omega_{mn})t}-1}{\omega-\omega_{nl}-\omega_{mn}} - \frac{e^{-i(2\omega-\omega_{mn})t}-1}{2\omega-\omega_{mn}} \right]
$$

$$
+ \frac{\hat{F}_{2mn}^+\hat{F}_{1nl}^+}{\omega - \omega_{nl}} \left[\frac{e^{-i(3\omega-\omega_{nl}-\omega_{mn})t}-1}{3\omega-\omega_{nl}-\omega_{mn}} - \frac{e^{-i(2\omega-\omega_{mn})t}-1}{2\omega-\omega_{mn}} \right] + \frac{\hat{F}_{0mn}\hat{F}_{1nl}}{\omega+\omega_{nl}} \left[\frac{e^{i(\omega+\omega_{nl}+\omega_{mn})t}-1}{\omega+\omega_{nl}+\omega_{mn}} - \frac{e^{i\omega_{mn}t}-1}{\omega_{mn}} \right]
$$

$$
+ \frac{\hat{F}_{0mn}\hat{F}_{1nl}^+}{\omega-\omega_{nl}} \left[\frac{e^{-i(\omega-\omega_{nl}-\omega_{mn})t}-1}{\omega-\omega_{nl}-\omega_{mn}} + \frac{e^{i\omega_{mn}t}-1}{\omega_{mn}} \right] + \frac{\hat{F}_{1mn}\hat{F}_{2nl}}{2\omega+\omega_{nl}} \left[\frac{e^{i(3\omega+\omega_{nl}+\omega_{mn})t}-1}{3\omega+\omega_{nl}+\omega_{mn}} - \frac{e^{i(\omega+\omega_{mn})t}-1}{\omega+\omega_{mn}} \right]
$$

$$
+ \frac{\hat{F}_{1mn}\hat{F}_{2nl}^+}{2\omega-\omega_{nl}} \left[\frac{e^{-i(\omega-\omega_{nl}-\omega_{mn})t}-1}{\omega-\omega_{nl}-\omega_{mn}} + \frac{e^{i(\omega+\omega_{mn})t}-1}{\omega+\omega_{mn}} \right] + \frac{\hat{F}_{0nl}\hat{F}_{1mn}}{\omega_{nl}} \left[\frac{e^{i(\omega+\omega_{nl}+\omega_{mn})t}-1}{\omega+\omega_{nl}+\omega_{mn}} - \frac{e^{i(\omega+\omega_{mn})t}-1}{\omega+\omega_{mn}} \right]
$$

$$
+ \frac{\hat{F}_{1mn}^+\hat{F}_{2nl}}{2\omega+\omega_{nl}} \left[\frac{e^{i(\omega+\omega_{nl}+\omega_{mn})t}-1}{\omega+\omega_{nl}+\omega_{mn}} + \frac{e^{-i(\omega-\omega_{mn})t}-1}{\omega-\omega_{mn}} \right] - \frac{\hat{F}_{1mn}^+\hat{F}_{2nl}^+}{2\omega-\omega_{nl}} \left[\frac{e^{-i(3\omega-\omega_{nl}-\omega_{mn})t}-1}{3\omega-\omega_{nl}-\omega_{mn}} - \frac{e^{-i(\omega-\omega_{mn})t}-1}{\omega-\omega_{mn}} \right]
$$

$$
- \frac{\hat{F}_{0nl}\hat{F}_{1mn}^+}{\omega_{nl}} \left[\frac{e^{-i(\omega-\omega_{nl}-\omega_{mn})t}-1}{\omega-\omega_{nl}-\omega_{mn}} - \frac{e^{-i(\omega-\omega_{mn})t}-1}{\omega-\omega_{mn}} \right]
$$

$$
+ \frac{1}{\hbar^3} \sum_{n,k} \left\{ -\frac{\hat{F}_{1mn}\hat{F}_{1nk}\hat{F}_{1kl}}{\omega+\omega_{kl}} \left[\frac{e^{i(3\omega+\omega_{nl}+\omega_{mn})t}-1}{(2\omega+\omega_{kl}+\omega_{nk})(3\omega+\omega_{kl}+\omega_{nk}+\omega_{mn})} \right. \right.
$$

$$
- \frac{e^{i(\omega+\omega_{mn})t}-1}{(2\omega+\omega_{kl}+\omega_{nk})(\omega+\omega_{mn})} - \frac{e^{i(2\omega+\omega_{nk}+\omega_{mn})t}-1}{(2\omega+\omega_{nk}+\omega_{mn})(\omega+\omega_{nk})}
$$

$$
\left. + \frac{e^{i(\omega+\omega_{mn})t}-1}{(\omega+\omega_{nk})(\omega+\omega_{mn})} \right] + \frac{\hat{F}_{1mn}\hat{F}_{1nk}\hat{F}_{1kl}^+}{\omega-\omega_{kl}} \left[\frac{e^{i(\omega+\omega_{kl}+\omega_{nk}+\omega_{mn})t}-1}{(\omega_{kl}+\omega_{nk})(\omega+\omega_{kl}+\omega_{nk}+\omega_{mn})} \right.
$$

$$
- \frac{e^{i(\omega+\omega_{mn})t}-1}{(\omega_{kl}+\omega_{nk})(\omega+\omega_{mn})} - \frac{e^{i(2\omega+\omega_{nk}+\omega_{mn})t}-1}{(\omega+\omega_{nk})(2\omega+\omega_{nk}+\omega_{mn})} + \frac{e^{i(\omega+\omega_{mn})t}-1}{(\omega+\omega_{nk})(\omega+\omega_{mn})} \right]
$$

$$
- \frac{\hat{F}_{1mn}\hat{F}_{1nk}^+\hat{F}_{1kl}^+}{\omega+\omega_{kl}} \left[\frac{e^{i(\omega+\omega_{kl}+\omega_{nk}+\omega_{mn})t}-1}{(\omega_{kl}+\omega_{mn})(\omega+\omega_{kl}+\omega_{nk}+\omega_{mn})} \right.
$$

$$
- \frac{e^{i(\omega+\omega_{mn})t}-1}{(\omega_{kl}+\omega_{nk})(\omega+\omega_{mn})} + \frac{e^{i(\omega_{nk}+\omega_{mn})t}-1}{(\omega-\omega_{nk})(\omega_{nk}+\omega_{mn})} - \frac{e^{i(\omega+\omega_{mn})t}-1}{(\omega-\omega_{nk})(\omega+\omega_{mn})} \right]
$$

$$
+ \frac{\hat{F}_{1mn}\hat{F}_{1nk}^+\hat{F}_{1kl}^+}{\omega-\omega_{kl}} \left[\frac{e^{-i(\omega-\omega_{kl}-\omega_{nk}-\omega_{mn})t}-1}{(2\omega-\omega_{kl}-\omega_{nk})(\omega-\omega_{kl}-\omega_{nk}-\omega_{mn})} \right.
$$

$$
\left. + \frac{e^{i(\omega+\omega_{mn})t}-1}{(2\omega-\omega_{kl}-\omega_{nk})(\omega+\omega_{mn})} + \frac{e^{i(\omega_{nk}+\omega_{mn})t}-1}{(\omega-\omega_{nk})(\omega_{nk}+\omega_{mn})} - \frac{e^{i(\omega+\omega_{mn})t}-1}{(\omega-\omega_{nk})(\omega+\omega_{mn})} \right]
$$

$$
-\frac{\hat{F}_{1mn}^{+}\hat{F}_{1nk}\hat{F}_{1kl}}{\omega+\omega_{kl}}\left[\frac{e^{i\left(\omega+\omega_{kl}+\omega_{nk}+\omega_{mn}\right)t}-1}{\left(\omega_{kl}+\omega_{nk}\right)\left(\omega+\omega_{kl}+\omega_{nk}+\omega_{mn}\right)}\right.
$$

$$
\left.+\frac{e^{-i\left(\omega-\omega_{mn}\right)t}-1}{\left(2\omega+\omega_{kl}+\omega_{nk}\right)\left(\omega-\omega_{mn}\right)}-\frac{e^{i\left(\omega_{nk}+\omega_{mn}\right)t}-1}{\left(\omega+\omega_{nk}\right)\left(\omega_{nk}+\omega_{mn}\right)}-\frac{e^{-i\left(\omega-\omega_{mn}\right)t}-1}{\left(\omega+\omega_{nk}\right)\left(\omega-\omega_{mn}\right)}\right]
$$

$$
-\frac{\hat{F}_{1mn}^{+}\hat{F}_{1nk}\hat{F}_{1kl}^{+}}{\omega-\omega_{kl}}\left[\frac{e^{i\left(\omega-\omega_{kl}-\omega_{nk}-\omega_{mn}\right)t}-1}{\left(\omega_{kl}+\omega_{nk}\right)\left(\omega-\omega_{kl}-\omega_{nk}-\omega_{mn}\right)}\right.
$$

$$
\left.-\frac{e^{-i\left(\omega-\omega_{mn}\right)t}-1}{\left(\omega_{kl}+\omega_{nk}\right)\left(\omega-\omega_{mn}\right)}+\frac{e^{i\left(\omega_{nk}+\omega_{mn}\right)t}-1}{\left(\omega+\omega_{nk}\right)\left(\omega_{nk}+\omega_{mn}\right)}+\frac{e^{-i\left(\omega-\omega_{mn}\right)t}-1}{\left(\omega+\omega_{nk}\right)\left(\omega-\omega_{mn}\right)}\right]
$$

$$
+\frac{\hat{F}_{1mn}^{+}\hat{F}_{1nk}^{+}\hat{F}_{1kl}}{\omega+\omega_{kl}}\left[\frac{e^{-i\left(\omega-\omega_{kl}-\omega_{nk}-\omega_{mn}\right)t}-1}{\left(\omega_{kl}+\omega_{nk}\right)\left(\omega-\omega_{kl}-\omega_{nk}-\omega_{mn}\right)}\right.
$$

$$
\left.-\frac{e^{-i\left(\omega-\omega_{mn}\right)t}-1}{\left(\omega_{kl}+\omega_{nk}\right)\left(\omega-\omega_{mn}\right)}+\frac{e^{-i\left(2\omega-\omega_{nk}-\omega_{mn}\right)t}-1}{\left(\omega-\omega_{nk}\right)\left(2\omega-\omega_{nk}-\omega_{mn}\right)}-\frac{e^{-i\left(\omega-\omega_{mn}\right)t}-1}{\left(\omega-\omega_{nk}\right)\left(\omega-\omega_{mn}\right)}\right]
$$

$$
+\frac{\hat{F}_{1mn}^{+}\hat{F}_{1nk}^{+}\hat{F}_{1kl}^{+}}{\omega-\omega_{kl}}\left[\frac{e^{-i\left(3\omega-\omega_{kl}-\omega_{nk}-\omega_{mn}\right)t}-1}{\left(2\omega-\omega_{kl}-\omega_{nk}\right)\left(3\omega-\omega_{kl}-\omega_{nk}-\omega_{mn}\right)}\right.
$$

$$
\left.-\frac{e^{-i\left(\omega-\omega_{mn}\right)t}-1}{\left(2\omega-\omega_{kl}-\omega_{nk}\right)\left(\omega-\omega_{mn}\right)}-\frac{e^{-i\left(2\omega-\omega_{nk}-\omega_{mn}\right)t}-1}{\left(2\omega-\omega_{nk}-\omega_{mn}\right)\left(\omega-\omega_{nk}\right)}+\frac{e^{-i\left(\omega-\omega_{mn}\right)t}-1}{\left(\omega-\omega_{nk}\right)\left(\omega-\omega_{mn}\right)}\right]
$$

15.2 The time reversal of transition probability amplitude of third order process of light's stimulated radiation and absorption

$$
a_{Tm}^{(3)}(t)=\frac{1}{\hbar^{2}}\sum_{n}\left\{\frac{\hat{F}_{2nl}'\hat{F}_{1mn}'}{\omega+\omega_{mn}}\left[\frac{e^{i\left(3\omega+\omega_{nl}+\omega_{mn}\right)t}-1}{3\omega+\omega_{nl}+\omega_{mn}}-\frac{e^{i\left(2\omega+\omega_{nl}\right)t}-1}{2\omega+\omega_{nl}}\right]\right.
$$

$$
-\frac{\hat{F}_{2nl}'\hat{F}_{1mn}'^{+}}{\omega-\omega_{mn}}\left[\frac{e^{i\left(\omega+\omega_{nl}+\omega_{mn}\right)t}-1}{\omega+\omega_{nl}+\omega_{mn}}-\frac{e^{i\left(2\omega+\omega_{nl}\right)t}-1}{2\omega+\omega_{nl}}\right]-\frac{\hat{F}_{2nl}'^{+}\hat{F}_{1mn}'}{\omega+\omega_{mn}}\left[\frac{e^{-i\left(\omega-\omega_{nl}-\omega_{mn}\right)t}-1}{\omega-\omega_{nl}-\omega_{mn}}-\frac{e^{-i\left(2\omega-\omega_{nl}\right)t}-1}{2\omega-\omega_{nl}}\right]
$$

$$
+\frac{\hat{F}_{2nl}'^{+}\hat{F}_{1mn}'^{+}}{\omega-\omega_{mn}}\left[\frac{e^{-i\left(3\omega-\omega_{nl}-\omega_{mn}\right)t}-1}{3\omega-\omega_{nl}-\omega_{mn}}-\frac{e^{-i\left(2\omega-\omega_{nl}\right)t}-1}{2\omega-\omega_{nl}}\right]+\frac{\hat{F}_{0nl}'\hat{F}_{1mn}'}{\omega+\omega_{mn}}\left[\frac{e^{i\left(\omega+\omega_{nl}+\omega_{mn}\right)t}-1}{\omega+\omega_{nl}+\omega_{mn}}-\frac{e^{i\omega_{nl}t}-1}{\omega_{nl}}\right]
$$

$$
-\frac{\hat{F}_{0nl}'\hat{F}_{1mn}'^{+}}{\omega-\omega_{mn}}\left[\frac{e^{-i\left(\omega-\omega_{nl}-\omega_{mn}\right)t}-1}{\omega-\omega_{nl}-\omega_{mn}}-\frac{e^{i\omega_{nl}t}-1}{\omega_{nl}}\right]+\frac{\hat{F}_{1nl}'\hat{F}_{2mn}'}{2\omega+\omega_{mn}}\left[\frac{e^{i\left(3\omega+\omega_{nl}+\omega_{mn}\right)t}-1}{3\omega+\omega_{nl}+\omega_{mn}}-\frac{e^{i\left(\omega+\omega_{nl}\right)t}-1}{\omega+\omega_{nl}}\right]
$$

$$
+\frac{\hat{F}_{1nl}'\hat{F}_{2mn}'^{+}}{2\omega-\omega_{mn}}\left[\frac{e^{-i\left(\omega-\omega_{nl}-\omega_{mn}\right)t}-1}{\omega-\omega_{nl}-\omega_{mn}}+\frac{e^{i\left(\omega+\omega_{nl}\right)t}-1}{\omega+\omega_{nl}}\right]+\frac{\hat{F}_{0mn}'\hat{F}_{1nl}'}{\omega_{mn}}\left[\frac{e^{i\left(\omega+\omega_{nl}+\omega_{mn}\right)t}-1}{\omega+\omega_{nl}+\omega_{mn}}-\frac{e^{i\left(\omega+\omega_{nl}\right)t}-1}{\omega+\omega_{nl}}\right]
$$

$$
+\frac{\hat{F}_{1nl}'^{+}\hat{F}_{2mn}'}{2\omega+\omega_{mn}}\left[\frac{e^{i\left(\omega+\omega_{nl}+\omega_{mn}\right)t}-1}{\omega+\omega_{nl}+\omega_{mn}}+\frac{e^{-i\left(\omega-\omega_{nl}\right)t}-1}{\omega-\omega_{nl}}\right]+\frac{\hat{F}_{1nl}'^{+}\hat{F}_{2mn}'^{+}}{2\omega-\omega_{mn}}\left[\frac{e^{-i\left(3\omega-\omega_{nl}-\omega_{mn}\right)t}-1}{3\omega-\omega_{nl}-\omega_{mn}}-\frac{e^{-i\left(\omega-\omega_{nl}\right)t}-1}{\omega-\omega_{nl}}\right]
$$

$$
\left.-\frac{\hat{F}_{0mn}'\hat{F}_{1nl}'^{+}}{\omega_{mn}}\left[\frac{e^{-i\left(\omega-\omega_{nl}-\omega_{mn}\right)t}-1}{\omega-\omega_{nl}-\omega_{mn}}-\frac{e^{-i\left(\omega-\omega_{nl}\right)t}-1}{\omega-\omega_{nl}}\right]\right\}
$$

$$+\frac{1}{\hbar^3}\sum_{n,k}\left\{-\frac{\hat{F}'_{1nl}\hat{F}'_{1mk}\hat{F}'_{1kn}}{\omega+\omega_{mk}}\left[\frac{e^{i(3\omega+\omega_{mk}+\omega_{kn}+\omega_{nl})\,t}-1}{(2\omega+\omega_{mk}+\omega_{kn})(3\omega+\omega_{mk}+\omega_{kn}+\omega_{nl})}\right.\right.$$

$$-\frac{e^{i(\omega+\omega_{nl})\,t}-1}{(2\omega+\omega_{mk}+\omega_{kn})(\omega+\omega_{nl})}-\frac{e^{i(2\omega+\omega_{kn}+\omega_{nl})\,t}-1}{(\omega+\omega_{kn})(2\omega+\omega_{kn}+\omega_{nl})}+\frac{e^{i(\omega+\omega_{nl})\,t}-1}{(\omega+\omega_{kn})(\omega+\omega_{nl})}\right]$$

$$-\frac{\left(\hat{F}'^{+}_{1nl}\right)^{*}\hat{F}'^{+}_{1mk}\left(\hat{F}'^{+}_{1kn}\right)^{*}}{\omega-\omega_{mk}}\left[\frac{e^{i(\omega+\omega_{mk}+\omega_{kn}+\omega_{nl})\,t}-1}{(\omega_{mk}+\omega_{kn})(\omega+\omega_{mk}+\omega_{kn}+\omega_{nl})}\right.$$

$$-\frac{e^{i(\omega+\omega_{nl})\,t}-1}{(\omega_{mk}+\omega_{kn})(\omega+\omega_{nl})}-\frac{e^{i(2\omega+\omega_{kn}+\omega_{nl})\,t}-1}{(\omega+\omega_{kn})(2\omega+\omega_{kn}+\omega_{nl})}+\frac{e^{i(\omega+\omega_{nl})\,t}-1}{(\omega+\omega_{kn})(\omega+\omega_{nl})}\right]$$

$$+\frac{\hat{F}'_{1nl}\hat{F}'_{1mk}\hat{F}'^{+}_{1kn}}{\omega+\omega_{mk}}\left[\frac{e^{i(\omega+\omega_{mk}+\omega_{kn}+\omega_{nl})\,t}-1}{(\omega_{mk}+\omega_{kn})(\omega+\omega_{mk}+\omega_{kn}+\omega_{nl})}\right.$$

$$-\frac{e^{i(\omega+\omega_{nl})\,t}-1}{(\omega_{mk}+\omega_{kn})(\omega+\omega_{nl})}+\frac{e^{i(\omega_{kn}+\omega_{nl})\,t}-1}{(\omega-\omega_{kn})(\omega_{kn}+\omega_{nl})}-\frac{e^{i(\omega+\omega_{nl})\,t}-1}{(\omega-\omega_{kn})(\omega+\omega_{nl})}\right]$$

$$-\frac{\hat{F}'_{1nl}\hat{F}'^{+}_{1mk}\hat{F}'^{+}_{1kn}}{\omega-\omega_{mk}}\left[\frac{e^{-i(\omega-\omega_{mk}-\omega_{kn}-\omega_{nl})\,t}-1}{(2\omega-\omega_{mk}-\omega_{kn})(\omega-\omega_{mk}-\omega_{kn}-\omega_{nl})}\right.$$

$$+\frac{e^{i(\omega+\omega_{nl})\,t}-1}{(2\omega-\omega_{mk}-\omega_{kn})(\omega+\omega_{nl})}+\frac{e^{i(\omega_{kn}+\omega_{nl})\,t}-1}{(\omega-\omega_{kn})(\omega_{kn}+\omega_{nl})}-\frac{e^{i(\omega+\omega_{nl})\,t}-1}{(\omega-\omega_{kn})(\omega+\omega_{nl})}\right]$$

$$-\frac{\hat{F}'^{+}_{1nl}\left(\hat{F}'^{+}_{1mk}\right)^{*}\hat{F}'^{+}_{1kn}}{\omega+\omega_{mk}}\left[\frac{e^{-i(\omega-\omega_{mk}-\omega_{kn}-\omega_{nl})\,t}-1}{(\omega_{mk}+\omega_{kn})(\omega-\omega_{mk}-\omega_{kn}-\omega_{nl})}\right.$$

$$-\frac{e^{-i(\omega-\omega_{nl})\,t}-1}{(\omega_{mk}+\omega_{kn})(\omega-\omega_{nl})}+\frac{e^{-i(2\omega-\omega_{kn}-\omega_{nl})\,t}-1}{(\omega-\omega_{kn})(2\omega-\omega_{kn}-\omega_{nl})}-\frac{e^{-i(\omega-\omega_{nl})\,t}-1}{(\omega-\omega_{kn})(\omega-\omega_{nl})}\right]$$

$$+\frac{\hat{F}'^{+}_{1nl}\left(\hat{F}'^{+}_{1mk}\right)^{*}\left(\hat{F}'^{+}_{1kn}\right)^{*}}{\omega+\omega_{mk}}\left[\frac{e^{i(\omega+\omega_{mk}+\omega_{kn}+\omega_{nl})\,t}-1}{(2\omega+\omega_{mk}+\omega_{kn})(\omega+\omega_{mk}+\omega_{kn}+\omega_{nl})}\right.$$

$$+\frac{e^{-i(\omega-\omega_{nl})\,t}-1}{(2\omega+\omega_{mk}+\omega_{kn})(\omega-\omega_{nl})}-\frac{e^{i(\omega_{kn}+\omega_{nl})\,t}-1}{(\omega+\omega_{kn})(\omega_{kn}+\omega_{nl})}\frac{e^{-i(\omega-\omega_{nl})\,t}-1}{(\omega+\omega_{kn})(\omega-\omega_{nl})}\right]$$

$$+\frac{\hat{F}'^{+}_{1nl}\hat{F}'^{+}_{1mk}\hat{F}'_{1kn}}{\omega-\omega_{mk}}\left[\frac{e^{-i(\omega-\omega_{mk}-\omega_{kn}-\omega_{nl})\,t}-1}{(\omega_{mk}+\omega_{kn})(\omega-\omega_{mk}-\omega_{kn}-\omega_{nl})}\right.$$

$$-\frac{e^{-i(\omega-\omega_{nl})\,t}-1}{(\omega_{mk}+\omega_{kn})(\omega-\omega_{nl})}+\frac{e^{i(\omega_{kn}+\omega_{nl})\,t}-1}{(\omega+\omega_{kn})(\omega_{kn}+\omega_{nl})}+\frac{e^{-i(\omega-\omega_{nl})\,t}-1}{(\omega+\omega_{kn})(\omega-\omega_{nl})}\right]$$

$$-\frac{\hat{F}'^{+}_{1nl}\hat{F}'^{+}_{1mk}\hat{F}'^{+}_{1kn}}{\omega-\omega_{mk}}\left[\frac{e^{-i(3\omega-\omega_{mk}-\omega_{kn}-\omega_{nl})\,t}-1}{(2\omega-\omega_{mk}-\omega_{kn})(3\omega-\omega_{mk}-\omega_{kn}-\omega_{nl})}\right.$$

16. References

[1] Mei Xiaochun, Science in China, Series G, Volume 51, Number 3, 2008; co-published with Springer-Verlag GmbH 1762-1799 (Print) 1862-2844 (On line).

[2] P. N. Butcher, D. Cotter, (1990), The elements of nonlinear Optics, Cambridge University Press.

[3] S. L. McCall, H. M. Gibbs and T. N. C. Venkateasn, (1975), J. Opt. Soc. Am. 65, pp. 1184. D. A. B. Miller, D. S. Chemla, T. C. Demen, C. Gossard, W. Wiegmann, T. H. Wood and C. A. Burrus, (1984), Appl. Phys. lette.45, pp. 83. T. Kobayashi, N. C. Kothari and H. Uchiki, (1983), Phys. Rev. A 29, pp. 2727; A.K.Kar, J.G.H. Mathew et al., (1984), Appl. Phys. Lett. 42, pp. 334.

[4] M. M. T. Loy and Y. R. Shen, (1971), Phys.Rev.Lett, 25, pp. 1333; (1970), Appl. Phys. Lett. 19, pp. 285; (1973), G. K. L. Wong and Y. R. Shen, Phys Rev. Lett, 32, pp. 527.

[5] W. Kaiser and C. G. B.Garret, (1961), Phy. Rev. Lett. 7, pp. 229

[6] N. A. Kurnit, I. D. Abella and S. R. Hartman, (1964), Phys. Rev. Lett, 13, pp. 567.

[7] Y. Wu. et al. (1977), Phys Rev. Lett. 42, 1077; K.J. Boller et al, (1991), Phys Rev Lett., 66 2591. Jinyue Gao et al. (1992), Opt.Comm, pp. 93, 323.

[8] Lei Shizhan, (1987), New World of Optics, Science Publishing Company, pp.25.

[9] Miao Dongsheng, Liu Huajie, (1993), Chaos Theory, Publishing House of Chinese People University, pp. 262.

[10] Luo Liaofu, (1990), Statistical Theory of Nonequivalence, Publishing House of Neimengu University, pp. 329.

5

The Electromagnetic Field in Accelerated Frames

J. W. Maluf and F. F. Faria

Instituto de Física, Universidade de Brasília, Brasília, DF
Brazil

1. Introduction

Maxwell's electromagnetic theory is more than a century old. It is a well established and understood theory. Usually the theory is presented in standard textbooks as a field theory in flat space-time (1; 2). The establishment of the theory in curved space-time (3) requires the understanding of how exactly the Faraday tensor couples with the gravitational field, and presently this is an open issue. In the ordinary description of electrodynamics in flat space-time one almost always assumes that the sources and fields are established in an inertial reference frame. Very few investigations (4) attempt to extend the analysis to accelerated frames. Such extension is mandatory because most frames in nature are, in one or another way, accelerated.

Until recently the attempts to describe the electromagnetic field in an accelerated frame consisted in performing a coordinate transformation of the Faraday tensor defined in an inertial frame in flat space-time. For this purpose one considers a coordinate transformation from the flat space-time cartesian coordinates to coordinates that describe a hyperbola in Rindler space, in case of uniform acceleration. This procedure is not satisfactory for two reasons. First, a coordinate transformations is not a frame transformation. A coordinate transformation is carried out on vectors and tensors on a manifold, and they just express the fact that (i) a point on the manifold may be labelled by different coordinates in different charts, and that (ii) one can work with any set of coordinates. On the other hand, a frame transformation is a Lorentz tranformation, it satisfies the properties of the Lorentz group and is carried out in the tangent space of the manifold.

The second reason is that by considering an accelerated frame as a frame obtained by a coordinate transformation, one cannot provide satisfactory answers to situations that are eventually understood as paradoxes, because the inertial and "accelerated" fields are described in different coordinates. One of these paradoxes is the following: are the two situations, (i) an accelerated charge in an inertial frame, and (ii) a charge at rest in an inertial frame described from the perspective of an accelerated frame, physically equivalent?

The procedure to be considered here consists, first, in assuming that the Faraday tensor and Maxwell's equations are abstract tensor quantities in space-time. Then we make use of tetrad fields to project the electromagnetic field either on an inertial or on a non-inertial frame, in the same coordinate system, in flat space-time. Tetrad fields constitute a set of four orthonormal vectors, that are adapted to observers that follow arbitrary paths in space-time. They constitute the local frame of these observers. Since the fields in the inertial frame and in the accelerated frame are defined in the same coordinate system, they can be compared with each other unambiguously.

Given any set of tetrad fields, we may construct the acceleration tensor, as we will show. This tensor determines the inertial (i.e., non-gravitational) accelerations that act on a given observer. For instance, a stationary observer in space-time undergoes inertial forces, otherwise it would follow a geodesic motion determined by the gravitational field. A given frame (or a given tetrad field) may be characterized by the inertial accelerations.

In this chapter we will obtain the general form of Maxwell's equations that hold in inertial or noninertial frames. The formalism ensures that the procedure for projecting electromagnetic fields in noninertial frames is mathematically and physically consistent, and allows the investigation of several paradoxes. It is possible to conclude, for instance, that the radiation of an accelerated charged particle in an inertial frame is different from the radiation of the same charged particle measured in a frame that is co-accelerated (equally accelerated) with the particle. Consequently, the accelerated motion in space-time is not relative, and the radiation of an accelerated charged particle is an absolute feature of the theory (5).

We will study in detail the description of plane and spherical electromagnetic waves in linearly accelerated frames in Minkowski space-time. We will show that (i) the amplitude, (ii) the frequency and the wave vector of the plane wave, and (iii) the Poynting vector in the accelerated frame vary (decrease) with time, while the light speed remains constant.

Notation:

1. Space-time indices $\mu, \nu, \ldots$ and Lorentz (SO(3,1)) indices $a, b, \ldots$ run from 0 to 3. Time and space indices are indicated according to $\mu = 0, i, \quad a = (0), (i)$.

2. The space-time is flat, and therefore the metric tensor in cartesian coordinates is given by $g_{\mu\nu} = (-1, +1, +1, +1)$

3. The tetrad field is represented by $e^a{}_\mu$. The flat, tangent space Minkowski space-time metric tensor raises and lowers tetrad indices and is fixed by $\eta_{ab} = e_{a\mu}e_{b\nu}g^{\mu\nu} = (-1, +1, +1, +1)$.

4. The frame components are given by the inverse tetrads $e_a{}^\mu$, although we may as well refer to $\{e^a{}_\mu\}$ as the frame. The determinant of the tetrad field is represented by $e = \det(e^a{}_\mu)$

2. Reference frames in space-time

The electromagnetic field is described by the Faraday tensor $F^{\mu\nu}$. In the present analysis we will consider that $\{F^{\mu\nu}\}$ are just *tensor components* in the flat Minkowski space-time described by *arbitrary* coordinates x^μ. The projection of $F^{\mu\nu}$ on inertial or noninertial frames yields the electric and magnetic fields E_x, E_y, E_z, B_x, B_y and B_z, which are the *frame components* of $\{F^{\mu\nu}\}$. The projection is carried out with the help of tetrad fields $e^a{}_\mu$. For instance, $E_x = -cF^{(0)(1)}$, where c is the speed of light and $F^{(0)(1)} = e^{(0)}{}_\mu e^{(1)}{}_\nu F^{\mu\nu}$. The study of the kinematical properties of tetrad fields is mandatory for the characterization of reference frames.

Tetrad fields constitute a set of four orthonormal vectors in space-time, $\{e^{(0)}{}_\mu, e^{(1)}{}_\mu, e^{(2)}{}_\mu, e^{(3)}{}_\mu\}$, that establish the local reference frame of an observer that moves along a trajectory C, represented by functions $x^\mu(s)$ (6–8) (s is the proper time of the observer). The tetrad field yields the space-time metric tensor $g_{\mu\nu}$ by means of the relation $e^a{}_\mu e^b{}_\nu \eta_{ab} = g_{\mu\nu}$, and $e^{(0)}{}_\mu$ and $e^{(i)}{}_\mu$ are timelike and spacelike vectors, respectively.

We identify the $a = (0)$ component of $e_a{}^\mu$ with the observer's velocity u^μ along the trajectory C, i.e., $e_{(0)}{}^\mu = u^\mu/c = dx^\mu/(cd\tau)$. The observer's acceleration a^μ is given by the absolute derivative of u^μ along C,

$$a^\mu = \frac{Du^\mu}{d\tau} = c\,\frac{De_{(0)}{}^\mu}{d\tau}. \tag{1}$$

The absolute derivative is constructed with the help of the Christoffel symbols. Thus $e_{(0)}{}^\mu$ and its absolute derivative determine the velocity and acceleration along the worldline of an observer adapted to the frame. The set of tetrad fields for which $e_{(0)}{}^\mu$ describes a congruence of timelike curves is adapted to a class of observers characterized by the velocity field $u^\mu = c\,e_{(0)}{}^\mu$ and by the acceleration a^μ. If $e^a{}_\mu = \delta^a_\mu$ everywhere in space-time, then $e^a{}_\mu$ is adapted to static observers, and $a^\mu = 0$.

We may consider not only the acceleration of observers along trajectories whose tangent vectors are given by $e_{(0)}{}^\mu$, but the acceleration of the whole frame along C. The acceleration of the frame is determined by the absolute derivative of $e_a{}^\mu$ along the path $x^\mu(\tau)$. Thus, assuming that the observer carries a frame, the acceleration of the latter along the path is given by (4; 9),

$$\frac{De_a{}^\mu}{d\tau} = \phi_a{}^b e_b{}^\mu, \tag{2}$$

where ϕ_{ab} is the antisymmetric acceleration tensor of the frame ($\phi_{ab} = -\phi_{ba}$). According to Refs. (4; 9), in analogy with the Faraday tensor we can identify $\phi_{ab} \equiv (\vec{a}/c, \vec{\Omega})$, where $\vec{a}$ is the translational acceleration ($\phi_{(0)(i)} = a_{(i)}/c$) and $\vec{\Omega}$ is the frequency of rotation ($\phi_{(i)(j)} = \epsilon_{(i)(j)(k)}\Omega^{(k)}$) of the spatial frame with respect to a nonrotating (Fermi-Walker transported (6; 8)) frame. It follows from Eq. (2) that

$$\phi_a{}^b = e^b{}_\mu \frac{De_a{}^\mu}{d\tau}. \tag{3}$$

Therefore given any set of tetrad fields for an arbitrary space-time, its geometrical interpretation may be obtained by suitably interpreting the velocity field $u^\mu = e_{(0)}{}^\mu$ and the acceleration tensor ϕ_{ab}.

Using the definiton of the absolute derivative, we can write Eq. (3) as

$$\begin{aligned}
\phi_a{}^b &= e^b{}_\mu \left(\frac{de_a{}^\mu}{d\tau} + \Gamma^\mu{}_{\lambda\sigma} \frac{dx^\lambda}{d\tau} e_a{}^\sigma \right) \\
&= e^b{}_\mu \left(\frac{dx^\lambda}{d\tau} \frac{\partial e_a{}^\mu}{\partial x^\lambda} + \Gamma^\mu{}_{\lambda\sigma} \frac{dx^\lambda}{d\tau} e_a{}^\sigma \right) \\
&= e^b{}_\mu u^\lambda \left(\frac{\partial e_a{}^\mu}{\partial x^\lambda} + \Gamma^\mu{}_{\lambda\sigma} e_a{}^\sigma \right) = e^b{}_\mu u^\lambda \nabla_\lambda e_a{}^\mu.
\end{aligned} \tag{4}$$

Following Ref. (7), we take into account the orthogonality of the tetrads and write Eq. (4) as $\phi_a{}^b = -u^\lambda e_a{}^\mu \nabla_\lambda e^b{}_\mu$, where $\nabla_\lambda e^b{}_\mu = \partial_\lambda e^b{}_\mu - \Gamma^\sigma{}_{\lambda\mu} e^b{}_\sigma$. Next we consider the identity

$\partial_\lambda e^b{}_\mu - \Gamma^\sigma{}_{\lambda\mu} e^b{}_\sigma + {}^0\omega_\lambda{}^b{}_c e^c{}_\mu = 0$, where ${}^0\omega_\lambda{}^b{}_c$ is the Levi-Civita spin connection given by Eq. (21) below, and express $\phi_a{}^b$ according to

$$\phi_a{}^b = u^\lambda e_a{}^\mu \left({}^0\omega_\lambda{}^b{}_c e^c{}_\mu \right) = c\, e_{(0)}{}^\mu ({}^0\omega_\mu{}^b{}_a). \tag{5}$$

Finally we make use of the identity ${}^0\omega_\mu{}^a{}_b = -K_\mu{}^a{}_b$, where $K_\mu{}^a{}_b$ is the contortion tensor defined by

$$K_{\mu ab} = \frac{1}{2} e_a{}^\lambda e_b{}^\nu (T_{\lambda\mu\nu} + T_{\nu\lambda\mu} + T_{\mu\lambda\nu}), \tag{6}$$

where

$$T^\lambda{}_{\mu\nu} = e_a{}^\lambda T^a{}_{\mu\nu} = e_a{}^\lambda \left(\partial_\mu e^a{}_\nu - \partial_\nu e^a{}_\mu \right), \tag{7}$$

is the object of anholonomity. Note that $T^\lambda{}_{\mu\nu}$ is also the torsion tensor of the Weitzenböck space-time. After simple manipulations we arrive at

$$\phi_{ab} = \frac{c}{2} \left[T_{(0)ab} + T_{a(0)b} - T_{b(0)a} \right], \tag{8}$$

where $T_{abc} = e_b{}^\mu e_c{}^\nu T_{a\mu\nu}$. The expression above is not covariant under local Lorentz (SO(3,1) or frame) transformations, but is invariant under coordinate transformations. The noncovariance under local Lorentz transformations allows us to take the values of ϕ_{ab} to characterize the frame.

In order to measure field quantities with magnitude and direction (velocity, acceleration, etc.), an observer must project these quantities on the frame carried by the observer. The projection of a vector V^μ on a particular frame is determined by

$$V^a(x) = e^a{}_\mu(x)\, V^\mu(x), \tag{9}$$

and the projection of a tensor $T^{\mu\nu}$ is

$$T^{ab}(x) = e^a{}_\mu(x)\, e^b{}_\nu(x)\, T^{\mu\nu}(x). \tag{10}$$

Note that the projections are carried out in the same coordinate system.

We consider now an accelerated observer that follows a worldline $\bar{x}^\mu(\tau)$ in Minkowski space-time and carries a tetrad $e^a{}_\mu$, such that $e_{(0)}{}^\mu = u^\mu/c$ and $De_a{}^\mu/d\tau = \phi_a{}^b e_b{}^\mu$. At each instant τ of proper time along the worldline there are spacelike geodesics orthogonal to the worldline that form a local spacelike hypersurface. The observer can assign local coordinates $x^a = \{x^{(0)}, x^{(i)}\} = \{c\tau, \vec{x}'\}$ to an event, which is also described by Cartesian coordinates $x^\mu = \{ct, \vec{x}\}$ belonging to this hypersurface, where

$$x^{(0)} = c\tau, \qquad x^{(i)} = [x^\mu - \bar{x}^\mu] e^{(i)}{}_\mu. \tag{11}$$

The inverse transformation reads

$$x^\mu = \bar{x}^\mu + e_{(i)}{}^\mu x^{(i)}. \tag{12}$$

If we differentiate both sides of this equation over the worldline, we find

$$dx^\mu = \left(\frac{1}{c} \frac{d\tilde{x}^\mu}{d\tau} + \frac{1}{c} \frac{de_{(i)}}{d\tau}^\mu x^{(i)} \right) dx^{(0)} + e_{(i)}{}^\mu dx^{(i)}$$

$$= \left(e_{(0)}{}^\mu + \frac{1}{c} \phi_{(i)}{}^a e_a{}^\mu x^{(i)} \right) dx^{(0)} + e_{(i)}{}^\mu dx^{(i)}. \tag{13}$$

Substituting Eq. (13) into the line element $ds^2 = \eta_{\mu\nu} dx^\mu dx^\nu$, we obtain the metric in the local coordinate system of an accelerated observer,

$$ds^2 = \left[-\left(1 + \frac{\vec{a} \cdot \vec{x}'}{c^2} \right)^2 + \frac{1}{c^2} \left(\vec{\Omega} \times \vec{x}' \right)^2 \right] (dx^{(0)})^2 + \left(\frac{2}{c} \vec{\Omega} \times \vec{x}' \right) dx^{(0)} dx^{(i)}$$

$$+ \eta_{(i)(j)} dx^{(i)} dx^{(j)}, \tag{14}$$

where we used $\phi_{(i)}{}^{(0)} x^{(i)} = (\vec{a} \cdot \vec{x}')/c$ and $\phi_{(j)}{}^{(i)} x^{(j)} = \left(\vec{\Omega} \times \vec{x}' \right)^{(i)}$.

We see from Eq. (14) that $\eta_{(0)(0)} \cong -1$ only in the regions of space-time where

$$|\vec{x}'| \ll \frac{c^2}{|\vec{a}|}, \quad \text{and} \quad |\vec{x}'| \ll \frac{c}{|\vec{\Omega}|}. \tag{15}$$

Furthermore, some $c\tau = $ constant surfaces will intersect each other if we extend the spatial local coordinates far away from the observer's worldline, which is not an admissible situation. Since we cannot assign two sets of coordinates for the same event, the local spatial coordinates have a limit of validity. In fact, the local coordinate system of Eq. (11) is valid only in those regions in the neighborhood of the observer's wordline in which Eqs. (15) hold. We call $c^2/|\vec{a}|$ the translational acceleration length and $c/|\vec{\Omega}|$ the rotational acceleration length. On the Earth's surface, for example, we have ($|\vec{a}| = 9,8\,\text{m/s}^2$, $|\vec{\Omega}| = \Omega_\oplus$)

$$\frac{c^2}{|\vec{a}|} = 9.46 \cdot 10^{15}\,\text{m} \approx 1\,\text{ly} \quad \text{and} \quad \frac{c}{|\vec{\Omega}|} = 4.125 \cdot 10^{12}\,\text{m} \approx 27.5\,\text{AU}. \tag{16}$$

Hence we can use the local coordinates x^a with confidence in most experimental situations in a laboratory on the Earth, where $|\vec{x}'|$ is negligible comparing to the acceleration lengths.

3. The formulation of Maxwell's theory in moving frames

The vector potential A^μ, the Faraday tensor $F_{\mu\nu} = \partial_\mu A_\nu - \partial_\nu A_\mu$ and the four-vector current J^μ are vector and tensor components in space-time. Space-time indices are raised and lowered by means of the flat space-time metric tensor $g_{\mu\nu} = (-1, +1, +1, +1)$. On a particular frame the electromagnetic quantities are projected and measured according to $A^a(x) = e^a{}_\mu(x) A^\mu(x)$ and $F^{ab}(x) = e^a{}_\mu(x) e^a{}_\nu(x) F^{\mu\nu}(x)$.

An inertial frame is characterized by the vanishing of the acceleration tensor ϕ_{ab}. A realization of an inertial frame in Minkowski space-time is given by $e^a{}_\mu(t, x, y, z) = \delta^a_\mu$. It is easy to verify that this frame satisfies $\phi_{ab} = 0$. More generally, all tetrad fields that are function of space-time *independent* parameters (boost and rotation parameters) determine inertial frames. Suppose that A^a are componentes of the vector potential in an inertial frame, i.e., $A^a = (e^a{}_\mu)_{in} A^\mu = \delta^a_\mu A^\mu$. The components of A^a in a noninertial frame are obtained by means of a local Lorentz transformation,

$$\tilde{A}^a(x) = \Lambda^a{}_b(x) A^b(x),\tag{17}$$

where $\Lambda^a{}_b(x)$ are space-time dependent matrices that satisfy

$$\Lambda^a{}_c(x)\Lambda^b{}_d(x)\eta_{ab} = \eta_{cd}.\tag{18}$$

In terms of covariant indices we have $\tilde{A}_a(x) = \Lambda_a{}^b(x)A_b(x)$. An alternative but completely equivalent way of obtaining the field components $\tilde{A}_a(x)$ consists in performing a frame transformation by means of a suitable noninertial frame $e^a{}_\mu$, namely, in projecting A^μ on the noninertial frame,

$$\tilde{A}^a(x) = e^a{}_\mu(x) A^\mu(x).\tag{19}$$

Of course we have $\Lambda^a{}_b \delta^b_\mu = \Lambda^a{}_b (e^b{}_\mu)_{in} = e^a{}_\mu$.

The covariant derivative of A_a is defined by

$$D_a A_b = e_a{}^\mu D_\mu A_b$$

$$= e_a{}^\mu (\partial_\mu A_b - {}^0\omega_\mu{}^c{}_b A_c),\tag{20}$$

where

$${}^0\omega_{\mu ab} = -\frac{1}{2}e^c{}_\mu(\Omega_{abc} - \Omega_{bac} - \Omega_{cab}),$$

$$\Omega_{abc} = e_{av}(e_b{}^\mu \partial_\mu e_c{}^v - e_c{}^\mu \partial_\mu e_b{}^v),\tag{21}$$

is the metric-compatible Levi-Civita connection considered in Eq. (5). Note that we are considering the flat space-time, and yet this connection may be nonvanishing. In particular, for noninertial frames it is nonvanishing. The Weitzenböck torsion tensor $T^a{}_{\mu v}$ is also nonvanishing. However, the curvature tensor constructed out of ${}^0\omega_{\mu ab}$ vanishes identically: $R^a{}_{b\mu v}({}^0\omega) \equiv 0$.

Under a local Lorentz transformation the spin connection transforms as

$$\widetilde{{}^0\omega}_\mu{}^a{}_b = \Lambda^a{}_c({}^0\omega_\mu{}^c{}_d)\Lambda_b{}^d + \Lambda^a{}_c\partial_\mu\Lambda_b{}^c.\tag{22}$$

It follows from eqs. (17), (21) and (22) that under a local Lorentz transformation we have

$$\tilde{D}_a\tilde{A}_b = \Lambda_a{}^c(x)\Lambda_b{}^d(x)\, D_c A_d.\tag{23}$$

The natural definition of the Faraday tensor in a noninertial frame is

$$F_{ab} = D_a A_b - D_b A_a.\tag{24}$$

In view of eq. (24) we find that the tensors F_{ab} and $\tilde{F}_{ab}$ in two arbitrary frames are related by

$$\tilde{F}_{ab} = \Lambda_a{}^c(x)\Lambda_b{}^d(x)F_{cd}. \tag{25}$$

The Faraday tensor defined by eq. (24) is related to the standard expression defined in inertial frames. By substituting (20) in (24) we find

$$F_{ab} = e_a{}^\mu(\partial_\mu A_b - {}^0\omega_\mu{}^m{}_b A_m) - e_b{}^\mu(\partial_\mu A_a - {}^0\omega_\mu{}^m{}_a A_m)$$

$$= e_a{}^\mu(\partial_\mu A_b) - e_b{}^\mu(\partial_\mu A_a) + ({}^0\omega_{abm} - {}^0\omega_{bam})A^m. \tag{26}$$

We make use of the *identity*

$$ {}^0\omega_{abm} - {}^0\omega_{bam} = T_{mab}, \tag{27}$$

where T_{mab} is given by eq. (7), and write

$$F_{ab} = e_a{}^\mu e_b{}^\nu(\partial_\mu A_\nu - \partial_\nu A_\mu) + T^m{}_{ab}A_m$$

$$+ e_a{}^\mu(\partial_\mu e_b{}^\nu)A_\nu - e_b{}^\mu(\partial_\mu e_a{}^\nu)A_\nu. \tag{28}$$

In view of the orthogonality of the tetrad fields we have

$$\partial_\mu e_b{}^\nu = -e_b{}^\lambda(\partial_\mu e^c{}_\lambda)e_c{}^\nu. \tag{29}$$

With the help of the equation above we find that the last two terms of eq. (28) may be rewritten as

$$e_a{}^\mu(\partial_\mu e_b{}^\nu)A_\nu - e_b{}^\mu(\partial_\mu e_a{}^\nu)A_\nu = -T^m{}_{ab}A_m. \tag{30}$$

Therefore the last three terms of (28) cancel each other and finally we have

$$F_{ab} = e_a{}^\mu e_b{}^\nu(\partial_\mu A_\nu - \partial_\nu A_\mu). \tag{31}$$

The equation above shows that given the abstract, tensorial expression of the Faraday tensor we can simply project it on any moving frame in Minkowski space-time. This is exactly the procedure adopted by Mashhoon (4) in the investigation of electrodynamics in accelerated frames. Mashhoon is interested in developing the non-local formulation of electrodynamics. However, if we restrict attention to the evaluation of total quantities, such as the integration of the Poynting vector and the total radiated power (and not to pointwise measurements), then the standard formulation suffices to arrive at qualitative conclusions.

We may obtain Maxwell's equations with sources from an action integral determined by the Lagrangian density

$$L = -\frac{1}{4}e F^{ab}F_{ab} - \mu_0 e A_b J^b, \tag{32}$$

where $e = \det(e^a{}_\mu)$, $J^b = e^b{}_\mu J^\mu$ and μ_0 is the magnetic permeability constant. Although in flat space-time we have $e = 1$, we keep e in the expressions below because it allows a straightforward inclusion of the gravitational field. Note that in view of eq. (31) we have

$$F^{ab}F_{ab} = F^{\mu\nu}F_{\mu\nu}. \tag{33}$$

Therefore L is frame independent, besides being invariant under coordinate transformations. The field equations derived from L are

$$\partial_\mu (e F^{\mu b}) + e F^{\mu c} \left({}^0\omega_\mu{}^b{}_c \right) = \mu_0 e J^b , \tag{34}$$

or

$$e_b{}^\nu [\partial_\mu (e F^{\mu b}) + e F^{\mu c} \left({}^0\omega_\mu{}^b{}_c \right)] = \mu_0 e J^\nu , \tag{35}$$

where $F^{\mu c} = e_b{}^\mu F^{bc}$. In view of eq. (33) it is clear that the equations above are equivalent to the standard form of Maxwell's equations in flat space-time.

The second set of Maxwell's equations is obtained by working out the quantity $D_a F_{bc} + D_b F_{ca} + D_c F_{ab}$, where the covariant derivative of $D_a F_{bc}$ is defined by

$$D_a F_{bc} = e_a{}^\mu D_\mu F_{bc} \tag{36}$$
$$= e_a{}^\mu (\partial_\mu F_{bc} - {}^0\omega_\mu{}^m{}_b F_{mc} - {}^0\omega_\mu{}^m{}_c F_{bm}).$$

Taking into account relations (27) and (29) we find that the source free Maxwell's equations in an arbitrary moving frame are given by

$$D_a F_{bc} + D_b F_{ca} + D_c F_{ab} = e_a{}^\mu e_b{}^\nu e_c{}^\lambda (\partial_\mu F_{\nu\lambda} + \partial_\nu F_{\lambda\mu} + \partial_\lambda F_{\mu\nu}) = 0 , \tag{37}$$

in agreement with the standard description.

We refer the reader to Ref. (5), where we consider an accelerated frame with velocity $v(t)$ with respect to an inertial frame, and describe Gauss law in the accelerated frame for the situations (i) in which the source is at rest in the inertial frame, and (ii) in which the source is at rest in the accelerated frame.

4. Plane electromagnetic waves in a linearly accelerated frame

In this section we consider an observer in Minkowski space-time that is uniformly accelerated in the positive x direction. The wordline and velocity of the observer in terms of its proper time τ are

$$\bar{x}^\mu = \left\{ \frac{c^2}{a} \sinh\left(\frac{a\tau}{c}\right), \frac{c^2}{a} \left[\cosh\left(\frac{a\tau}{c}\right) - 1\right], 0, 0 \right\}, \tag{38}$$

and

$$u^\mu = \frac{d\bar{x}^\mu}{d\tau} = \left\{ c \cosh\left(\frac{a\tau}{c}\right), c \sinh\left(\frac{a\tau}{c}\right), 0, 0 \right\}, \tag{39}$$

respectively.

A simple form of tetrad fields adapted to the observer with velocity u^μ, i.e., for which $e_{(0)}{}^\mu = u^\mu/c$ and $e^a{}_\mu e_{a\nu} = \eta_{\mu\nu}$, is given by

$$e^a{}_\mu = \begin{pmatrix} \cosh(a\tau/c) & -\sinh(a\tau/c) & 0 & 0 \\ -\sinh(a\tau/c) & \cosh(a\tau/c) & 0 & 0 \\ 0 & 0 & 1 & 0 \\ 0 & 0 & 0 & 1 \end{pmatrix}. \tag{40}$$

If we substitute the tetrad fields and the inverses into Eq. (4), we see that the only nonvanishing component of ϕ_{ab} is

$$\phi_{(0)}{}^{(1)} = \frac{a}{c}. \tag{41}$$

The frame described by Eq. (40) is moving with uniform acceleration a in the positive x direction, and its axes are oriented along the global Cartesian frame. In view of Eqs. (12), (38) and (40), it follows that

$$
\begin{aligned}
t &= \frac{c}{a}\left(1 + \frac{ax'}{c^2}\right)\sinh\left(\frac{a\tau}{c}\right), \\
x &= \frac{c^2}{a}\left(1 + \frac{ax'}{c^2}\right)\cosh\left(\frac{a\tau}{c}\right) - \frac{c^2}{a}, \\
y &= y', \\
z &= z'.
\end{aligned}
\tag{42}
$$

We note that Eq. (39) can be given alternatively in terms of the time coordinate t of the inertial frame by

$$u^\mu(t) = \{c\gamma(t),\, c\beta(t)\,\gamma(t),\, 0,\, 0\}, \tag{43}$$

where

$$\gamma(t) = \sqrt{1 + a^2 t^2/c^2}, \qquad \text{and} \qquad \beta(t)\gamma(t) = at/c .$$

In terms of the coordinates (t, x, y, z) adapted to the inertial frame, the Faraday tensor for a plane electromagnetic wave that propagates in the positive x direction reads

$$
F^{\mu\nu} = \begin{pmatrix}
0 & 0 & -E_y/c & 0 \\
0 & 0 & -B_z & 0 \\
E_y/c & B_z & 0 & 0 \\
0 & 0 & 0 & 0
\end{pmatrix}. \tag{44}
$$

where

$$E_y(t, \vec{x}) = E_0 \cos(kx - \omega t), \tag{45}$$

$$B_z(t, \vec{x}) = \frac{E_0}{c}\cos(kx - \omega t). \tag{46}$$

In these expressions k is the wave number and ω is the frequency of the wave, which are related by $k = |\vec{k}| = \omega/c$. The speed of propagation of the electromagnetic wave is

$$v_p = \frac{\omega}{|\vec{k}|} = c . \tag{47}$$

The expression of the electromagnetic field in the inertial frame is formally obtained out of Eqs. (45) and (46) by means of the tetrad field $e^a{}_\mu = \delta^a_\mu$. However we will consider that (45) and (46) do represent the fields in the inertial frame.

In view of the expressions above we see that the only nonzero component of the Poynting vector

$$\vec{S} = \frac{1}{\mu_0}\vec{E} \times \vec{B}, \tag{48}$$

is given by

$$S_x = \frac{(E_0)^2}{\mu_0 c} \cos^2 (kx - \omega t), \tag{49}$$

where μ_0 is the magnetic permeability constant. Thus the energy flux of the electromagnetic wave points in the same direction of the wave propagation.

In order to obtain the electric and magnetic field components of the electromagnetic wave in the uniformly accelerated frame, we insert Eqs. (44) and (40) into $F^{ab} = e^a{}_\mu e^b{}_\nu F^{\mu\nu}$. We obtain

$$E_{(y)} = \cosh \left(\frac{a\tau}{c} \right) E_y - c \sinh \left(\frac{a\tau}{c} \right) B_z, \tag{50}$$

$$B_{(z)} = -\frac{1}{c} \sinh \left(\frac{a\tau}{c} \right) E_y + \cosh \left(\frac{a\tau}{c} \right) B_z, \tag{51}$$

where E_y and B_z are given by (45) and (46), respectively.

In Eqs. (50) and (51) E_y and B_z are expressed in terms of the coordinates (t, x). In order to present the electric and magnetic fields in terms of the coordinates $(\tau, \vec{x}')$ of the accelerated frame we make use of Eq. (42). We arrive at

$$E_{(y)}(\tau, \vec{x}') = E_0 e^{-a\tau/c} \cos \left[k \left(e^{-a\tau/c} \right) x' - \frac{\omega c}{a} \left(1 - e^{-a\tau/c} \right) \right], \tag{52}$$

$$B_{(z)}(\tau, \vec{x}') = \frac{E_0}{c} e^{-a\tau/c} \cos \left[k \left(e^{-a\tau/c} \right) x' - \frac{\omega c}{a} \left(1 - e^{-a\tau/c} \right) \right], \tag{53}$$

where we used

$$e^{-a\tau/c} = \cosh(a\tau/c) - \sinh(a\tau/c).$$

The only nonzero component of the Poynting vector is

$$S_{(x)} = \frac{(E_0)^2}{\mu_0 c} e^{-2a\tau/c} \cos^2 \left[k \left(e^{-a\tau/c} \right) x' - \frac{\omega c}{a} \left(1 - e^{-a\tau/c} \right) \right], \tag{54}$$

We see that the density of energy flux decreases in time by a factor $e^{-2a\tau/c}$ in a frame that is uniformly accelerated in the same direction of the propagation of the electromagnetic wave.

The amplitudes in Eqs. (52) and (53) may be written as

$$E_{(0)} = E_0 e^{-a\tau/c}, \tag{55}$$

$$B_{(0)} = \frac{E_0}{c} e^{-a\tau/c} = \frac{E_{(0)}}{c}. \tag{56}$$

The identification of the wave number and of the frequency of the wave in the accelerated frame is made by means of a projection of the wave vector $k_\mu = (-\omega/c, k, 0, 0)$ from the inertial to the accelerated frame, according to $k^a = e^a{}_\mu k^\mu$. We recall that this procedure is equivalent to performing a local Lorentz transformation where the coefficients $\Lambda^a{}_b$ of the transformation satisfy $e^a{}_\mu = \Lambda^a{}_b (e^b{}_\mu)_{in} = \Lambda^a{}_b \delta^b_\mu$. Thus we have

$$k' = k e^{-a\tau/c}, \tag{57}$$

$$\omega' = \omega e^{-a\tau/c}. \tag{58}$$

We conclude that the amplitude, wave number and frequency of the electromagnetic wave decrease in proper time by a factor $e^{-a\tau/c}$ in a frame that is uniformly accelerated in the same direction of the wave propagation. We note that the observer will never reach the speed of light. Considering Eqs. (57) and (58) we see that the speed of propagation of the electromagnetic wave in the uniformly accelerated frame is

$$v'_p = \frac{\omega'}{k'} = \frac{\omega e^{-a\tau/c}}{k e^{-a\tau/c}} = c. \tag{59}$$

Therefore the speed of the electromagnetic wave is independent of the observer's acceleration.

5. Spherical waves in a radially accelerated frame

We will repeat the analysis carried out in the previous section and consider the measurement of spherical electromagnetic waves, produced in an inertial frame, in a radially accelerated frame. A spherical wave in an inertial frame may be characterized by the following expressions for the electric and magnetic fields,

$$\vec{E}(t,r,\theta,\phi) = E_0 \frac{\sin\theta}{r} \left[\cos(kr - \omega t) - \frac{1}{kr} \sin(kr - \omega t) \right] \hat{\phi}, \tag{60}$$

$$\vec{B}(t,r,\theta,\phi) = -\frac{E_0}{c} \frac{\sin\theta}{r} \left[\cos(kr - \omega t) - \frac{1}{kr} \sin(kr - \omega t) \right] \hat{\theta}, \tag{61}$$

where the unit vectors $\hat{\phi}$ and $\hat{\theta}$ are defined in terms of the cartesian unit vectors by

$$\hat{\phi} = -\sin\phi\,\hat{x} + \cos\phi\,\hat{y},$$

$$\hat{\theta} = \cos\theta\cos\phi\,\hat{x} + \cos\theta\sin\phi\,\hat{y} - \sin\theta\,\hat{z}. \tag{62}$$

A set of tetrad fields in spherical coordinates, adapted to an observer that undergoes uniform acceleration in the radial direction, is given by

$$e^a{}_\mu(t,r,\theta,\phi) = \begin{pmatrix} \gamma & -\gamma\beta & 0 & 0 \\ -\gamma\beta & \gamma & 0 & 0 \\ 0 & 0 & r & 0 \\ 0 & 0 & 0 & r\sin\theta \end{pmatrix}, \tag{63}$$

where

$$\gamma = \sqrt{1 + \frac{a^2 t^2}{c^2}}, \qquad \gamma\beta = \frac{at}{c}. \tag{64}$$

The inverse components of Eq. (63) are such that $e_{(0)}{}^\mu(t,r,\theta,\phi) = (\gamma, \beta\gamma\,0, 0)$. Therefore the frame is accelerated along the radial direction.

We start with the Faraday tensor in cartesian coordinates,

$$F^{\mu\nu}(t,x,y,z) = \begin{pmatrix} 0 & -E_x/c & -E_y/c & -E_z/c \\ E_x/c & 0 & -B_z & B_y \\ E_y/c & B_z & 0 & -B_x \\ E_z/c & -B_y & B_x & 0 \end{pmatrix}. \tag{65}$$

The electric and magnetic field components in the expression above are obtained out of Eqs. (60), (61) and (62). We must consider the expression above in spherical coordinates. So we perform the coordinate transformation

$$F'^{\alpha\beta}(t,r,\theta,\phi) = \frac{\partial x'^{\alpha}}{\partial x^{\mu}} \frac{\partial x'^{\beta}}{\partial x^{\nu}} F^{\mu\nu}(t,x,y,z). \tag{66}$$

After some algebra we obtain

$$F'^{01} = 0,$$
$$F'^{02} = 0,$$
$$F'^{03} = -\frac{E_0}{c} \frac{1}{r^2} \left[\cos(kr - \omega t) - \frac{1}{kr} \sin(kr - \omega t) \right],$$
$$F'^{12} = 0,$$
$$F'^{23} = 0,$$
$$F'^{31} = \frac{E_0}{c} \frac{1}{r^2} \left[\cos(kr - \omega t) - \frac{1}{kr} \sin(kr - \omega t) \right]. \tag{67}$$

The quantities in Eq. (67) represent both the abstract tensor components of the Faraday tensor in spherical coordinates, and the components of the Faraday tensor in an inertial frame. Next we project these tensor components on the accelerated frame defined by Eq. (63). We arrive at

$$F'^{(0)(1)} = 0,$$
$$F'^{(0)(2)} = 0,$$
$$F'^{(0)(3)} = -\frac{E_0}{c}(\gamma - \gamma\beta)\frac{\sin\theta}{r} \left[\cos(kr - \omega t) - \frac{1}{kr} \sin(kr - \omega t) \right],$$
$$F'^{(1)(2)} = 0,$$
$$F'^{(2)(3)} = 0,$$
$$F'^{(3)(1)} = \frac{E_0}{c}(\gamma - \gamma\beta)\frac{\sin\theta}{r} \left[\cos(kr - \omega t) - \frac{1}{kr} \sin(kr - \omega t) \right]. \tag{68}$$

Note that the factor $(\gamma - \gamma\beta)$ may be rewritten as

$$\gamma - \gamma\beta = \sqrt{\frac{1-\beta}{1+\beta}}. \tag{69}$$

In order to verify how Eqs. (60) and (61) are modified in the accelerated frame we just compare the structure of Eqs. (67) and (68), and indentify (60) and (61) in the latter expression. We obtain

$$\vec{E}(t,r,\theta,\phi) = E_0(\gamma - \beta\gamma)\frac{\sin\theta}{r}\left[\cos(kr - \omega t) - \frac{1}{kr}\sin(kr - \omega t)\right]\hat{\phi}, \tag{70}$$

$$\vec{B}(t,r,\theta,\phi) = -\frac{E_0}{c}(\gamma - \beta\gamma)\frac{\sin\theta}{r}\left[\cos(kr - \omega t) - \frac{1}{kr}\sin(kr - \omega t)\right]\hat{\theta}, \tag{71}$$

in the inertial frame coordinates.

By comparing Eqs. (70) and (71) with (60) and (61) we see that the major qualitative difference between these expressions is the emergence, in the former pair of equations, of the time dependent Doppler factor $(\gamma - \beta\gamma)$ given by Eq. (69). If the accelerated frame is at the radial position (r, θ) at the instant t, then the measured amplitude of the wave in the accelerated frame will be smaller by a factor $(\gamma - \beta\gamma)$ than if the frame were at rest at the same position. Thus the amplitude of the spherical wave in the accelerated frame varies with time, and approaches zero in the limit $t \to \infty$, since in this limit $\beta \to 1$.

6. Final comments

The tetrad field and its interpretation as a frame adapted to arbitrary observers in space-time allow the formulation of electrodynamics in accelerated frames. The idea is to project the electromagnetic vectorial and tensorial quantities in any moving frame by means of the tetrad field. Specific issues regarding electromagnetic radiation were discussed in ref. (5).

Of course all the results derived from the procedure adopted in this chapter are valid as long as the very concept of tetrad field and its interpretation are also valid. The justification behind the usage of tetrad fields for this purpose is given by principle of locality (10). The idea is the following. A physical measurement is considered to be reliable if it is performed in an inertial reference frame. Normally it is admitted that the observer is standing in an inertial frame. Measurements in accelerated frame are, in general, not easily performed. When an electromagnetic field quantity is projected in a frame by means of the tetrad field, it is assumed that this tetrad field is, at each instant of time, physically equivalent (identical) to another frame that is inertial and momentarily co-moving with the accelerated frame. The worldline of the two frames, the accelerated and the inertial, coincide at that instant of time. To a certain extent, the hypothesis of locality, together with the concept of tetrad field, extends the principle of relativity, since it relates inertial and non-inertial frames.

An interesting consequence of the present analysis is the following. Let us suppose that an accelerated observer in the context of section 4 measures the frequencies ω'_1 and ω'_2 at the instants of proper time τ_1 and τ_2, respectively,

$$\omega'_1 = \omega e^{-a\tau_1/c}, \qquad \omega'_2 = \omega e^{-a\tau_2/c}, \tag{72}$$

according to eq. (58). By dividing the two frequencies of the electromagnetic waves we obtain

$$a = \frac{c}{\Delta\tau}\ln\left(\frac{\omega'_1}{\omega'_2}\right), \tag{73}$$

where $\Delta\tau = \tau_2 - \tau_1$. Therefore the accelerated observer may determine the value of its own acceleration provided the luminosity of the source is constant and the acceleration is uniform. This formula may be useful in the evaluation of the acceleration of the solar system, for instance, with respect to the distant supernovas, provided it is verified that in the interval $\Delta\tau$ the luminosity of the supernova is not substantially changed. Of course the resulting value will provide just the order of magnitude of the acceleration of the expansion of the universe.

The final expressions of the electric and magnetic fields in the accelerated frames, Eqs. (52-53), and (70-71), for plane and spherical waves, respectively, are related to the expressions in the inertial frame by means of simple time dependent functions. The simplicity of the final expressions ensures that the present tehcnique is correct, and suggests that all manifestations of electrodynamics may be investigated in any moving frame.

7. References

[1] *The Classical Theory of Fields*, L. D. Landau and E. M. Lifshitz (Pergamon, Oxford, 1908).

[2] *Classical Electrodynamics*, J. D. Jackson (Wiley, New York, 1999).

[3] *Foundations of Classical Electrodynamics*, F. W. Hehl and Y. N. Obukhov (Birkhäuser, Boston, 2003).

[4] B. Mashhoon, Ann. Phys. (Berlin) 12, 586 (2003) [hep-th/0309124]; Phys. Rev. A 70 (2004) 062103 [hep-th/0407278]; Phys. Rev. A 79 (2009) 062111 [arXiv:0903.1315].

[5] J. W. Maluf and S. C. Ulhoa, Ann. Phys. (Berlin) 522, 766 (2010) [arXiv:1009.3968].

[6] F. H. Hehl, J. Lemke and E. W. Mielke, "Two Lectures on Fermions and Gravity", in *Geometry and Theoretical Physics*, edited by J. Debrus and A. C. Hirshfeld (Springer, Berlin Heidelberg, 1991).

[7] J. W. Maluf, F. F. Faria and S. C. Ulhoa, Class. Quantum Grav. 24 (2007) 2743 [arXiv:0704.0986].

[8] J. W. Maluf and F. F. Faria, Ann. Phys. (Berlin) 17 (2008) 326 [arXiv:0804.2502].

[9] B. Mashhoon, Ann. Phys. (Berlin) 11, (2002) 532 [gr-qc/0206082].

[10] B. Mashhoon, Phys. Lett. A 143, 176 (1990); 145, 147 (1990).

6

Neoclassical Theory of X-Ray Scattering by Electrons

V.V. Aristov

Institute of Microelectronics Technology and High Purity Materials RAS,
Russia

1. Introduction

Many successes in physics of 20th century associated with quantum mechanics in the statement, which was given by the N. Bohr, W. Heisenberg, M. Born, W. Pauli and others. It is known as "Copenhagen interpretation". It bases on the statement that the purpose of quantum theory is the description of results of observation, instead of getting knowledge about reality and the processes occurring in material systems in an interval between the first observation and the next. Moreover: "Any attempt to find such a description would lead to contradictions" (Heisenberg, 1958). Canonization of this opinion considerably changed a view at the purpose of researches, led to the confidence accepted today by the majority of physicists that "the quantum mechanics has rejected usual classical mechanics determinism in behavior of microobjects", and that "the aim of quantum mechanics to give a method for finding probability distributions for various physical values in various states of microobjects" (Akhiezer & Polovin, 1973). Postulation a fundamental nature of quantum mechanics "obscurity" has led to the statement on completeness of quantum mechanics, i.e. that there are no parameters (probably hidden) which define physical sense of processes in quantum system and give fuller description of the nature, than are based on probability functions or a density matrix.

This radical change of the research concept was denied by many of those who stood at the basis of the quantum theory – M. Planck, A. Einstein, E. Schrödinger, M. von Laue, and L. de Broglie. In second half of 20th century, dispute over reality and search of physical reasons, causing the quantum phenomena, has led to various formulations of quantum mechanics - the theories containing hidden variables, describing behavior and evolution of quantum systems in space and time. Herein D. Bohm mechanics, based on de Broglie idea about a "pilot wave" directing electron movement (Bohm, 1952), and the semiclassical or neoclassical electrodynamics theory developed by E. Jaynes and his colleagues (Crisp & Jaynes, 1969), based on Schrödinger idea (supported by E. Fermi) that square of wave function describes not probability, but actual charge density distribution in atom (Schrödinger, 1926). In the same years J. Bell demonstrated, that contrary to belief of the majority of physicists about impossibility of hidden variables existence, the nonlocality of quantum mechanics reflects nonlocality of hidden variables (Bell, 2004). Nevertheless, today, in spite of enough numerous researches directed on revision of view of Copenhagen interpretation founders, the disputes round the purposes of physical researches and the essence of hypotheses and postulates, containing in the theory, majority of physicists are perceived as the philosophical debates representing only historical interest. "...truth does not triumph by convincing its opponents and making them see the light, but rather because its opponents eventually die..." (M. Planck). Such point of view is grounded usually by impossibility to offer any significant experiment distinguishing this realistic theory from canonical quantum mechanics. This opinion is deeply erroneous. Quantum mechanics from the very beginning developed as an empirical science. Its formulations, models were formed by only trial and error method. Any discrepancies of quantum models with experimental results were removed by imposing of "exclusion principle" for any processes, which should be observed according to the logic of classical physics, were removed by acceptance of postulates on principles of matter behavior in quantum world and hypotheses about a structure and properties of quanta. As a result of such "development" was creation of purely mathematical theory with the logic which is initially not reduced to the logic of classical physics. Such theory necessarily should meet difficulties in definition of physical sense some the phenomenon: "Quantum electrodynamics is not a perfectly consistent theory" (Dirac, 1965). "It is plagued by divergences, some of which are carried over from the classical theory of electromagnetic fields, and some of which are introduced by the procedure of quantizing the electromagnetic field. Quantizing a field that has an infinite number of degrees of freedom seems to lead unavoidably to an infinite amount of energy in the zero-point oscillations. Furthermore, the usual QED derivation of spontaneous decay, the Lamb shift, and the calculation of the anomalous moment of the electron seem to require these zero-point oscillations. Nevertheless, it is difficult to rationalize that these zero-point oscillations actually exist in nature" (Crisp, 1990).

Other fundamental difficulty of quantum mechanics in explaining of an interference of particles - photons, electrons etc. R. Feynman wrote about an interference problem on two slits: "Impossible, absolutely impossible to explain it any classical way ... has in the heart of QM. Really, it contains the only mystery" (Feynman, 1965).

Usually considering, the mysticism inherent in quantum description of some experiments is eliminated by replacement of the Copenhagen interpretation most popular in quantum physics by one of equivalent interpretations. It is considering, that all formulations (and them more than 10) despite "differ dramatically in mathematical and conceptual overview, yet each one makes identical predictions for all results" (Styer et al., 2002).

Perhaps, firm belief in correctness of quantum mechanics conclusions is connected with considering in quantum mechanics simple enough phenomena isolated from others. We will below analyze complex experiments in X-ray physics where the various effects are the consequence of the same process but for its explanation till now theory formulates assumptions, not agreed among themselves, and by that we will prove, all mysticism of behavior of quantum particles is connected only with the theory. Important, that a lot of troubles and "obscurities" inherent in quantum physics, are concentrated in physics of X-ray scattering.

In this chapter, results of experimental research of hard x-ray radiation scattering will be compared to those theories, which have arisen for their explanation. It is interesting to note, already in the early twenties of last century the common theory of scattering of electromagnetic radiation could be created. Instead of it, unique process of radiation scattering by atoms of matter began to be considered as consisting of the several independent phenomena described by various mechanisms of scattering. Theories describing the scattering are:

Classical (Thomson) theory: electromagnetic wave scattering by point electrons. All electrons of atom scatter incoherently with each other. Frequency of a scattering wave coincides with incident frequency. Possibility of existence of the "big" electron having the certain size and the form is supposed.

Quantum (Compton) theory: the photon interact with the point electron in atom in the state, defined at the moment of scattering in coordinate and momentum corresponding to the solution of Schrödinger wave equation. This scattering named as Compton or incoherent because frequency of scattering radiation (new photon frequency) depends on a scattering angle and an initial momentum of an electron which it has at the moment of collision with a primary photon.

Neoclassical theory of scattering (NCT) - base for X-ray diffraction theory and X-ray crystallographic analysis: electromagnetic wave scattering by electron distribution in atom. This distribution is set by Schrödinger function, which is understood not as function of probability distribution of point electron in atom but as the real charge density. It is coherent (Bragg and Rayleigh) scattering without frequency changing.

The listed theories - classical, quantum and NCT are applied not to the description of scattering of X-ray radiation by atomic electrons, but for the description of separate fragments of scattering spectrum in various models of an atom structure and radiation properties.

From here follows singlevalued, though also a paradoxical conclusion. In the modern physics, there is no theory of scattering of hard electromagnetic radiation not only by bound, but also by free electrons.

Situation with treatment of x-ray radiation scattering on matter reminds the story about an elephant and the group of blind men. They touch an elephant at different parts to learn what it is like. As a result each of them formed his own "the theory of an elephant". Disputes what theory more correct are useless. Everyone reflects only a part of the general essence. It is necessary to see "elephant" entirely. For this purpose, in the case of X-ray scattering phenomena, it is necessary to rethink available experimental data, recognizing, that all of them reflect uniform process of scattering, and all theories existing for today describing its separate fragments are not perfectly correct, and provide some misrepresentation about a matter structure.

Basic concept of quantum electrodynamics is the concept about a photon and electron as a point-like particle. The most convincing, but actually, the first and unique proof of photon existence is Compton effect: changing and angular dependence of the scattered (secondary) x-ray radiation wavelength in comparison to the incident radiation.

For Compton effect explaining by photons scattering it is necessary, that the electron in atom: 1) was point-like, 2) had a certain momentum - instant velocity and a direction of movement in atom (theory of impulse approximation), explaining broadening of Compton line.

On the other hand, as noted above, X-ray diffraction method bases on the assumption that electromagnetic wave scatters not on separate point electrons, but on a charge distributed in atom. This mechanism of scattering is put in a basis of methods of atoms and molecules structures definition, research of crystal real structure. Clearly, that it cannot be realized from quantum mechanics point of view. Imagine, that the distance between slits (electrons in atoms) quickly changes in time under some probability law in a big area defined by Schrödinger function, then named Feynman "mystical" process of an interference of photons on two slits becomes inconceivable. Thus it is clear, in experiments on x-ray diffraction (i.e. at coherent scattering), X-rays, scattered by electrons of a crystals, is electromagnetic radiation, and the charge occupies certain volume in atoms, instead of consists of point electrons. However, if it so, what conclusion should be made about plausibility of quantum origin of Compton effect, and other effects arising at X-ray scattering?

For the answer to this question, at the analysis of results of X-ray radiation scattering experiments we will base on representations about mechanisms of scattering, that are used in the neoclassical theory. We will see, both effects of scattering, "coherent" (Rayleigh) and "incoherent" (Compton), represent in a spectrum only phenomenon: scattering of electromagnetic radiation on a volumetric electron charge which structure is deformed in the field of atomic forces. In such understanding of scattering process Thomson theory plays the role of the theory of wave scattering on the elementary charge contained in small volume dV of electronic density.

2. Classical theory of scattering (1903-1923)

To develop new theory on x-ray scattering we are used the technology suggested by J. Dodd at a classical treatment of Compton effect (Dodd, 1983): «I should like to include a little fantasy: to presented that we stand near the beginning of the 20th century and attend to discover the lands for the interaction between light and matter using the classical theory of the day being guiding by experiments which in principle could be performed near that time».

2.1 Basic conceptions and models of scattering used at first two decades in 20th century

The view of physics at the time we consider is dominated by theories of J. Maxwell (explanation of electromagnetic fields) and H. Lorentz (force equation explained how electrical charges and current interact). We have learned, that light propagation as a wave of electromagnetic fields described by these equation. Einstein's "lichtquanten" hypothesis did not obtain wide acceptance from the physicists at the time. Von Laue, for example, was opposed the light quanta and suggested that quantization resided in matter, not radiation (at least in x-ray wave region). Though, Einstein in 1918 wrote "I do not doubt any more, the reality of radiation quanta", in 1921 he complained to P. Ehrenfest that "problem of quanta was enough to drive him to the madhouse" (Kidd et al., 1989).

Discussions of size and shape of electron followed from point of view of different physical properties such as magnetic properties, possibility of electron to be compressed because external force to produce Lorentz radiation force for it compensation. The most popular model was electric charge compressed to shape of sphere of radius is equal to

$$r_e = \frac{e^2}{mc^2} = 2.82 \cdot 10^{-15} m$$

Later Dirac suggests mathematic point like model of electron for quantum electrodynamics theory. Results of some experiments on the X and γ rays led to assumption that electron may be flexible ring or spherical shape distributed charge of electricity with radius order

$$\Lambda_e = \frac{h}{mc} = 2.42 \cdot 10^{-12} m$$

(Bergman, 2004). From point of view of investigation of any real body structure and form, it is logical to use electromagnetic radiation with length of a wave, comparable with assumptive size of structure details of investigated object. Usual X-ray radiation satisfies to such criterion for atomic structure of matter: in X-ray experiments typical investigation radiation with wave length 0.5Å - 2Å are used. From this point of view, for the theory of X-ray scattering it was reasonable to start with assumption about point electron.

2.2 Thomson theory — Scattering by point elementary charge

In 1903 J. Thomson published the theory of scattering of electromagnetic radiation by point charge of an electricity. As noted above, the model of a point electron with radius r_e was offered along with model of an electron with the size of an order Λ_e

In this case, it is necessary, that "elementary charges", parts of the "big" electron, scatter coherently and independently, and relation of charge quantity to mass in volume dV in each part remains constant and equal to e/m, where e and m charge and mass of electron (Compton, 1919a). Assume charge distribution in a free electron[1] $\rho(r) = e|\psi(r)|^2$, mass - $m|\psi(r)|^2$.

In the classical theory of scattering is assumed, that electron can have the sizes of an order of $2 \cdot 10^{-12}$ m and can consist of elementary charges in which the relation e/m is constant. It means, that

$$\frac{de}{dm} = \frac{e}{m}, de = e|\psi|^2 dV, dm = m|\psi|^2 dV, \int |\psi|^2 dV = 1 \qquad (1)$$

On Fig. 1 the scheme of scattering plane electromagnetic wave with wavelength λ_0 and frequency ν_0 on an electron consisting of elementary charges. Any "elementary" volume dV<< (h/mc)³ inside distribution of $\rho(r)$ looks as elementary electronic charge with constant charge to mass relation.

---

[1] In QM, for free electron ψ(r)=δ(r). For the electron in atom probability of its position in any point are calculated and presented by Schrödinger function | ψ (r)|². Acceptable distribution of electron density in NCT is under discussion in the paragraph 3.

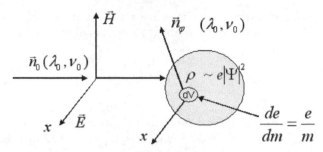

Fig. 1. To the theory of electromagnetic wave scattering by electron. Charge electron density ρ is presented by distribution described by function $|\psi|^2$.

Each charge oscillates in a direction of electric field $\vec{E}$ under the influence of force:

$$F = -de\frac{1}{c}\frac{\partial A_0}{\partial t} \tag{2}$$

where

$$\frac{1}{c}\frac{\partial \vec{A}_0}{\partial t} = -\vec{E}_0 ,$$

$\vec{A}$ - vector potential of a field. The movement equation of an elementary charge is:

$$\ddot{x} = \frac{e}{m}\vec{E}_0(t) \text{ or } \vartheta(t) = -\frac{e}{mc}\vec{A}_0(t) \tag{3}$$

Here $\vartheta(t)$ - velocity of fluctuations of an electron. Thus along an electron in direction $\vec{n}_0$ of distribution of an electromagnetic field current $J(r,t)$ flows. Thus:

$$J(r,t) = e|\psi(r)|^2 \vartheta(t) \tag{4}$$

Further we consider only spherical symmetric functions $\psi(r)$ for which it is convenient to use polar co-ordinates - radius vector $\vec{r}$ and polar angles: φ - an angle between vectors $\vec{n}_0$ and - $\vec{n}_\varphi$, $\vec{n}_\varphi$ - a direction of propagation of a scattering wave, a - an angle between vectors $\vec{E}$ and $\vec{n}_\varphi$, in this case $dV = dxdydz = r^2\sin\alpha\,drd\alpha\,d\varphi = r^2dr d\Omega$

Let's define parameters of scattering radiation for a point electron when $|\psi(r)|^2$ is δ-function. This condition is satisfied at radiation wavelengths λ>10⁻²Å, most widely used in X-ray crystallographic analysis. Besides, the equation (3) is valid in small fields $\vec{A}$, with oscillation amplitude $x_{max} < \Lambda_e$.

Easy to show, scattering of a polarised wave by a point charge leads to dipole waves, intensity equation

$$d|A|^2 = (\frac{e^2}{mc^2})^2|A_0|^2 \sin^2\alpha d\Omega$$

Then differential cross section $d\sigma_0$ and cross section σ_0 is:

$$\frac{d\sigma_0}{d\Omega} = r_e^2 \sin^2 \alpha; \quad \sigma_0 = \frac{8}{3}\pi r_e^2 \tag{5}$$

Notice one essential circumstance. At making above assumptions of structure and properties of the "big" electron, the value of scattering cross section is proportional to r_e.

Therefore, for the "big" charge easy to write the equation for amplitude and intensity scattered in direction n_φ in limit far field diffraction zone $R \gg r_e$ (R - distance from an electron to a point of observation). Assume big electron scattering function in a direction n_φ:

$$f_e(k_0 - k_\varphi) = \int |\psi(r)|^2 \, e^{2\pi i(k_0 - k_\varphi)r} \, dV \tag{6}$$

$$k_0 - k_\varphi = \frac{n_0 - n_\varphi}{\lambda_0} = \vec{H}, \quad |\vec{H}| = \frac{2\sin\frac{\varphi}{2}}{\lambda_0}$$

$\vec{H}$ - vector in space of reciprocal wave vectors.

In this case scattering cross section:

$$\frac{d\sigma(\vec{H})}{d\Omega} = r_0^2 \sin^2\varphi \cdot |f_e(\vec{H})|^2, \, d\sigma(\vec{H}) = d\sigma_0 |f_e(\vec{H})|^2 \tag{7}$$

According to Rutherford atomic model each electron in atom scatter short-wave radiation independently (incoherently) from other electrons. Thomson model also suggest, that intensity of scattering by each atom is proportional to amount of atomic electrons Z:

$$I = \sigma_0 Z \tag{8}$$

In his experimental researches, C. Barkla demonstrates that electrons actually scatter independently and incoherently (Barkla, 1911), at least, in atoms of light elements (the neat result of this researches was definition the number of electrons in atom). Barkla also discovered polarization of X-ray radiation and validity Maxwell theory for calculation of X-rays scattering, thus identified it with the electromagnetic waves, having very short length in comparison with a visible region. P. Evald took for the basis of the wave propagation theory in crystals and assumed that wavelengths of X-ray radiation are comparable with interplane distances. This hypothesis was the basis for Laue experiment (performed by W.Friedrich and P. Knipping in 1912) for discovering X-rays diffraction on crystals. A year later, the father and son Braggs created the theory of Bragg diffraction, and the device for spectral analysis of scattering radiation. Discovering of diffraction of X-ray radiation has allowed to draw a remarkable but paradoxical conclusion that, in our opinion, was a forerunner of the subsequent "obscurity" the quantum physics. In spite of the fact that the majority of experiments shows, that electrons in each atom scatter independently from each other, the lattice atoms in a crystal, under certain conditions (Bragg conditions), become coherent scatterers for X-ray radiation.

2.3 Compton theory – Scattering by finite size electron

Among other important experimental results it is necessary to mention one more result, that did not become discovery but should become. D. Florance showed, that the angular diagram of X-rays scattering $I(\varphi)$ is asymmetric. $I(0^0)>I(180^0)$, and dependence $I(180°)/I(0°)$ rapidly decreases with λ_0 decreasing (Florance, 1910).

These results allow to suppose that the electron has the size comparable to x-ray radiation wavelengths. These years Mie and Debay already created the theory of light scattering on dielectric and metallic small diameter spherical particles, with diameter comparable to radiation wavelength (Mie, 1908; Debye, 1909). Rigorous solution problem of diffraction plane wave on homogeneous sphere of any diameter and structure was obtained. The next years many authors analyzed various aspects of this problem. In present paragraph, consider such researches in x-ray wavelength region.

In 1918-1919 Compton dedicate theoretical researches to analysis of experimental data of x-ray scattering on electrons (Compton, 1919a,b). Fig.2 represents experimental measurement (circles) of relative intensity radiation scattered at different angles when the hard X-ray (λ =0.09 Å) from radium bromide traverse a plate of iron (Florance, 1910), and calculated curves of angular scattering dependence of unpolarized wave (Compton, 1919a). In figure, calculated curves for "big" spherical shell electron with radius r =2·10⁻¹⁰cm - internal continuous line; for an electron in the form of a ring (ring electron, r =2·10⁻¹⁰cm) - dashed line; for point charge - an external continuous line.

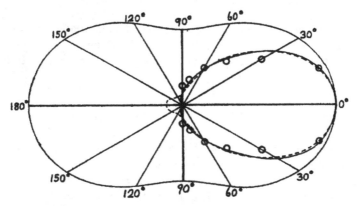

Fig. 2. Experimental (Florance, 1910) and theoretical (Compton, 1919a) angular dependencies integral scattering intensity from point-like and "big" electron with different charge distribution $\rho(r)$ (spherical shell and ring electron with radius 2·10⁻¹²m), see text above.

Compton for the definition of distribution $\rho(r)$ has analyzed not only results of measurements of angular diagram of scattering. Analyzing measurements of dependence an absorption coefficient by various materials, he demonstrated, that μ_a - factor of atomic absorption of energy in an electron distributed in some volume, is defined by the sum consisting of fluorescent absorption, and scattering σ, depending on wavelength λ_0:

$$\mu_a = kZ^4\lambda_0^3\,\Theta(r,\lambda_0) + \sigma(r,\lambda_0) \tag{9}$$

k=const, r–electron radius, functions Θ and σ calculated for flexible ring electricity, are expressed approximately by the equations

$$\Theta = 14.8(\frac{r}{\lambda_0})^2 + 93.6(\frac{r}{\lambda_0})^4 + ..., \quad \sigma = \sigma_0(1 - 26.6(\frac{r}{\lambda_0})^2 + 524(\frac{r}{\lambda_0})^4...)$$

Formula (9) demonstrated, that experimental data well coincide with calculated at $r \approx 0.8 \cdot 10^{-10}$ cm.

Absorption measurements were less precise, than scattering measurements, but values of the calculated radiuses of an electron according to these various methods of measurements showed, that the electron is a particle with radius of an order 10^{-12} m. Such coincidence of experimental values of scattering and absorption, obtained at determination of electron radius, is especially important in view of dispute, continued more than centuries on character of interaction of radiation with matter. In conclusion review of first two decades results obtained by theory of scattering hard electromagnetic radiation, it is necessary to discuss in brief a problem of absorption and radiation by the charged particle of electromagnetic impulse and energy. This problem has been solved in classical electrodynamics more than 100 years ago. Compton in 1923 has offered the new decision, in fact reducing a problem of an exchange of energy and an impulse between field and electron to a mechanical problem of particles collision. His suggestion leads to dramatic consequences, especially for physics of X-ray radiation scattering.

2.4 Absorption of an impulse of an electromagnetic field, Lorentz forces – Impulse and energy conservation laws

Consider scattering of an electromagnetic wave by an elementary charge. Force F_S, acting from a field on an electron, is equal to average value of an impulse absorbed in unit of time.

$$F_S = \frac{8\pi}{3}r_e^2\frac{E^2}{4\pi}\,\vec{n} \tag{10}$$

A charge particle on being accelerated by the force (10) recover electromagnetic energy and itself losses energy. H. Lorentz point to occurrence of braking force of an electron at radiation of X-ray electromagnetic wave (Lorentz, 1905). Loss of energy is interpreted by him as caused by a force F_L acting on the particle given in value equal to F_S and opposite to it:

$$F_L = -F_S \tag{11}$$

The sense of the equations (11) and (10) that the elementary charge in the electromagnetic field gets and loses energy and an impulse simultaneously. Therefore neither an electron at rest, nor a moving electron with constant velocity, does not change the kinetic energy at radiation scattering. Equality of forces F_L and F_S – is a consequence of equality of energy of the waves absorbed and reradiated by an electron at scattering by it of an electromagnetic wave. Presence of Lorentz force was checked experimentally though the physical sense of it origin remains till now not clear. In 1945 discussing this problem J. Wheeler and R. Feynman wrote: "The origin the force of radiative reaction has not been really so clear as its existence". Lorentz, explaining the mechanism radiative reaction forces found the cause in final size of electron and particle elasticity. Under the influence of external forces the sizes of a particle change, there are internal pressure, resistance to external forces. Dirac offered method of its calculation, come from point particle model and not trying to explain physical origin of radiation damping. On base of Dirac solution Wheeler and Feynman offered the scheme of creation braking forces in that the electron simultaneously radiates a spherical wave (out, retarded wave) and accepts another wave (in, advanced wave). In this case the semidifference of forces acting on an electron is equal to Lorentz force of any elementary charge (1). Therefore, equation (11) also as (7), is universal and correct at any form of an electron. The electron scatter the energy proportional to its cross section, and electromagnetic field loses the energy equal to the same value. The energy lost by a field comes back in the form of the sum dipole electromagnetic waves radiated by elementary charges of the big electron. The electron, absorbing energy, absorbs also proportional amount of an impulse that leads to occurrence of Lorentz force equal and opposite F_S.

Note fundamental meaning of a conclusion that follows from the above consideration. The equation (3) of charge transverse oscillation describes the mechanism of energy extraction from primary wave field and simultaneously returning back this energy by radiation dipole waves. The equation (11) confirms: absorption of an impulse of a field at energy absorption causes force acting on an electron in opposite direction at energy radiation.

2.5 Classical theory: Short "Golden age" and swift break-down

In conclusion of the section we will underline some important statements:

- X-rays are electromagnetic radiation, the hypothesis about radiation quanta was discussed, but did not find support in the scientific community up to 1923;
- Electron is a main reradiating matter. Cross section of scattering proportional to the size $r_e = \dfrac{e^2}{mc^2}$ named classical radius;
- The main part of radiation scattered by individual electrons in atom are incoherent;
- Electrons in the atoms from a crystal lattice is a coherent scatter under certain conditions (Bragg diffraction conditions);
- Hard X-ray radiation scattering experiments shows the possibility for electron to have a radius in order to h/mc.

These years were developed methods of the x-ray analysis of molecules and crystals structure, were outlined methods electronic structure analysis of atoms and electrons structure itself. Development of common theory of hard radiation scattering by electrons has been interrupted in 1923. Compton, and, independently, Debay, offered the theory explaining effect "softening" by scattering of particles of light (photons) on point electrons (Compton, 1923; Debye, 1923). "Softening" effect (found out earlier many authors (Eve, 1904; Barkla, 1904)) was: "back" scattering radiation (on large angles φ) absorbs more strongly, than "forward" scattering radiation (on small angles). In experiments Gray and Compton, executed independently, has been shown, that this effect is caused by scattering radiation wavelength increasing with increase in scattering angle φ (Gray, 1920; Compton, 1922). This effect is known today as Compton effect.

Fig. 3 presents Compton spectra of synchrotron radiation (Bessy-2) λ_0 = 0.88 Å scattered on carbon crystal.

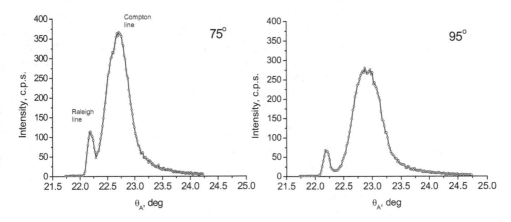

Fig. 3. Experimental scattering spectra 14 keV on diamond at φ = 75°, 95°. Center of Compton line at φ = 95° approximately corresponds to $\Delta\lambda = \Lambda_e$ (see equations (12)). θ_A – angle of Ge crystal analyzer.

At scattering by electrons of carbon atoms, the spectrum contains a weak narrow line with wavelength which occurrence was expected according to the classical theory of scattering. It is surprising, that intensity of this unmodified line (named coherent or Rayleigh line), makes some percent from full intensity of scattering. This part of spectra is responsible for diffraction phenomena and widely used for the structural analysis, X-ray optics and interferometry. This line responsible also for formation of atomic scattering factor and refraction. Much more intensive part of scattering spectrum contains the wavelengths $\lambda > \lambda_0$. This is "Compton" part of scattering spectra, considered in quantum theory as incoherent. Experimentally, with very high accuracy, it has been established, that line centre (its wavelength and frequency) determinate by equations:

$$\lambda'(\varphi) = \lambda_0 + \Lambda_e(1 - \cos\varphi) \quad \nu'(\varphi) = \frac{\nu_0}{1 - \frac{\nu_0}{\nu_e}(1 - \cos\varphi)}$$

$$\frac{\Delta\lambda'}{\lambda_0} = \frac{\Lambda_e}{\lambda_0}(1 - \cos\varphi), \quad \frac{\Delta\nu'}{\nu'} = -\frac{\nu_0}{\nu_e}(1 - \cos\varphi) \tag{12}$$

$$\Lambda_e = \frac{h}{mc}, \quad \nu_e = \frac{mc^2}{h}, \quad \frac{\nu_0}{\nu_e} = \frac{\Lambda_e}{\lambda_0}$$

The Compton line is very wide $\delta\lambda \approx \Delta\lambda'$ and from this point of view it is partially coherent. Value of $\delta\lambda$ of "incoherent" Compton line is only 1-2 orders more than it is for coherent Rayleigh line.

Main part of the radiation energy, scatter by an electron, is contained in a wide spectrum with wavelengths $\lambda' > \lambda_0$. Contrary to the standard opinion, the explanation of this remarkable fact is not found till now. The evidence to that of hundred publications in which various variants of the theory of effect under discussed up to day. Theoretical and experimental investigations of the Compton scattering mechanism, attempts to understand physical sense, proceed already almost century. This is unique phenomenon, since the effect has not obtained any significant practical application. In the theory of x-ray radiation scattering, this effect mention only with necessity to correct calculated atomic and structural scattering factors (along with other just incoherent effects - absorption, diffuse scattering, etc).

Such attention to Compton effect among experts is caused by two factors. The first: Compton effect has made the enormous influence on formation of quantum mechanics and quantum electrodynamics (Glauber, 2006). As it was already mentioned in introduction, it became a turning point in physics development, therefore any attempt to revise quantum physics inevitably should lead to attempts of revision of the accepted standard theory of Compton effect. The second factor. An explanation "softening" effect by collision of particles of point electron with photons broke the describing X-ray scattering theory of into "puzzles". Whole "image of an elephant" has broken up to images of its "foot", "trunk" and "body". Instead of common theory of scattering three various theories which are only formally agreed among themselves have come. Actually, in them are used various representations about entity of matter, various theories of atom structure, various mathematical apparatus. In the next paragraphs we will consider the theory based on de Broglie representation about the electron occupies all space and Schrödinger wave ψ-function as real fields. We will name this theory neoclassical (Crisp, 1990). This name reflects its essence more successfully than semiclassical theory, used various authors.

3. Neoclassical theory of coherent X-ray scattering by bound electrons

For an explanation of results of electromagnetic scattering on the bound electrons, Schrödinger, for the first time, interpreted the wave function ψ found him as the function defining density of a charge in atom. According to his "electrodynamic" hypothesis, the charge density in atom in stationary condition is defined by the formula $\rho = e |\psi(r)|^2$. The elementary charges representing an electron in atom, oscillate under the influence of electromagnetic radiation. At such understanding of a charge, scattering on an electron in atom is reduced to a problem defining the function of scattering $f(H)$ for the big electron (7). Because all electrons in atom are in the central field and all charges density distributions are symmetric about centre, it is obvious, that scattering of each electron is coherent with others. The factor of atomic scattering $f_a = \sum_{i=1}^{Z} f_i$, where Z – atomic number. Usual opinion: there is exist experimental prove that all electrons scatter coherently according to Schrödinger $|\psi(r)|^2$ representation for each electron. This opinion is widely used in X-ray structure analysis. At the same time, as already was point above, it has been shown, that electrons in atoms of easy elements scatter incoherently with each other, and their size much less than defined for an electron by function $|\psi(r)|^2$.

Actually, as we can see from analysis of the Fig. 3, all electrons simultaneously participate both in Rayleigh, and in Compton scattering. It is possible to consider approximately, that factor of coherent scattering by electrons in atoms of light elements is $R_{ch} = I_{ch}/I_{th}$, where I_{ch} and I_{th} intensity of coherent and Thomson scatterings.

Evidently, the same electrons participate in forming both part of spectra. It is obvious, that charges in atom scatter coherently should remain motionless and occupy great volume of atom. So there are no alternatives of function interpretation $|\psi(r)|^2$ except real distribution of charge density.

It is necessary to assert, that electrons in atom remain motionless. Therefore standard model of atom, with electrons are point particles having an impulse corresponding to velocity $10^{-2}c$ - $10^{-1}c$ where c - velocity of light, is incorrect from point of view of X-ray diffraction experiment. Otherwise, coherent, unmodified part of spectrum would not be observed at all for moving electrons. In the case of standard model, separate atoms even united in a crystal lattice, will give smoothed diffraction picture (like diffuse thermal scattering), as for it formation it is necessary synchronous moving electrons in atoms of a crystal lattice. But also in this case the atomic scattering factor will be differs from calculated from formula (6), because instead of addition scattering amplitude it is necessary to sum up intensity of it scattering from points, defined by function[2] $|\psi(r)|^2$. "Electrodynamic interpretation" gives some distribution of electronic density in atom, but does not explain loss more than 90 % of a scattering matter in atoms of light elements (intensity of "incoherent" scattering much higher than "coherent" one). Above noticed, the condition (1) - basic condition both valued for classical electrodynamics and NCT. Hence, only a part of the electronic mass, located in an electron, provide Rayleigh scatterings. Designate this mass m_a. It is obvious, that it makes a small part of a lump of an electron. Other part, we name "dark matter" m', scatter "incoherent", forming Compton spectrum.

Quantum theory of scattering explain existence this "dark matter" in atom only that the part of electrons is pulled out from the position determined by function $|\psi|^2$ for very short time, an order t $\approx 1/v_0$ (Cooper, 1985). Other electrons "do not notice" this loss and keep the previous position. The pulled out electron scatter radiation on the mechanism of incoherent scattering, which we will discuss in the next paragraph. We will notice only, that this electron "recollects" an impulse, which ostensibly was in atom. This is one more "ad hoc" hypothesis of quantum theory of x-ray scattering.

Let's quote Compton reasoning about the discrete electrons scattering concept which reflects the standard point of view on the mechanism coherent scatterings: "According to the wave-mechanics theory, under the influence of the field of the incident electromagnetic wave the characteristic functions for higher energy states of an atom assume finite values and the radiation which it emits has the frequencies described by $hv_0 = hv + W_i - W_f$ (where W_i and W_f are the initial and final energies). If the final state of the atom is identical with the initial state i, the frequency is unchanged, and coherent radiation is emitted. In calculating this part of the scattering, only the ψ functions of the normal state 0 of the atom are therefore concerned. That is, the coherent scattering is identical with that from an atom having a

---

[2]The similar situation considered Bosanac in article where analyzed application consequences semiclassical theories for the analysis of Compton lines (Bosanac, 1998).

continuous distribution of electric charge of density $\rho = -e\psi_0\psi_0^*$ " (Compton, 1935). Such reasoning naively reduced to false statement, that the probability of a finding of a point electron in volume dV is equivalent to distribution of a scattering charge. From the point of view of the quantum theory of a structure of atom, neither orbital movement, nor any another movement of electrons in atom outwardly should not be shown, as the electron cannot lose energy, being in a stationary condition. From experiments on coherent scattering it is possible to draw a conclusion, that movement of electrons in atom actually is not exist and invisible. We will show now, how it is possible to "remove" electron movements in atom, having real, explicit physical sense "nonobservability" of them.

Schrödinger to find ψ-function placed the electron in a force field of atom, and electron get an impulse

$$p = \sqrt{2m(\varepsilon - V)},$$

where V - a force field potential.

Lets the stationary condition has some energy E_i. Electron-binding energy E_i defines an impulse $p_i = \sqrt{2mE_i}$ or de Broglie wavelength

$$\Lambda_{Bi} = \frac{h}{p_i} = \frac{h}{m\vartheta}$$

We will consider, that "coherently" scattering "part" of an electron, mass m_a, is proportional to R_{ch}. It is natural to suppose, that this relation is equal to

$$R_{ch} = \frac{m_a}{m} = \frac{p_i}{mc} = \frac{\Lambda_e}{\Lambda_{Bi}} = \sqrt{\frac{2E_i}{mc^2}} = \alpha_i,$$

For "incoherent" ("dark") mass m', forming a Compton spectrum part, it is possible to obtain, using the same logic, the result: the bound electron gets impulse $p_i = \sqrt{2m\,E_i}$. Compton wavelength Λ_e and mass m of a free electron changes on the values Λ'_e and m' in the manner:

$$\Lambda'_e = \frac{h}{mc + p_i} = \Lambda_e(1 - \alpha_i), \quad m' = m(1 + \alpha_i) \tag{13}$$

Try to estimate factor of scattering by atoms of easy elements:

$$R = \frac{I_{ch} + I_{in}}{I_{th}} = \alpha^2 \left| Z - \sum_{i=1}^{Z} \frac{\alpha_i}{\alpha} f_i \right|^2 + (Z - \alpha^2 \sum_{i=1}^{Z} \left| 1 - \frac{\alpha_i}{\alpha} f_i \right|^2),$$

$$\alpha = \sum_{i=1}^{Z} \alpha_i \tag{14}[3]$$

---

[3] The value R was estimated by Compton and other authors (Compton, 1935) from the consideration of quantum derivation for "incoherent" part scattering spectra. Their results rather differ from ours. Significance and sense of this fact will be explain in the next paragraphs.

This estimation is valid only for the scattering angles outside of Bragg conditions. In the case of Bragg diffraction on crystal structure only coherent radiation in φ direction is observed.

Thus, coherent Rayleigh scattering of electromagnetic radiation is possible only by bound electrons. Bound energy of electrons in atoms of light elements is small in comparison with mc^2 and it is enough to localise near to atom only small part of electronic mass of each electron. Interpretation $|\psi(r)|^2$ as a real function lead to necessity to replace momentum representation $\vec{p}_i$ on de Broglie wavelength Λ_{Bi} representation. Distribution $\chi(p)$ - momentum distribution in atom, is a Fourier transformation of $\psi(r)$ function (Dirac, 1926), also necessary replace by de Broglie wavelength distribution:

$$\chi(\vec{p}) = \chi(\frac{h}{\Lambda_B}) = \left(\frac{h}{2\pi}\right)^{\frac{3}{2}} \int \psi(r) \exp\left(-i2\pi\frac{r}{\Lambda_B}\right) dr \tag{15}$$

This equation gives clear physical sense to why "momentum" of electrons being in atom is not visible at coherent scattering of x-ray radiation. Such understanding of "coherent" x-ray scattering give us possibility to regard "incoherent" scattering actually as partly coherent and changes a view at structure of a free electron. We will consider this questions in details.

4. Wave-particle model of free electron in "electrodynamic interpretation" as real wave packet electrical charge density

In the previous section we pointed to the contradiction between an explanation of coherent scattering in " electrodynamic interpretation » $|\psi(r)|^2$ function with concept of point electron. We made a choice in favour of Schrödinger $|\psi(r)|^2$ interpretation as the main concept, in spite of usually auxiliary considering, helping to understand physical sense of an event. Even Crisp, one of authors of the neoclassical theory, writes: "Schrödinger's interpretation of ψ was found to be seriously flawed when used to explain the behavior of a free particle. The general spreading of a particle's wave packet made it too transient to be as stable as particles that are found in nature. Furthermore, the application of this interpretation of quantum mechanics to a scattering experiment suggest a splitting, or division, of particles which is not experimentally observed" (Crisp, 1990).

The true reason of seeming "splitting" of an electron in experiments on x-ray radiation scattering lays in misunderstanding of the free electron state description. Schrödinger equations can be received in nonrelativistic approach at $\nu_B \ll \nu_e$, (or $\Lambda_B \gg \Lambda_e$), where

ν_B – any frequency of function $\psi(r,t)$. In this case it is possible to consider, that ψ function is slow envelop function of high-frequency relativistic function $\exp(i2\pi\nu_e)$.

It is known, that each electronic de Broglie wave satisfies Klein-Gordon equation

$$\nabla^2\Phi - \frac{1}{c^2}\frac{\partial^2\Phi}{\partial t^2} = (\frac{mc}{h})^2\Phi \tag{16}$$

Plain wave decision of this equation is

$$\Phi = \Phi_0(r,t) = \exp 2\pi i(k_0 x - v_0 t),$$
$$k_0 = 0, v_0 = v_e$$

Let's $\Phi(r,t) = \psi(r,t)\exp 2\pi i v_e$, then at the conditions $v_B \ll v_e$, we can neglect member $\ddot{\psi}$ and write down approximately:

$$ih\dot{\psi} = -\frac{h^2}{2m}\nabla^2\psi + (\frac{h^2}{mc^2}\ddot{\psi}) \approx \frac{h}{2m}\nabla^2\psi \qquad (17)$$

Let's pay attention to that difficulty ("serious flaw") in Schrödinger interpretation connected with interpretation ψ as a real wave of electronic density at excitation of an electron at rest by an electromagnetic wave. The frequency v_e in this case are not localised, but synchronous everywhere. In such interpretation of the wave process inherent to a rest electron, de Broglie wave of a moving electron automatically become waves of probability, which describe electron movement at a big distance R from it (R>>Λ_e) in a far zone, and do not give the information on its real position. At movement of an electron with a velocity ϑ in a direction r in fixed coordinate system, the de Broglie wave is:

$$\sin 2\pi v_e \left(\frac{t - \frac{\vartheta}{c^2}r}{(1-\beta^2)^{\frac{1}{2}}}\right) \equiv \sin 2\pi \frac{r}{(1-\beta^2)^{\frac{1}{2}}}\left(\frac{1}{\Lambda_e} - \frac{1}{\Lambda_B}\right) \qquad (18)$$

Phase velocity of process of field propagation V, de Broglie wavelength Λ_e and β are:

$$V = \frac{c^2}{\vartheta}, \Lambda_B = \frac{h}{m\vartheta}, \beta = \frac{\vartheta}{c},$$

By analogy with paragraph 3, replace time fluctuations with frequency v_e a of a rest electron, on the spatial fluctuations formed by real waves. For this purpose it is necessary to image, that two counter electronic waves - outgoing and ingoing, entering in a point $r=0$, and formed a standing wave of electronic density. If the length of each of this waves is equal to $2\Lambda_e$, the electronic density for a free electron at rest is described by set of plane waves. In any direction $\vec{r}$ is a wave of charge density:

$$\rho(R)d\tau = e|\psi(R)|^2 R^2 dRd\Omega,$$
$$R = \frac{r}{\Lambda_e}, d\tau = \frac{dV}{\Lambda_e^3} \qquad (19)$$

where

$$|\psi(R)|^2 = \frac{\sin^2(\frac{\pi r}{\Lambda_e})}{(\frac{\pi r}{\Lambda_e})^2} = \frac{\sin^2 \pi R}{\pi^2 R^2} = \frac{(1-\cos 2\pi R)}{2\pi^2 R^2} \qquad (20)$$

It means, that in any direction $\vec{r}$ at r >> Λ_e the wave de Broglie a rest electron with spatial frequency $\vec{k} = \Lambda_e^{-1}$ are exist. At electron movement with a velocity ϑ wave process is described by both representation in the formula (18), but physical sense has only real wave:

$$\sin 2\pi \frac{r}{\sqrt{1-\beta^2}} (\frac{1}{\Lambda_e} - \frac{1}{\Lambda_B})$$

This movement are described by standing wave of electronic density movement and by "in" and "out" waves moving with relative velocity $\vartheta = \beta c$ each of them receive usual Doppler shift (Wolf, 1998). Phase velocity in a movement direction in a far zone corresponds to usual de Broglie waves as in (18), but now concrete sense is given to these waves. Electron movement in this consideration should be described not by single plane wave spreading in x direction, but by all plane waves, in a point $x = \vartheta_x$ t, y =0, z=0. In such representation both moving and rest electrons as de Broglie waves, Klein-Gordon equation gets sense of the equation of a real plane waves. It is obvious, that all wave structure of an electron scatter an electromagnetic field under the complex law, and leads to a two-part spectrum. "Coherent" part is formed as a result of deformation phase periodic structure of an electron. This elastic deformation leads to appearance of small homogenous electron charge distribution described by Schrödinger equation. In the next paragraph, we try to consider Compton effect as effect of unusual coherent scattering. At the same time, wave structure of atomic electron is deformed according (15).

5. Compton effect as effect of coherent scattering

Fig 4 represents the scheme, suggested by Compton, explaining effect as incoherent scattering. The primary photon with energy $h\nu_0$ is absorbed by a point electron. The new particle, possessing total energy and an impulse of a photon and an electron is formed. For a short time interval (it is sometimes supposed about time of an order $1/\nu_0$) this quasiparticle breaks up on two - a new photon with energy $h\nu'$ <$h\nu_0$ and an electron with some kinetic energy. Writing down the energy and impulse conservation laws it is easy to connect energy and impulses of the new particles. In case of scattering by free electron formula (12) describes relationship between frequencies and wavelengths of the primary and secondary photons. Formula (12) allow to define only centre of line, explanation of wide width of scattered lines, not envisioned in Compton scheme, offered Jauncey (Jauncey, 1925). In 1933 DuMond has published the theory is known as impulse approximation in which the width of the line is compared with those impulses $\vec{p}$, which the electron has in atom in the bound state, as he guess (DuMond, 1933).

In his theory, Doppler shift provide line spreading. In common, most physicists grounded in classical electrodynamics were a grudging acceptance for Compton explanation of "incoherent" scattering effect. However, the twentieth of 20 century was turbulent period in the evolution of physics. Compton effect was "turning point of physics" (Stuewer, 1975) and pull the trigger of a general acceptance of quantum ideas. It is possible to agree with this estimation of Compton effect, its treatments by Compton, but possible to ask a question: whether on a correct way physics development, since this point has gone. For this purpose once again we will analyse Compton, DuMond and other founders of the theory "incoherent" scattering effect arguments from the neoclassical theory point of view. It is necessary "to combine all puzzles" experimental results X-ray radiation scattering from Barkla to Compton in a whole picture.

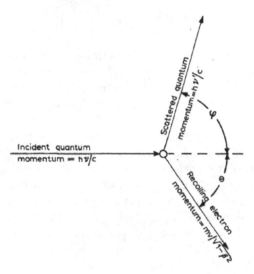

Fig. 4. An X-ray photon is deflected through an angle φ by an electron, which in turn recoils at an angle θ, taking a part of the energy of the photon. (Compton, 1923, 1927)

5.1 Scattering by a free electron (early models of Compton effect)

Behind seeming simplicity and grace of the modern Compton effect theory, stand some "ad hoc" hypotheses, which has put difficulties, contradictions and "obscurity" the quantum physics, mentioned in introduction. Was used the hypothesis: reduction or a collapse of wave function (in this case, an x-ray electromagnetic wave), i.e. instant transformation wave in a particle. It contradicted to previous physical experience: this particle-photon get not only the energy received by it at a "birth" but impulse $\bar{p}$, directed from a source to a concrete electron. Clearly, already at this analysis stage theory need to add two new, very strong new hypotheses to the earlier quantum hypotheses: Planck (atom radiates quanta of energy $E = h\nu_0$) and Einstein (absorption of energy by matter occurs also discretely to energy $h\nu_0$, equal some resonant energy).

First from new hypotheses: absorption of energy is accompanied by absorption of impulse quantum.

Second hypothesis: electron radiate just new photon, but not radiation, and wave function of the photon is only the function describing probability of its occurrence in concrete point of space - a collapse at a "birth".

Compton theory is a remarkable example of replacement in quantum electrodynamics of the physical sense in interaction of an electromagnetic field with charged particles, by the mathematical fictions, and hypotheses. Really, in Compton theory electromagnetic interaction between photon an electron is considered only at determination of probability photon scattering by an electron which is proportional to the scattering cross section υ_0, calculated in the classical theory. Wonderfully, in Compton theory there is no consideration of the processes occurring in a time interval between the beginning of interaction of radiation with an electron and its finishing. In introduction it was noticed, ignoring of transient process typical for all quantum electrodynamics and is very important feature influencing on interpretation of particles interaction process. In the case of Compton effect, removed the process occurring at interaction of radiation with electrons, we removed such physical processes, as: electron acceleration, appearance of Lorentz force radiating braking. Their account in classical electrodynamics leads to a conclusion that the electron should remain motionless at scattering of a wave.

Schrödinger criticized the Compton scheme of formation radiation spectrum with the shifted wavelength and pointed out the discrepancy with the electromagnetic theory (Schrödinger, 1927). He offered model in which the electromagnetic wave scatters on moving de Broglie real wave. In this model, at correctly chosen velocity of an electron, frequency of a primary wave gets the shift corresponding to the formula (12). Later Dodd investigated this problem as the problem of plane electromagnetic wave scattering on a moving electron (Dodd, 1983).

According to the Dodd model of Compton scattering, the electron moves with a velocity

$$\vartheta = \vartheta_0 \left(\frac{c}{c + \vartheta_0}\right), \text{ where } \vartheta_0 = \frac{h\nu_0}{m_e c}$$

The plane electromagnetic wave with frequency ν_0 propagate in a direction parallel to movement of electron. Lorentz transformations at scattering on such electron lead to frequencies and wavelengths value equal to (12). Neither Schrödinger, nor Dodd, did discuss reason of electron movement. They only postulated necessity of electron movement to receiving angular distribution of frequency of scattering wave.

As we demonstrate above, scattering on a free electron should not lead to absorption - neither energy, nor an impulse. The free electron has no possibility to take impulse from electromagnetic radiation, it should remain motionless, making under the influence of radiation only periodic oscillations. All experience of X-ray structural analysis shows, electrons in atoms keep the position at scattering. The photoeffect, other effects of incoherent scattering at which there is a displacement of electrons from stationary position, are connected with interatomic absorption. These secondary effects are absent at scattering on free electrons and are small at scattering of hard radiation on atoms of easy elements. Compton effect is effect of scattering on a free electron and dominates at scattering on atoms of easy elements, but nevertheless looks like effect being in close connection with point electron movement. From that consideration we have to confirm: Compton effect have to be regarded evidently as a experimental argument in favor of conclusion about real wave particle model of free electron, taking some volume in space. Linear sizes and details of it structure are comparable to X-ray radiation wavelengths.

5.2 Neoclassical theory of scattering by free electrons

Let's note once again: the base for Thomson's classical theory of X-ray scattering on a free electron is the assumption about physical pointness an electron as particle with the size r_e.

The theory admits also possibility of existence of the "big" electron with the sizes of order Λ_e, with relation of a charge to mass in each elementary volume equal to e/m (1). Quantum electrodynamics starts with mathematical model of a point electron, though the condition Schrödinger equation is derived limits the size of "point" equal to Λ_e.

This restriction is known as a Heisenberg uncertainty principle, and it reflects not character of basic impossibility to define coordinate of an electron (more precisely, than this size), but fundamental statement, that the charge and mass of an electron which occupies all space (as represented still de Broglie) can mainly take place in the particle with diameter Λ_e.

In the previous paragraph we represent the function $\cos 2\pi R$ as function for describing of electronic density distribution in a free electron in any direction $\vec{r}$. Let electromagnetic wave propagates along periodic distribution of electronic density in direction $\vec{n}_0$. In this case, oscillation along electronic structure moves with a velocity c followed electromagnetic wave in direction $\vec{n}_0$. Such scheme is completely equivalent Dodd scheme, but electron movement replaced by propagation electronic waves along electronic structure. It means that along standing wave electron density moves toward each other with a velocity of light two waves. Phase shift between them is equal to $4\pi\Lambda_e/\lambda_0$.

In other words, along an electron two transverse waves (excitations) propagate with a velocity of light, that causes two dipole electromagnetic waves, shift of phases between them in directions $\vec{n}_\varphi$ leads to angular change of wavelength (the same shift received in theories Compton, Dodd and Schrödinger):

$$\Delta\lambda = c\Lambda_e\left(\frac{1}{c} - \frac{\cos\varphi}{c}\right) \tag{21}$$

Note, at derivation an equation (21) we explain both Compton shift and also physical sense of mechanism of formation Lorentz radiating force (11). Along an electron with periodically distributed electronic density the phase wave do not transfer an impulse - the electron remains motionless. Fig.5a presents model of an electron (19) - a standing wave of electronic density with spherical symmetric periodic distribution. Let's now the plane electromagnetic wave $\vec{n}_0$ propagate under any angle φ to the plane electronic wave in direction $\vec{n}_\varphi$ (Fig. 5b). It causes in direction $\pm n_\varphi$ two opposite transverse electronic waves and two sets of dipole electromagnetic waves. Phase shift between in-out dipole waves formula (21) describes.

Thus, the electronic structure with distribution of electronic density (19) radiates unusual "composite" dipole wave. The wavelength of this radiation depends on an angle φ. The width of spectrum $\delta\lambda = \Lambda_e(1 - \cos\varphi)$, that is coherency, is defined by a spectrum of spatial frequencies of function sinc $2\pi r$ at any direction φ.

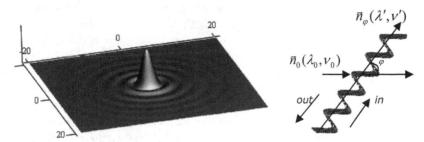

Fig. 5. a) Schematic radial distribution of charge density in electron $\rho(r)$ (19), (20). De Broglie wavelength in any direction $\bar{n}_\varphi$ is equal to Λ_e ; b) wave scattering in one of direction $\bar{n}_\varphi$. In-out electronic waves run along $\bar{n}_\varphi$ direction. Phase velocity of propagation of electron excitation is equal to $c/\cos\varphi$.

5.3 Neoclassical theory of scattering by bound electrons in light atoms

Theoretical distribution of intensity Compton line for free electron at the angle $\varphi = 90°$ and angular interval $0° < \varphi < 180°$ is presented on Fig.6a,b.

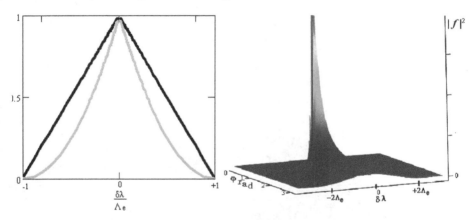

Fig. 6. a) Black line: Fourier spectra for $|\psi|^2$; color line: intensity distribution $|f(\varphi, \delta\lambda)|^2$ for "incoherent" scattering by free electron at $\varphi = 90°$; b) Theoretical dependence $|f(\varphi, \delta\lambda)|^2$ for "incoherent" line in angular range $0° < \varphi < 180°$

On Fig.6a black line is Fourier spectra of $|\psi(r)|^2$ function for free electron (20):

$$F\left[|\psi(r)|^2\right] = f\left(\frac{\delta\lambda}{\Delta\lambda'(\varphi)}\right);$$

at $\varphi = 90°$, $\Delta\lambda'(\varphi) = \Lambda_e$; color line is intensity of this spectra $|f(\varphi)|^2$.

Integral intensity of scattering at $\lambda_0 \gg \Lambda_e$ equal to classical one (scattering factor $R=1$) both for point-like and wave-particle (20) electron, and independent of angle φ. Fig.6b presents calculated distribution $|f(\varphi, \delta\lambda)|^2$.

At calculations we assumed, that $\lambda_0 \gg \Lambda_e$ (we compare calculation to experiments results with $\lambda_0 = 0.88\text{Å}$). A case $\lambda \geq \Lambda_e$ we consider later. Let's remind, since work DuMond, width of Compton lines someone try to explain as electron (has been beaten out from atom) save velocity that it had in the bound state, i.e. impulse $\vec{p}$ from distribution $\chi(\vec{p})$ (15). New photon, forming by scattering on such free electron has Doppler shift of frequency. Since $p_{max} = \sqrt{2E_i m}$ the width of a line should be defined by size $\alpha_i \lambda_0$. This width in case of scattering on carbon comparable to the width of the Compton line for free electron. Therefore, DuMond, theory seemed plausible, and experimental checks of "impulse approximation" continue in the same spirit till now. Actually, the change of width of Compton line from bound electron comparable with a width of line from free electron. The difference have to be very small and is order to $\delta\lambda = \Lambda_e - \Lambda'_e \approx \alpha_i \Lambda_e$ (See (13)). This value is much less than in the quantum theory accepted today. In the case of scattering on carbon atoms, even for K-electron ($E_i \approx 280$ eV), $\alpha_i \approx 3 \cdot 10^{-2}$. It means, that the width and the spectrum form of Compton line on carbon is defined by scattering on free electron within several percent accuracy. Fig. 7 represents the spectrum of scattering on a diamond monocrystal in angular interval $10°$ - $160°$. We see, line form qualitatively coincides with the calculated (Fig 6b).

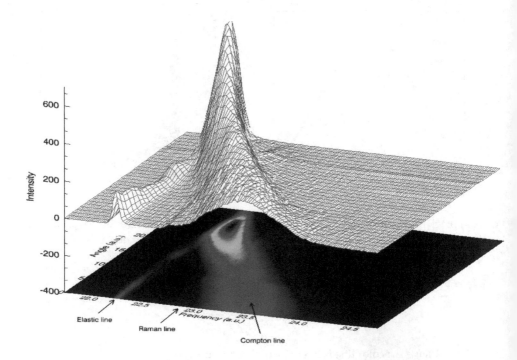

Fig. 7. Experimental spectra at $\lambda_0 = 0.88\text{Å}$ on a diamond single crystal, at various angles φ from $10°$ to $160°$. Among "coherent" (elastic) and "incoherent" (Compton) lines, is weak Raman (J-line; about reason for term $h\nu_J$ for Raman-line more detailed in paragraph 6).

Fig. 8a presents dependences: theoretical (color line) and experimental (points) full width at half intensity of a line $|f(\varphi, \delta\lambda)|^2$. Fig.8b presents theoretical (color line) and experimental results (points) intensity in a Compton lines maximum $|f(\varphi, 0)|^2$.

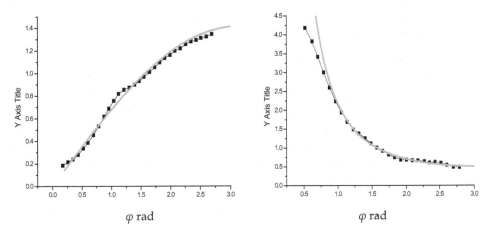

φ rad φ rad

Fig. 8. a) Full width at half intensity of a line $|f(\varphi, \delta\lambda)|^2$ of "incoherent" line (for Fig.7) (in arbitrary unit). Points – experimental date; color line theoretical $\delta\lambda \sim (1-\cos\varphi)^{1/2}$; b) Intensity dependence in maximum (center) of "incoherent" line (for Fig.7). Points –experimental date; color line – theoretical intensity dependence $I_m \sim (1-\cos\varphi)^{-1}$. Y-axis in arbitrary units.

It is clear, these dependences well coincide. However, similar measurements on monocrystals Si show much less coincidence from calculated for a free electron (for k-electrons in Si, $\alpha_i \approx 1.2 \cdot 10^{-1}$).

Theory of impulse approximation evaluate some displacement (Ross & Kirkpatrick, 1934) for centre of Compton line on value

$$\Delta\lambda = -\frac{\lambda_0^2}{hc}\frac{1-n}{n}E_i \; ; n \approx \frac{1}{2}$$

This estimation show, that impulse approximation theory predicts for displacement of Compton shift

$$-\frac{\Delta\lambda}{\lambda_0} = \frac{\Delta\nu}{\nu_0} = \frac{\nu_i}{\nu_0}$$

This value derived under assumption of short collision of photon with electron. From another point of view, according to energy conservation law this line displacement have to be equal to usual equation for position of Raman X-ray line (J-line) (Raman, 1928; Pimpale & Mande, 1984) and do not observed as Compton line shift. In this case frequency change:

$$h\nu_J = h\nu_0 - E_i = h(\nu_0 - \nu_i) \; ; \Delta\nu = \nu_i(\nu_0 - \nu_J) \tag{22}$$

At the analysis of Compton scatterings consider, that frequency change Δv (φ) should be much less v_i. This condition is necessary to explain experimental fact, that the shift of a line given by the formula (22), not observed. Experimentally we see J-line (Raman line) inside the Compton scattering spectra (see Fig.7). It means, at all angles, Δv (φ) comparable to v_i. Frequency v_i for K -electron in carbon corresponds to energy excitation 280 eV. Frequency change Δv at φ=90°, corresponds to energy change approximately equal to 380 eV.

According to table data (Alexandropoulos et al., 2004) under the experimental condition incoherent scattering factor is more than 5 in wide angle φ interval. It means, that all 6 electrons in carbon should give the contribution to "incoherent" scattering. Nevertheless, the line centre is not displaced, that once again to get support to discussed model of scattering.

The mechanism of formation of a Compton spectrum, considered above, has been offered for the first time by V.Aristov in 2008 (Aristov, 2009b) and discussed in a conferences (Aristov, 2010, 2011). At first sight, suggested mechanism is unusual. However, in optics of visible light, radio optics, it is possible to find examples of formation of the angular wavelength radiation dependence. The nearest analogy to Compton effect - Purcell-Smith effect (Smith & Purcell, 1953), and especially inversed Purcell-Smith effect, was suggested for electron accelerators (Kim, 1993). Purcell-Smith effect is a radiation caused from electron movement with a velocity ϑ along a dielectric lattice with the period b. Angular dependence radiated electromagnetic wavelength is:

$$\lambda = \frac{b}{\beta}(1 - \beta \cos \varphi) \qquad (23)$$

The wavelengths difference between direction φ =0, and all others is equal to:

$$\lambda(\varphi) - \lambda_0 = \Delta \lambda = b(1 - \cos \varphi) \qquad (24)$$

5.4 Compton scattering cross section

One of the achievement of the quantum theory of x-ray radiation scattering is the theory explaining angular dependence of scattering factors from wavelength, based on works Klein and Nishina, Tamm (Klein & Nishina, 1929; Tamm, 1930).

This dependence was discussed already in paragraph 2 (Fig 2). There is nothing surprising that results of the quantum theory correspond to that have been received by Compton in 1919 in frame of classical scattering theory. The quantum theory operates in space of impulses. In the kinematical theory of scattering (in the first Born approximation) the interference is described in space of reciprocal wave vectors also.

In the neoclassical theory scattering in each directions φ is defined by independent oscillation by elementary charges. Each of them scatters as in classical electromagnetic theory. As was mentioned above, integral intensity at any angle φ is constant at $\Lambda_e << \lambda_0$, as well as in case of a point charge. In approximation of $\Delta v(\varphi) << v_i$ it is possible to conclude that classical theory is true, an electron is point, and the scattering factor is proportional r_0^2. Moreover, "core" of an electron in the interval between the two first zero of function $\mathrm{sinc}^2 \pi R$

forms an angular spectrum of quasi-monochromatic radiation in any direction. As a result of an interference this spectrum leads to angular dependence of integral intensity, as though in classical theory for "big" electron. Radius of "big" electron have an order of value Λ_e. Possible to calculate a scattering spectrum using method, that Compton suggested in 1919, using various models of a big electron. Note, at large angles φ "coherent" Rayleigh scattering it is possible to neglect and consider that all electrons in atom scatter as free, because $|\psi(r)|^2$ area distribution much larger than Λ_e.

6. About true X-ray incoherent scattering

The special section of physics is devoted to this topic and its consideration is not task of this chapter. However, we suspect the Compton "recoil" low-energy electrons do not connected with Compton effect, but with real absorption, i.e. with "true" incoherent scattering. In this chapter, under true incoherent scattering we consider X-ray Raman effect.

According to Einstein's hypothesis, the photoeffect reason is absorption of energy of light quantum $h\nu$, then electron overcome energy barrier that equal to binding energy and gets kinetic energy

$$E_k = \frac{m\vartheta^2}{2} = h\nu_0 - E_i \tag{25}$$

In this equation there are no suggestions about concrete mechanisms of absorption and radiation. In this sense it nothing differs from the equations of conservation of energy and an impulse, written by Compton. In connection with such extremely superficial similarity "Compton" and "Einstein" effects are considered together not only in textbooks, but also in serious researches. Such identification of effects is incorrect. Photoelectric resonance absorption is responsible for photoelectrons. From this point of view, the photoeffect – is effect of classical electrodynamics (Lamb & Scully, 1959). However, till now there is a debatable question how accumulation of energy quantum is carried out (Aristov, 2009a).

Compton theory is a theory for an explanation of effects of radiation scattering by free electron. However, any variants, except direct collision between point particles to transfer energy from one to another do not considered (even at scattering on the bound electrons). Instead of primitive billiard-like model, there is more complex and beautiful physical effect - scattering by infinite size electron with spherically symmetric distribution charge density. In this case, resonant absorption of radiation by bound electron is a more complicated phenomena, that is considered in classical and quantum electrodynamics. Excitation of electron gas produced due to photoelectric absorption lead to various secondary processes, including photoemission "Einstein-electrons" and "Compton-electrons". Cross section of scattering and absorption by bound electron is determined by radiation braking factor

$$\gamma = \frac{2e^2}{3mc^3}\nu_i^2 \; , \; \nu_i = \frac{E_i}{h}$$

In this case, instead of (3) movement equation of an electron:

$$V(t) = -\frac{e}{mc}\vec{A}(t)\frac{1}{(v_0^2 - v_i^2) - i\,v_0\gamma}$$ (26)

The imaginary part of the equation (26) determines the photoabsorption, real part – scattering cross section. Both scattering and absorption depends on frequencies v_0 and v_i and from their difference $(v_0 - v_i)$, that is equal of frequency of Raman scattering. Mechanism formation of this radiation only formally reminds optical Raman effect. We named this line as J-line. In our meaning, this line of radiation was observed Barkla and is named him a J-line (Barkla, 1917) by analogy to lines K and M characteristic radiation of atoms. It is remarkable, that Barkla and his opponents, including Compton, connected J-phenomenon with Compton scattering (Compton, 1924; Alexander, 1930).

J radiation arises because of energy absorption at frequency $v_0 - v_i = v_J$ (22), and not because of exchange of energies between a photon hv_0 and quantum of orbital movement of an electron (such movement does not exist). The equation (26) has been written at assumption that electron is point-like. In our case, it is necessary to replace v_0 on $v'(\varphi)$ (see (12)). It means, that absorption occurs at frequencies $v'(\varphi) - v_i = [(v_0 - \Delta v'(\varphi)) - v_i] = v'_J$.

In this case, for J-line, conservation energy law equation, combining (22) and (25), is:

$$h v_J{}' = h v_0 - E_i + \frac{m\vartheta^2}{2}$$ (27)

Here ϑ – velocity of low energy "Compton" electrons, their appearance caused by incoherent J'-line. Finally, it is possible to write down the equations for a J-line and low-energy electrons born with it:

$$v_0(1 - \frac{2\alpha}{1 - 2\alpha}) < (v_J + v_i) \le v_0, \quad \alpha = \frac{\Lambda_e}{\lambda_0}$$

$$0 \le \frac{m\vartheta^2}{2} \le mc^2(\frac{2\alpha^2}{1 - 2\alpha})$$ (28)

$$m\vec{\vartheta} = -\alpha mc\frac{\vec{H}}{|\vec{k}_0|}$$

values $\vec{H}, \vec{k}_0$ from (6). The equations (28) formally remind Compton equations. But physical sense considerably differ from them. The form of a J-line does not depend on an angle φ. It has the maximum energy at frequency $v_J = v_0 - v_i$.

The J-line is wide, and its width $\Delta\lambda$ remains in good approach by a constant, and approximately equal to $2\Lambda_e$. With a J-line corresponding to K-electrons the lines corresponding to L, M-electrons should be observed.

The spectrum of low-energy electrons according to the formula (28) is qualitatively close predicted by Compton. Let's pay attention, that in formulae (25) - (28) constant h is used only in the Planck sense. The bound electrons can radiate and absorb an electromagnetic wave according to classical electrodynamics laws.

7. Conclusion

In the chapter, we formed and proved general conceptions of the neoclassical theory of interaction of electromagnetic radiation with matter on the free and weak bound electrons in atom. Neoclassical theory unifies all experimental scattering fragments in whole picture by means model infinite size electron. This model - standing spherical wave electron density with wavelength Λ_e.

For such understanding all electromagnetic radiation scattered by an electrons, is coherent, that allows to suggest new methods for X-ray analysis and electronic spectroscopy. In thin films, can be realised methods of research similar Bragg diffraction, when incident radiation with λ_0 wavelength is reflects by the crystal under an angle ϕ with $\lambda_0 + \Delta\lambda(\varphi)$ wavelength (Aristov & Shulakov 2010). It is interesting to measure Compton spectra (not only intensity) in schemes of standing waves. It is necessary to return, certainly, to Compton idea: investigation of the form and internal structure of an electron using hard radiation scattering diagrams (Compton, 1919). Possible to measure of a phase of the atomic scattering factor for integrated intensity of Compton spectra, as can see from (14).

8. Acknowledgment

This work was supported by RFBR, grant № 09-02-12090. The author also thank S.N. Yakunin for experimental results, and A.A. Despotuli for assistance.

9. References

Alexander, N. (1930). The J-phenomenon in X-rays. *Proceedings of the Physical Society*, Vol.42, No.2, (December 1929), pp.82-96

Alexandropoulos, N., Cooper, M., Suortti, P. & Willis, B. (2004). Correction of systematic errors, In: *International Tables for Crystallography. Vol. C.*, E. Prince, (Ed.), pp.653-661, Kluwer Academic Publishers, ISBN 1-4020-19009, Dordrecht, Netherlands

Akhiezer A.I. & Polovin R.V. (1973). Why it is impossible to introduce hidden parameters into quantum mechanics, *Uspekhi Fizicheskikh Nauk*, Vol.15, (April 1973), pp. 500-512, ISSN 0042-1294

Aristov, V. (2009a). The photoelectric effect in the semiclassical theory. *Doklady Physics*, Vol.54, No.4 (July 2008), pp. 171 –173, ISSN 1028-3358

Aristov, V. (2009b). Scattering of an electromagnetic wave on a free electron in the semiclassical approximation. *Doklady Physics*, Vol.54, No.4 (July 2008), pp. 187 -189, ISSN 1028-3358

Aristov, V. (2010) Fundamental problem of hard X-ray scattering (in russian), In: *"X-ray optics". Working Conference. Chernogolovka-2010*, pp. 208-211, IMT RAS, Available from http://purple.iptm.ru/xray/xray2010/files/CHGXRAY2010_BOOK.pdf

Aristov, V. & Shulakov, E. (2010) Bragg scattering due to changing of wavelength (in russian), In: *"X-ray optics". Working Conference. Chernogolovka-2010*, pp. 57-59, IMT RAS, Available from http://purple.iptm.ru/xray/xray2010/files/CHGXRAY2010_BOOK.pdf

Aristov, V. (2011). New Compton effect solution as the way to general x-ray scattering theory, *Asia Pacific Academy of Materials-APAM. Abstract book.* Taiwan, August, 2011

Barkla, C.(1904).LXI. Energy of secondary Röntgen radiation. *Philosophical Magazine Series 6*, Vol.7, No.41, (May 1904), pp.543-560

Barkla, C. (1911). LXXVI. Note on the energy of scattered X-radiation. *Philosophical Magazine Series 6*, Vol.21, No.125, (May 1911), pp. 648-652

Barkla, C. (1918). Bakerian Lecture: On X-Rays and the Theory of Radiation. *Philosophical Transactions of the Royal Society*, A, Vol.217, No.549, (January 1918), pp. 315-360

Bell, J. (2004). *Speakable And Unspeakable In Quantum Mechanics* (second edition), Cambridge University Press, ISBN 0 521 81 86 2 1, United Kingdom

Bergman, D. (2004). Observation of the Properties of Physical Entities. Shape and Size of Electron & Neutron, In: *Common Sense Science*, May 2004, Available from http://CommonSenseScience.org

Bohm, D. (1952) A Suggested Interpretation of the Quantum Theory in Terms of "Hidden" Variables. I. *Physical Review*, Vol.85, No.2, (January 1952), pp. 166–179

Bosanac, S. (1998). Semiclassical theory of Compton and photoelectric effects. *The European Physical Journal D*, Vol.1, (January 1998), pp. 317-327

Compton, A. (1919a). The Size and Shape of the Electron I. *Physical Review* , Vol.14, No.1, (March, 1919), pp. 20–43 Compton, A. (1919b). The Size and Shape of the Electron II. *Physical Review*, Vol.14, No.3, (May, 1919), pp. 247–259

Compton, A. (1922). Secondary Radiations Produced by X-rays, and Some of Their Applications to Physical Problems. *Bulletin of the National Research Council*, Vol.20, No.19, (October 1922), pp. 1-56

Compton, A. (1923) The Spectrum of Scattered X-Rays. *Physical Review*, Vol.22, No.5, (November 1923), pp. 409-413

Compton, A. (1924). Scattering of X-ray Quanta and J-phenomena. *Nature*, Vol.113, No.2831, (February 1924), pp.160-161

Compton, A. (1935). Incoherent Scattering and the Concept of Discrete Electrons, *Physical Review*, Vol.47, (March 1935), pp. 367-370

Compton, A. (1927). X-rays as a branch of optics, *Nobel Lectures*.

Cooper, M. (1985). Compton scattering and electron momentum determination. *Reports on Progress in Physics*, Vol.48, No.4, (April 1985), pp. 415-481

Crisp, M. & Jaynes E. (1969). Radiative Effects in Semiclassical Theory. *Physical Review*, Vol.179, No.5, (March 1969), pp.1253–1261

Crisp, M. (1990) Self-fields in semiclassical radiation theory. *Physical Review A*, Vol.42, No.7, (October 1990), pp.3703–3717

Debye, P. (1909). Der Lichtdruck auf Kugeln von beliebigem Material. *Annalen der Physik*, Vol.30, No.1, pp 57-136

Debye, P. (1923) Zerstreuung von Rontgenstrahlen und Quantentheorie. *Zeitschrift für Physik*, Vol.24, (March 1923), pp. 161-166

Dirac, P. (1926) Relativity Quantum Mechanics with an Application to Compton Scattering. *Royal Society's Proceedings A*, Vol.111, (June 1926), pp. 405-423

Dirac, P. (1938). Classical Theory of Radiating Electrons. *Royal Society's Proceedings A*, Vol.167, (August 1938), pp. 148-169

Dirac, P. (1965). Quantum Electrodynamics without Dead Wood. *Physical Review*, Vol.139, No.3B, (August 1965), pp. B684–B690

Dodd, J. (1983). Compton effect – a classical treatment. *European Journal of Physics*, Vol.4, (November, 1983), pp.205-213

DuMond, J. (1933). The Linear Momenta of Electrons in Atoms and in Solid Bodies as Revealed by X-Ray Scattering. *Review of Modern Physics*, Vol.5, No.1, (January 1933), pp. 1-33

Eve, A. (1904). On the secondary radiation caused by the beta and gamma rays of radium. *Philosophical Magazine*, Vol.8, No.48, (December, 1904), pp. 669-685

Feynman, R. (1965). *The Feynman Lectures on Physics*, Addison-Wesley, ISBN 0-8053-9045-6, New York, USA

Florance, D. (1910). CIV. Primary and secondary γrays. *Philosophical Magazine*, Vol.20, No.120, (December 1910), pp. 921-938

Glauber, R. (2006). Nobel Lecture: One hundred years of light quanta. *Review of Modern Physics*, Vol.78, No.4, (November 2006), pp. 1267–1278

Gray, J. (1920). The scattering of X- and γ-rays. *Journal of The Franklin Institute*, Vol.190, No.5, pp. 633-655

Heisenberg, W. (1958). *Physics and Philosophy: The Revolution in Modern Science*, Harper & Brothers, ISBN: 0061305499, New York, USA

Jauncey, G. (1925) Quantum Theory of the Unmodified Spectrum Line in the Compton Effect, *Physical Review*, Vol.25,No.3, (March 1925), pp. 314–321

Kidd, R., Ardini, J. & Anton, A. (1989) Evolution of the modern photon. *American Journal of Physics*, Vol.57, No.1,(January, 1989), pp.27-34

Kim, S. (1993) Principle of Random Wave-Function Phase of the Final State in Free-Electron Emission in Wiggler. APEIRON, Vol.17, pp.13-17

Klein, O. & Nishina, Y (1929). Über die Streuung von Strahlung durch freie Elektronen nach der neuen relativistischen Quantendynamik von Dirac. *Zeitschrift für Physik*, Vol.52, No.11, pp. 853-868

Lorentz, H. (1916). *The Theory of Electrons*, Teubner, Leipzig Lamb W. & Scully, M. (1969). The Photoelectric Effect Without Photons, In: *Polarization, Matter and Radiation*, Presses Universitaires de France, pp.363-369, Paris

Mie, C. (1908). Beiträge zur Optik trüber Medien, speziell kolloidaler Metallösungen. *Annalen der Physik*, Vol.25, No.3, pp. 377-445

Pimpale, A. & Mande, Ch. (1984). Raman effect in x-ray region. *Pramana, A Journal of Physics*, Vol.23, No.3, (September 1984), pp.279-295.

Raman, G. (1928) A new radiation. *Indian Journal of Physics*, Vol.21, (March 1928), pp.368-377

Ross P. & Kirkpatrick P. (1934). The Constant in the Compton Equation. *Physical Review*, Vol.45, No.3, (February 1934), pp. 223–223

Schrödinger, E. (1926). Quantizierung als Eigenwertproblem (Erste Mitteilung). Part I. *Annalen der Physik*, Vol.79, No.4, pp. 361-376

Schrödinger, E. (1927). Uber den Comptoneffekt. *Annalen der Physik*, Vol.82, No.4, (January 1927), pp. 257-264

Smith, L. & Purcell, E. (1953). Visible Light from Localized Surface Charges Moving across a Grating, *Physical Review*, Vol.92, No.4, (November 1953), pp. 1069–1069

Stuewer, R. (1975). *The Compton effect: turning point in physics*, Science History Publications, ISBN 9780882020129, New York, USA

Styer, D., Balkin, M., Becker, K., Burns, M., Dudley, C., Forth, S., Gaumer, J., Kraer, M., Oertel, D., Park, L., Rinkoski, M., Smith, C. & Wotherspoon, D. (2002). Nine formulations of quantum mechanics. *American Journal of Physics*, Vol.73, No.3, (November 2001), pp.288-297

Tamm, I. (1930). Uber die Wechselwirkung der freien Elektronen mit der Strahlung nach der
    Diracschen Theorie des Elektrons und nach der Quantenelektrodynamik. *Zeitschrift
    fur Physik*, Vol.62, No.7, pp.545-568

Wheeler, J. & Feynman, R. (1945). Interaction with the Absorber as the Mechanism of
    Radiation. *Review of Modern Physics*, Vol.17, No.2,3, (April 1945)

Wolf, M. (n. d.). Beyond the Point Particle – a Wave Structure for the Electron, In: *Galilean
    Electrodynamics 6, No. 5*, (May1998), < http://mwolff.tripod.com/point.html >

7

Shift-Scale Invariance
of Electromagnetic Radiation

Nikolay Smolyakov
National Research Center "Kurchatov Institute" Moscow, Russia

1. Introduction

Electromagnetic radiation generated by relativistic beams in electron storage rings is of considerable current use. Today's third-generation synchrotron radiation facilities are designed with a number of straight sections to maximize the use of so-called insertion devices (wigglers and undulators). Wigglers and undulators are magnetic devices producing a spatially periodic (or slightly nonperiodic) field variation that cause a charged electron beam to emit electromagnetic radiation with special properties. A wiggler magnet is a succession of alternating polarity magnetic poles, each of which bends the electron beam through an angle large compared with the natural opening angle of the synchrotron radiation $1/\gamma$, where $\gamma = electron\ energy/mc^2$ is the electron reduced energy, see Fig. 1.

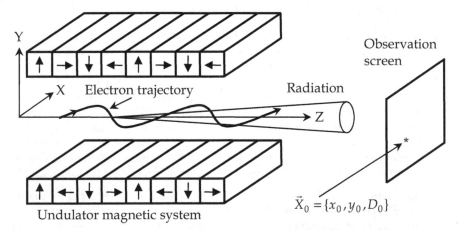

Fig. 1. Schematic of planar undulator and coordinate system. Arrows show magnetization directions.

Wigglers provide a strong magnetic field resulting in broadband emission of a fan-shaped beam of photons. Wiggler radiation is similar to standard synchrotron radiation produced by an individual bend magnet, but $2N$ times as intense due to $2N$ repetitive emission over the length of a $2N$ pole wiggler. The design principle of undulator magnets is basically the same as of wiggler magnets. The difference comes from the magnetic field strength. Undulators, having relatively weak magnetic field, cause small beam deflection. In this case the photons emitted by an individual electron at the various poles in the magnet array interfere coherently. Due to the constructive interference, the undulator radiation beam's opening angle is reduced by $\sqrt{N}$ and thus radiation intensity from one electron per solid angle goes as N^2. A great body of publications is dedicated to undulator/wiggler radiation.

Over the last two decades considerable attention has been focused on the edge radiation as a bright source in infrared – ultraviolet spectral range. Edge radiation is produced when a relativistic charged particle passes through a region of rapid change of magnetic fields at the edges of storage ring bending magnets, see Fig. 2.

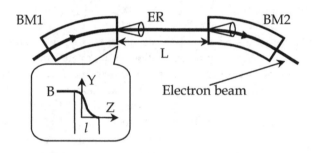

Fig. 2. Edge radiation setup: BM1, BM2 – bending magnets, ER – edge radiation, L – straight section length, l – length of the bending magnet fringe field.

It was discovered independently at the SPS proton synchrotron at CERN (Bossart et al., 1979, 1981) and the Tomsk electron synchrotron "Sirius" (Nikitin et al., 1980, 1981). Notice that the physics of edge radiation in proton storage rings (Coisson, 1979; Smolyakov, 1985, 1986) differs essentially from that in electron storage rings (Bessonov, 1981; Chubar & Smolyakov, 1993; Bosch, 1997; Geloni et al., 2009) though some similarities in their properties can be found. An important property of edge radiation from an electron beam is that its intensity substantially exceeds the standard synchrotron radiation intensity (from the regular bending magnetic field) in the long-wave spectral range, where its wavelength is much more than the corresponding synchrotron radiation critical wavelength. This feature has been confirmed experimentally (Shirasawa et al., 2003; Smolyakov & Hiraya, 2005).

The thorough simulations of radiation characteristics in real experiments call for the special-purpose computer codes. This chapter deals with the problem of how to calculate incoherent electromagnetic radiation, generated by a relativistic electron beam in external magnetic field.

Practically the measured magnetic field data, electron beam emittance, its energy spread, near-field effects (Hirai et al., 1984; Walker, 1988) as well as the finite width of the band-pass filter should be taken properly into account. We consider here the electron beam as an incoherent source, since it is assumed that electrons radiate independently from each other. Therefore, overall spectral intensity is determined by the sum of all the individual intensities. Generally speaking, the spectral-angular distribution of energy radiated by one electron is dependent upon the following nine parameters: the electron initial transversal coordinates and velocities (both horizontal and vertical), its energy, radiation wavelength and the horizontal, vertical and longitudinal coordinates of the observation point. Let us briefly review numerical methods used for simulation of a single-electron spectrum.

It has been suggested that the ideal undulator has a perfect sinusoidal magnetic field, as with the planar (Alferov et al., 1976), helical (Alferov et al., 1976; Kincaid, 1977), elliptical (Yamamoto & Kitamura, 1987), two-harmonic (Dattoli & Voykov, 1993) and figura-8 (Tanaka & Kitamura, 1995) undulators. In this case, the spectral-angular density of undulator radiation in the far-field region involves a series of Bessel functions. This method is used in the computer codes SMUT (Jacobsen & Rarback, 1986; Rarback et al., 1988), URGENT (Walker, 1989; Walker & Diviacco, 1992) and US, which is incorporated into the software toolkit XOP (Sanchez del Rio & Dejus, 1998).

Radiation from an ideal undulator can also be calculated by going into a drift frame where an electron, on average, is at rest (Anacker et al., 1989). Both of these approaches have a fast speed of computation as their main advantage; however, they cannot be used for real undulators with imperfect magnetic fields, nor can they take into account near-field effects.

The most recently developed computer codes are able to use a real magnetic field map and, hence, compute the Fourier transformation of the radiation field numerically, using one of three ways. The spectral integrals may be evaluated by the saddle point method (Leubner & Ritsch, 1986a, 1986b; Steinmuller-Nethl et al., 1989). This method is used in the computer code RADID (C. Wang & Xian, 1990; C. Wang et al., 1994). This approach provides formulae similar to those for standard synchrotron radiation and, by its asymptotic nature, is applicable to insertion devices with strong magnetic fields (C. Wang & Jin, 1992; Walker, 1993). The second method involves the calculation of the radiation field in the time domain followed by the Fourier analysis in order to find the radiation spectrum. Such an approach, with minor modifications, is employed in the codes B2E (Elleaume & Marechal, 1991, 1997), UR (Dejus, 1994; Dejus & Luccio, 1994), YAUP (Boyanov et al., 1994) and SPECTRA (Tanaka & Kitamura, 2001) and is also integrated in the code RADID (C. Wang & Jin, 1992; C. Wang & Xiao, 1992; C. Wang, 1993). It gives direct insight into the physics of radiation and enables the calculation of radiation in a wide spectral range simultaneously, although requiring a huge amount of memory for angular distribution computation (Boyanov et al., 1994). The third method involves the direct integration of the Lienard-Wiechert retarded potentials in the frequency domain. This method is used in the computer codes SpontLight (Geisler et al., 1994), SRW (Chubar & Elleaume, 1998), SMARTWIG (Smolyakov, 2001), WAVE (Bahrdt et al., 2006) and in a number of others programs (Tatchyn et al., 1986; Chapman et al., 1989; Elleaume & Marechal, 1991, 1997; Ch. Wang et al., 1992; Yagi et al., 1995). The main problem of this approach comes from the fast oscillating factor in the integrand, requiring the use of extreme care in integration routines. Comparative studies of these algorithms can be found in (C. Wang & Jin, 1992; Dattoli et al., 1994).

Emittance effects are usually simulated by one of two ways: the Monte Carlo technique or the off-axis approximation method (Jacobsen & Rarback, 1986; Rarback et al., 1988), also known as the shift-invariant property of the radiation pattern (Chapman et al., 1989).

The Monte Carlo method, generally considered to be the most accurate, is employed in a number of computer codes, such as RADID (C. Wang & Xian, 1990; C. Wang & Jin, 1992; C. Wang & Xiao, 1992; C. Wang, 1993; C. Wang et al., 1994), UR (Dejus, 1994; Dejus & Luccio, 1994) and SpontLight (Geisler et al., 1994) (see also (Tatchyn et al., 1986; Yagi et al., 1995)). The problem with this approach is that the large number of individual computations for single-electron radiation can sometimes be too time consuming and impractical (Tatchyn et al., 1986; C. Wang, 1993; Lumpkin et al., 1995). Indeed, it needs generally the 5-dimantional sampling (two for electron position, two for electron deflection and one for its energy). It should be also mentioned that usually the simulations are carried out for a large number of observation points (or spectra). As a result, a total amount of individual simulations is huge.

The off-axis approximation method is based on the concept that spatial distributions of radiation from different electrons are essentially identical and are related to each other by the simple coordinate shifts arising due to angular and spatial electron spreads in the beam. If this is so, there is no need to compute the radiation for each electron separately and, hence, the computational task is simplified considerably. To our knowledge, paper (Nikitin & Epp, 1976) was the first to pioneer the application of this method. Owing to computational speed, this method is used extensively in a number of computer codes such as SMUT (Jacobsen & Rarback, 1986; Rarback et al., 1988), URGENT (Walker, 1989; Walker & Diviacco, 1992), B2E (Elleaume & Marechal, 1991, 1997), YAUP (Boyanov et al., 1994), US (Sanchez del Rio & Dejus, 1998), SRW (Chubar & Elleaume, 1998), SPECTRA (Tanaka & Kitamura, 2001) and SMARTWIG (Smolyakov, 2001) (see also (Anacker et al., 1989; Chapman et al., 1989; Ch. Wang et al., 1992)). The validity of the off-axis approximation method has, however, been proven in the far-field region only (Jacobsen & Rarback, 1986; Rarback et al., 1988; Chapman et al., 1989), since the equations controlling the pattern of radiation in the near-field region are rather cumbersome. Because of this, it has been argued that in a real-life geometry, when the distance to the observer is limited and does not tends to infinity, the radiation does not have the scale properties of standard synchrotron radiation (Chapman et al., 1989) and thus, the off-axis approximation method fails in the near-field region (C. Wang & Xiao, 1992; C. Wang, 1993). This point is particular important for edge radiation simulation (Chubar & Elleaume, 1998; Shirasawa et al., 2003; Smolyakov & Hiraya, 2005) since the straight section length usually is of the same order as the distance to the observer (edge radiation is generated by ultrarelativistic charged particles in the region of magnetic field change at bending magnets edges).

In fact, as it is demonstrated in this chapter, this statement is wrong. The off-axis approximation is valid in near-field region. It has been additionally proven in this chapter that electromagnetic radiation generated by relativistic particle in arbitrary transversal magnetic field (bending magnet fringe fields for edge radiation, undulator or wiggler field and so on) offers the radiation scale property. It means the following. The derived expressions clearly show that the electron energy does not appear explicitly in the formulas for radiation intensity distributions, but implicitly only, through a set of three new variables. These three variables can be called as reduced angles (horizontal and vertical) and reduced wavelength of radiation. By this is meant that the electron energy variation leads to variation of these listed above three reduced parameters only. These variations can be effectively reduced to the proper variations of observation point position and radiation wavelength. As a result, we can make the following statement. The variations of the eight of the nine parameters (usually the distance to observer is constant) can be reduced to the correspondent variations of three parameters only: observation point position (horizontal and vertical) and radiation wavelength. This property simplifies considerably the simulation of electromagnetic radiation in a real situation. We did not employ any particular features of the external magnetic field in this study, except for taking the field to be uniform in the transversal directions. Therefore, the presented results are very general in nature.

We use the Gaussian unit system in this chapter.

2. Shift invariance of electron trajectories

Let us consider the motion of a single electron in the external transversal magnetic field $\vec{B}$. We choose the right-hand coordinate system in the usual fashion (Fig. 1): Z-axis is aligned with the electron beam propagation (straight section axis, undulator axis etc.), the X-axis is directed horizontally and the Y-axis is directed upwards. For simplicity the origin of the reference system is placed at the starting point of the magnetic field. In the vicinity of the Z-axis, the magnetic field can be approximated as:

$$\vec{B}(z) = \left\{ B_x(z), B_y(z), 0 \right\} \tag{1}$$

It follows from (1) that the magnetic field is presumed to be homogeneous in the transversal plane (i.e., its components do not depend on the coordinates x and y). The longitudinal component of the magnetic field B_z is ignored too. This means that the focusing properties of the undulator magnetic field are excluded from our analysis. This simplification is well suited for high-energy electron beams with relatively small transversal sizes and is a working standard for the simulation of spontaneous radiation. All the computer codes cited above in the introduction employ such approximations (or even a much higher degree of the undulator fields' idealization). Among the approximations that have been used in this study, the most radical limitations arise from expression (1). At the same time, there are no additional assumptions about the field profile so that its transversal components $B_{x,y}(z)$ can be considered to incorporate the correction fields at the ends of the undulator. These functions may also include magnetic system fabrication errors.

The electron motion in the magnetic field is governed by the Lorentz force equation:

$$\frac{d\vec{\beta}(t)}{dt} = \frac{e}{mc\gamma} \left[\vec{\beta}(t) \times \vec{B}(\vec{r}(t)) \right] \tag{2}$$

where c is the speed of light, e, m, $\vec{r}(t)$, $c\vec{\beta}(t)$ and γ are the electron's charge, mass, trajectory, velocity and reduced energy respectively: $\gamma = 1/\sqrt{1 - \beta^2}$. We assume here that the radiated energy is negligible as compared to the electron kinetic energy, hence leaving γ constant. It readily follows from Eqs. (1) and (2) that:

$$\frac{d\beta_{x,y}(t)}{dt} = \frac{\pm Q}{\gamma} B_{y,x}(z(t)) \frac{dz(t)}{dt} \tag{3}$$

where $Q = (-e)/(mc^2)$. Remember, an electron has the negative charge so Q is positive. Since the magnetic field is given as a function of the longitudinal coordinate z, it is natural to express the electron trajectory in terms of its longitudinal position as well. Integrating (3) with respect to time and using z as an independent variable instead of t, we get the following exact formulae:

$$\tilde{\beta}_x(z) = Q \int_0^z B_y(z')dz' \tag{4}$$

$$\tilde{\beta}_y(z) = -Q \int_0^z B_x(z')dz' \tag{5}$$

$$\beta_{x,y}(z) = \beta_{x,y}(0) + \frac{1}{\gamma}\tilde{\beta}_{x,y}(z) \tag{6}$$

$$\beta_z(z) = \sqrt{\beta^2 - \beta_x^2(z) - \beta_y^2(z)} \tag{7}$$

Similarly, we can now express the horizontal and vertical components of the electron trajectory $\vec{r}(z) = \{r_x(z), r_y(z), z\}$ as functions of its longitudinal position z:

$$r_{x,y}(z) = r_{x,y}(0) + \int_0^z \frac{\beta_{x,y}(z')}{\beta_z(z')}dz' \tag{8}$$

The time-dependence of the electron longitudinal position is implicitly governed by the equation:

$$t(z) = \int_0^z \frac{dz'}{c\beta_z(z')} \tag{9}$$

Let us suppose that the electron beam and the external magnetic field satisfy the following conditions:

i.   The beam is relativistic with reduced energy $\gamma \gg 1$.
ii.   The angular spread of the electron beam is small and the magnetic field slightly deflects an electron from its initial direction, i.e. $\left|\beta_{x,y}(z)\right| \ll 1$. For wiggler or undulator, this condition is equivalent to $K/\gamma \ll 1$, where K is the undulator deflection parameter.

Notice that this set of requirements is of a general nature and is fulfilled in practically all modern electron storage rings.

Expanding the function $\beta_z(z)$ in terms of the small quantities γ^{-2}, $\beta_{x,y}^2(z)$, we will get from (8) and (9):

$$r_{x,y}(z) = r_{x,y}(0) + \beta_{x,y}(0) \cdot z + \frac{1}{\gamma}\tilde{r}_{x,y}(z) \tag{10}$$

$$t(z) = \frac{z}{c}\left(1 + \frac{1}{2\gamma^2}\right) + \frac{1}{2c}\int_0^z \left(\beta_x^2(z') + \beta_y^2(z')\right)dz' \tag{11}$$

$$\tilde{r}_{x,y}(z) = \int_0^z \tilde{\beta}_{x,y}(z')dz' \tag{12}$$

Since the functions $\tilde{\beta}_{x,y}(z)$ and $\tilde{r}_{x,y}(z)$ are independent of the initial conditions $\beta_{x,y}(0)$, $r_{x,y}(0)$ and γ, they are the same for all electrons from the beam. The Eqs. (4) - (6), (10) - (12) express the electron trajectory in terms of the longitudinal coordinate z. These formulae relate all electron trajectories to a reference (ideal) orbit through a simple linear transformation, thus displaying the shift invariant property of electron trajectories. It is also important to note that the transversal components of the velocity, which are of fundamental importance to the density of electromagnetic radiation, are given by exact expressions. It should be noted that the simplicity of the equations defining the trajectory is a result of the fact that the magnetic field heterogeneity in transversal directions as well as its longitudinal component B_z are negligible. This leads to the separation of the problem into two independent equations, which can be integrated explicitly and independently from each other by the change of variables from t (time) to z (longitudinal position).

3. Spectral and angular distributions of radiations

Let us consider the radiation field of a moving electron, which is seen by the observer at time τ and at the observation point $\vec{X}_0$ with coordinates $\vec{X}_0 = \{x_0, y_0, D_0\}$. Notice that D_0 is the longitudinal distance between the starting point of the magnetic field and the observer, and is generally comparable to the magnetic field length L, although keeping in mind that $D_0 > L$. The electrical component of the radiation field is given by the following exact expression:

$$\vec{E}(\tau) = \frac{e}{cR(t)} \cdot \frac{[\vec{n}(t) \times [(\vec{n}(t) - \vec{\beta}(t)) \times \dot{\vec{\beta}}(t)]]}{(1 - (\vec{n}(t) \cdot \vec{\beta}(t)))^3} + \frac{e}{R^2(t)\gamma^2} \cdot \frac{\vec{n}(t) - \vec{\beta}(t)}{(1 - (\vec{n}(t) \cdot \vec{\beta}(t)))^3} \tag{13}$$

Here, the vector $\vec{R}(t) = \vec{X}_0 - \vec{r}(t)$ with absolute value $R(t)$ represents the distance between the emission point and the observer, the unit vector $\vec{n}(t) = \vec{R}(t)/R(t)$ points from the instantaneous position of the electron to the observer, and acceleration $c\dot{\vec{\beta}}(t)$ is calculated by Eq. (2). The quantities $\vec{n}(t)$, $\vec{\beta}(t)$, $\dot{\vec{\beta}}(t)$ and $R(t)$ on the right-hand side of Eq. (13) are to be evaluated at the retarded time t which must obey the equation:

$$c\tau = ct + R(t) \tag{14}$$

Differentiating the last relation, we find:

$$\frac{d\tau}{dt} = 1 - (\vec{n}(t) \cdot \vec{\beta}(t)) \tag{15}$$

Upon integrating (15) with respect to the retarded time we obtain the alternative relationship between observer time τ and retarded time t :

$$\tau(t) = \tau(0) + \int_0^t (1 - (\vec{n}(t') \cdot \vec{\beta}(t'))) dt' \tag{16}$$

The first term in (13) arises from the electron acceleration and varies as R^{-1}. The second term in (13) is the so-called velocity field, which is independent of acceleration and essentially arises from static fields falling off as R^{-2}. At large distances this term is negligibly small and is often ignored. However in some cases of the long wavelength radiation simulation both terms in Eq. (13) should be considered (Roy et al., 2000). Thus we will include the velocity term into our analysis and furthermore will prove that its inclusion does not violate the shift-scale properties of radiation.

Let us consider the electromagnetic radiation with wavelength λ passing through an infinitesimal surface area ds located at the observation point $\bar{X}_0$ perpendicular to the Z - axis, as it shown in Fig. 1. The general expression for the number of photons $dN_{x,y}$ with horizontal (x) and vertical (y) polarization, emitted by one electron in its passage through the magnetic field per relative bandwidth $d\lambda/\lambda$ and per area ds can be written as follows:

$$dN_{x,y} = \frac{ds}{D_0^2} \left(\frac{d\lambda}{\lambda} \right) \frac{\alpha c^2 \gamma^2}{4\pi^2 e^2} \left| \tilde{E}_{x,y}(\lambda) \right|^2 \tag{17}$$

where α is the fine structure constant and $\tilde{E}_{x,y}(\lambda)$ is proportional to the Fourier-transform of the radiation field:

$$\tilde{E}_{x,y}(\lambda) = \frac{D_0}{\gamma} \int_{-\infty}^{\infty} \exp(i\frac{2\pi c}{\lambda}\tau) E_{x,y}(\tau) d\tau \tag{18}$$

A comprehensive study of electromagnetic radiation includes the simulation of its polarization, which is usually calculated via the Stokes parameters:

$$S_0 = A\left(\tilde{E}_x \tilde{E}_x^* + \tilde{E}_y \tilde{E}_y^* \right),$$
$$S_1 = A\left(\tilde{E}_x \tilde{E}_x^* - \tilde{E}_y \tilde{E}_y^* \right),$$
$$S_2 = A\left(\tilde{E}_x \tilde{E}_y^* + \tilde{E}_y \tilde{E}_x^* \right),$$
$$S_3 = -iA\left(\tilde{E}_x \tilde{E}_y^* - \tilde{E}_y \tilde{E}_x^* \right) \tag{19}$$

where

$$A = \frac{ds}{D_0^2} \left(\frac{d\lambda}{\lambda} \right) \frac{\alpha c^2 \gamma^2}{4\pi^2 e^2},$$

thus normalizing S_0 to the photon flux density (17).

One additional comment is necessary. The far-field approximation, which is extensively used in radiation simulations, implies that the distance between the source and the observer is vastly larger than the magnetic field length, so that the source of radiation is considered point-like. At this level of approximation, the variables $\vec{n}$ and R in (13) may be considered as constants, which uniquely determine the position of the observation point. The radiant energy in this case can be expressed conveniently in terms of angular coordinates of the observation point n_x and n_y. However, as it is shown in (Walker, 1988), the near-field effects that are caused by the finite distance to the observer can significantly change the properties of undulator radiation and hence should be properly taken into consideration. It means that the time-dependence of the unit vector $\vec{n}$ must be taken into account. That is why we specify the observer position by its Cartesian coordinates and consider the photon flux per small surface area rather than per a small solid angle.

In order to calculate the Fourier transform of the field in the frequency domain, we substitute the expressions (13) and (16) for electric field $\vec{E}(\tau)$ and observer time τ respectively, in Eq. (18). Furthermore, we change the variable of integration from retarded time t to the particle longitudinal coordinate z with the help of Eq. (9), thereby obtaining the following exact expression:

$$\tilde{E}_{x,y}(\lambda) = \frac{-2eQ}{c}\int_0^L \exp(i\Phi(z))\left(f_{x,y}(z)B_x(z) + g_{x,y}(z)B_y(z) - \frac{h_{x,y}(z)}{Q(D_0 - z)}\right)\frac{D_0}{D_0 - z}dz, \tag{20}$$

$$\vec{f}(z) = \frac{[\vec{n}(z)\times[(\vec{n}(z)-\vec{\beta}(z))\times[\vec{\beta}(z)\times\vec{i}]]]}{2\gamma^2\beta_z(z)(1-(\vec{n}(z)\cdot\vec{\beta}(z)))^2}\cdot\frac{D_0 - z}{R(z)}, \tag{21}$$

$$\vec{g}(z) = \frac{[\vec{n}(z)\times[(\vec{n}(z)-\vec{\beta}(z))\times[\vec{\beta}(z)\times\vec{j}]]]}{2\gamma^2\beta_z(z)(1-(\vec{n}(z)\cdot\vec{\beta}(z)))^2}\cdot\frac{D_0 - z}{R(z)}, \tag{22}$$

$$\vec{h}(z) = \frac{\vec{n}(z)-\vec{\beta}(z)}{2\gamma^3\beta_z(z)(1-(\vec{n}(z)\cdot\vec{\beta}(z)))^2}\cdot\left(\frac{D_0 - z}{R(z)}\right)^2, \tag{23}$$

$$\Phi(z) = \frac{2\pi c}{\lambda}\tau(0) + \frac{2\pi}{\lambda}\int_0^z(1-(\vec{n}(z')\cdot\vec{\beta}(z')))\frac{dz'}{\beta_z(z')}, \tag{24}$$

where $\vec{i}$ and $\vec{j}$ are the unit vectors along the axis X and Y respectively. Since the first term $(2\pi c/\lambda)\tau(0)$ in phase (24) is z-independent, it can be dropped out because the constant phase factor leaves the physical quantities (17) and (19) unaltered.

It should be noted that in many papers the other expression is used instead of (20), see (Rarback et al., 1988; Chapman et al., 1989; Boyanov et al., 1994; Dattoli et al. 1994; Chubar & Elleaume, 1998; Tanaka & Kitamura, 2001). This expression has a more simple analytical form and can be derived from (20) as a result of integration by parts accompanied by the omitting of the boundary terms. It has been pointed out, however, that this simplification is not generally valid (Tatchyn et al., 1986) and ignoring the boundary terms may introduce considerable errors in computer results (Walker, 1989). Thus we will use the expressions (20) - (24) here.

We will consider the radiation at small observation angles: $|n_x(z)| << 1$ and $|n_y(z)| << 1$. If so, let us expand R, n_z and β_z in power series of the small quantities γ^{-1}, $n_{x,y}$ and $\beta_{x,y}$:

$$R(z) = D_0 - z$$

$$n_z(z) = 1 - 0.5 \cdot (n_x^2(z) + n_y^2(z))$$

$$\beta_z(z) = 1 - 0.5 \cdot (\gamma^{-2} + \beta_x^2(z) + \beta_y^2(z))$$

Using these expansions and keeping only the leading terms, it is easily found that:

$$1 - (\vec{n} \cdot \vec{\beta}) = 0.5 \cdot \left(\gamma^{-2} + (n_x - \beta_x)^2 + (n_y - \beta_y)^2 \right) \tag{25}$$

Expanding the triple cross products in (21) and (22), using (25) and keeping again the leading terms, after a rather long computation we finally get:

$$f_x(z) = \left(\vec{f}(z) \cdot \vec{i} \right) = \frac{2u_x(z)u_y(z)}{\left(1 + u_x^2(z) + u_y^2(z) \right)^2} \tag{26}$$

$$f_y(z) = \left(\vec{f}(z) \cdot \vec{j} \right) = \frac{1 + u_x^2(z) - u_y^2(z)}{\left(1 + u_x^2(z) + u_y^2(z) \right)^2} \tag{27}$$

$$g_x(z) = \left(\vec{g}(z) \cdot \vec{i} \right) = \frac{1 - u_x^2(z) + u_y^2(z)}{\left(1 + u_x^2(z) + u_y^2(z) \right)^2} \tag{28}$$

$$g_y(z) = \left(\vec{g}(z) \cdot \vec{j} \right) = -f_x(z) \tag{29}$$

$$h_{x,y}(z) = \frac{2u_{x,y}(z)}{\left(1 + u_x^2(z) + u_y^2(z) \right)^2} \tag{30}$$

$$\Phi(z) = \frac{\pi}{\lambda \gamma^2} \int_0^z \left(1 + u_x^2(z') + u_y^2(z') \right) dz' \tag{31}$$

$$u_{x,y}(z) = \gamma n_{x,y}(z) - \gamma \beta_{x,y}(z) \tag{32}$$

The values of $f_{x,y}$, $g_{x,y}$ and $h_{x,y}$ are rapidly varying functions of the instantaneous electron angles $\beta_{x,y}(z)$ and the observation angles $n_{x,y}(z)$. The peak values are 0.25 for f_x and g_y, 1.0 for f_y and g_x, and $0.375\sqrt{3}$ for $h_{x,y}$. The values of these functions differ noticeably from zero only within a small angle $\pm 1/\gamma$ from the electron velocity direction, which is common to electromagnetic radiation emitted by a relativistic charged particle. More detailed calculations show that correction terms for the expressions (26) - (31) are γ^2 times smaller than the leading terms and hence may be safely discarded.

All the functions $f_{x,y}(z)$, $g_{x,y}(z)$ and $h_{x,y}(z)$ in (20) are of the same order of magnitude and have a similar scale of variation. Thus the relative contribution of the velocity term $h_{x,y}(z)$ is determined by the ratio between magnetic field amplitude and factor $1/(QD_0)$. For an undulator with a sinusoidal magnetic field, the ratio is equal to $\lambda_w/(2\pi K D_0)$, where λ_w is the undulator period length and K is the undulator deflection parameter. Usually this quantity is negligibly small and thus the velocity term can be dropped out. However, at the level of approximation used in this paper, only those correction terms, which are proportional to the γ^{-2}, $n_{x,y}^2$ and $\beta_{x,y}^2$, are disregarded. Thus the velocity term has been retained here for generality.

4. Shift-scale invariance of radiation

Detailed theoretical investigations of electromagnetic radiation are certain to include simulation of radiation intensity for a great number of observation points and photon wavelengths. This means that we should repeat the individual calculations for a wide range of the following radiation's three parameters: its wavelength λ and transversal coordinates of the observation points x_0 and y_0. The longitudinal coordinates of the observation points D_0 are essentially the same since the observation plane is aligned perpendicularly to the Z - axis. To include the multi-electron effects into simulation, one must then repeat the calculation of single-electron radiation for a huge number of different electrons, each moving in its own trajectory. The trajectory is determined uniquely by the following five parameters: initial transversal positions of the electron $r_{x,y}(0)$, its initial angles $\beta_{x,y}(0)$ and its reduced energy γ. So, we have to compute single-electron radiation for a wide variety of values for each of just listed eight parameters by itself. Though this straightforward way is conceptually satisfying (and it is used in Monte-Carlo sampling), it is computationally formidable since the needed number of individual calculations is multiplicatively accumulated and the simulation quickly becomes too time consuming. That is why the reduction in number of independent variables is of fundamental importance.

Let us now proceed to analyze how the radiated fields $\tilde{E}_{x,y}$ are controlled by the following eight parameters: $r_{x,y}(0)$, $\beta_{x,y}(0)$, γ, x_0, y_0 and λ. It should be noted that the radiated fields depend on these eight parameters only implicitly through the variable $\Lambda = \lambda\gamma^2$ and two functions, $u_x(z)$ and $u_y(z)$ (see Eqs. (20) and (26) - (31)).

In the case of small observation angles $|n_{x,y}(z)| \ll 1$, we can write:

$$n_x(z) = \frac{x_0 - r_x(z)}{D_0 - z} \tag{33}$$

$$n_y(z) = \frac{y_0 - r_y(z)}{D_0 - z} \tag{34}$$

By putting these relations and Eqs. (6) and (10) for the electron trajectory into Eqs. (32) for $u_x(z)$ and $u_y(z)$, we have after some simple algebraic manipulations:

$$u_{x,y}(z) = \Theta_{x,y}\frac{D_o}{D_o - z} - v_{x,y}(z) \tag{35}$$

$$v_{x,y}(z) = \frac{\tilde{r}_{x,y}(z)}{D_o - z} + \tilde{B}_{x,y}(z) \tag{36}$$

$$\theta_x = \frac{x_o - r_x(0)}{D_o} - \beta_x(0) \tag{37}$$

$$\theta_y = \frac{y_o - r_y(0)}{D_o} - \beta_y(0) \tag{38}$$

where $\Theta_{x,y} = \gamma\theta_{x,y}$. The quantities $\theta_{x,y}$ are the angles between the initial electron's velocity and the direction of observation. It is clear that the functions $v_{x,y}(z)$ are the same for all electrons in the beam being dependent only on the magnetic field $B_{x,y}(z)$ and longitudinal coordinate of the observation points D_o (see Eqs. (4), (5) and (12)). By substituting Eqs. (35) into the integral (31) and performing integration, we have:

$$\Phi(z) = \Phi_0(z) + \Phi_1(z), \tag{39}$$

$$\Phi_0(z) = \frac{\pi}{\Lambda}\left[z\left(1 + \Theta_x^2 + \Theta_y^2\right) - 2\Theta_x \tilde{r}_x(z) - 2\Theta_y \tilde{r}_y(z) + \int_0^z \left(\tilde{B}_x^2(z') + \tilde{B}_y^2(z')\right)dz'\right], \tag{40}$$

$$\Phi_1(z) = \frac{\pi}{\Lambda(D_o - z)}\left[\left(z\Theta_x - \tilde{r}_x(z)\right)^2 + \left(z\Theta_y - \tilde{r}_y(z)\right)^2\right] \tag{41}$$

where $\Lambda = \lambda\gamma^2$. These three relations show explicitly that the phase $\Phi(z)$ depends on the electron initial parameters, its energy and radiation parameters only through the following four variables: $\Theta_{x,y}$, Λ and D_o. The first term in (39) is explicitly independent of D_o although implicitly it is D_o-dependent through the variables $\Theta_{x,y}$.

The foregoing allows us to express the main result of this study through the following compact relation:

$$\tilde{E}_{x,y}\left(r_x(0), r_y(0), \beta_x(0), \beta_y(0), \gamma, x_o, y_o, \lambda, D_o; \{B_x, B_y\}\right) = \tilde{E}_{x,y}\left(\Theta_x, \Theta_y, \Lambda, D_o; \{B_x, B_y\}\right) \tag{42}$$

where $\{B_x, B_y\}$ denotes the appropriate set of parameters, which uniquely determines the external magnetic field along. In the general case, this is an experimentally measured magnetic field mesh. For the case of ideal planar undulator with a sinusoidal field, it may be the following three parameters: number of the undulator periods, the length of period and magnetic field amplitude (or an undulator deflection parameter).

Expressions (37) and (38) for $\theta_{x,y}$ show explicitly that photon density (17) as well as the Stokes parameters (19) possess the shift-invariant property. It means that any changes in the electron initial parameters $r_{x,y}(0)$ and $\beta_{x,y}(0)$ may be reduced to the corresponding shift of the observer transversal coordinates x_o and y_o. Hence there is no need to repeatedly simulate the radiation from a great number of electrons with different trajectories. The desired radiation distribution for any electron can be obtained from the corresponding distribution of radiation that has been calculated for the electron with the same energy but following the equilibrium trajectory.

One can see from expression (42) that the functions $\tilde{E}_{x,y}$ are also scale-invariant relative to the reduced energy γ. Let us consider changing the electron energy $\gamma \to k\gamma$ such that the restriction $\gamma \gg 1$ remains intact. Such a change of energy results in the following variations of arguments: $\Lambda \to k^2\Lambda$ and $\Theta_{x,y} \to k\Theta_{x,y}$. These variations may be thought of as rescaling of the radiation wavelength $\lambda \to k^2\lambda$ and the observation angles $\theta_{x,y} \to k\theta_{x,y}$, which in turn can be boiled down to the corresponding shifts of the observer coordinates x_o and y_o. It means again that the simulation of radiation from electrons of differing energy may effectively be reduced to the simulation of radiation from a single electron with energy and trajectory in equilibrium, but with different radiation wavelengths and at different observation points.

It follows from the expressions (17) and (42) that photon density is also scale-invariant, which is to say that the number of photons is invariant under the following transformation:

$$\gamma \to k\gamma ,$$
$$\theta_{x,y} \to k^{-1}\theta_{x,y} ,$$
$$\lambda \to k^{-2}\lambda ,$$
$$d\lambda \to k^{-2}d\lambda ,$$
$$ds \to k^{-2}ds \tag{43}$$

These transformations leave the Stokes parameters also unchanged.

Since the parameters

$$\Theta_x = \gamma\theta_x = \gamma\left(\frac{x_o - r_x(0)}{D_o} - \beta_x(0)\right), \quad \Theta_y = \gamma\theta_y = \gamma\left(\frac{y_o - r_y(0)}{D_o} - \beta_y(0)\right) \text{ and } \Lambda = \lambda\gamma^2$$

are of primary importance in describing the shift-scale invariant properties of electromagnetic radiation, we can call these variables "reduced angles" and "reduced wavelength" respectively.

Assuming the electron proceeds along an equilibrium trajectory which passes through the origin (i.e. $r_{x,y}(0) = 0$), the correction term $\Phi_1(z)$, which is primarily responsible for near-field effects, may be written as:

$$\Phi_1(z) = \frac{\pi}{\Lambda(D_o - z)}\left[\left(z\frac{\gamma x_o}{D_o} - \gamma r_x(z)\right)^2 + \left(z\frac{\gamma y_o}{D_o} - \gamma r_y(z)\right)^2\right] \tag{44}$$

For undulators with deflection parameter K and period length λ_w, the functions $\gamma r_{x,y}(z)$ oscillate with the amplitudes of the order of $K\lambda_w/(2\pi)$, what is typically much less than the undulator length L, which is the maximum value for z. If so, we can safely omit the terms $\gamma r_{x,y}(z)$ in (44). Let us consider the radiation at a large distance $D_o \gg L$, which enables us to substitute D_o for $(D_o - z)$ in (44). As a result, the phase term (44) can be approximated by:

$$\Phi_1(z) = \frac{\pi}{\lambda D_o^3}\left(x_o^2 + y_o^2\right)z^2 \tag{45}$$

This is in agreement with the correction term derived by (Walker, 1988).

To get the corresponding expressions in far-field approximation, we should proceed to limit $D_o \to \infty$, $x_o \to \infty$ and $y_o \to \infty$ while the ratios

$$\frac{x_o}{D_o} \text{ and } \frac{y_o}{D_o}$$

are kept constant and become new transversal angular coordinates of the observation point. Then the ratio

$$\frac{ds}{D_o^2}$$

is turned to the infinitesimal solid angle $d\Omega$. In that event the initial transversal positions of the electron $r_{x,y}(0)$ are irrelevant and the correction term $\Phi_1(z)$ in (39) tends to zero.

It is notable that the shift-scale invariance of electromagnetic radiation, which is given by Eq. (42), is identical to those of standard synchrotron radiation. Indeed, the spectral – angular density of synchrotron radiation in far – field region depends on the ratio λ_c/λ, where $\lambda_c = 4\pi\rho/(3\gamma^3)$ is the critical wavelength and ρ is the bending radius. However, as Eq. (42) shows, bending magnetic field B_0 should appear in this ration explicitly, rather than implicitly through ρ. Using the relation $eB_0\rho = mc^2\gamma$, we get:

$$\frac{\lambda_c}{\lambda} = \frac{4\pi mc^2}{3eB_0}\frac{1}{\lambda\gamma^2} \tag{46}$$

thus displaying the Λ - dependence of synchrotron radiation spectra.

A comprehensive theoretical analysis of edge radiation in sharp-edge approximation has been recently done (Geloni et al., 2009). Sharp-edge approximation means that the magnetic field at the bending magnet edge has supposedly a stepwise change from its regular value B_0 to zero. In other words, the fringe field length $l = 0$, see Fig. 2. In this case we have two parameters completely defining the magnetic system: magnetic field amplitude B_0 and straight section length L, or $\{B_x, B_y\} = (B_0, L)$ in terms of Eq.(42). It has been shown analytically and verified numerically that in this case the edge radiation distributions have the property of similarity. Two dimensionless parameters

$$\delta \equiv \sqrt[3]{\frac{\rho^2 \lambda}{2\pi L^3}}$$

and

$$\phi \equiv \frac{2\pi L}{\lambda \gamma^2}$$

were defined, where ρ is the radius of the trajectory bend. The property of similarity means that data for different sets of edge radiation parameters corresponding to the same values of δ and ϕ reduce to a single curve when properly normalized (Geloni et al., 2009).

This result is in complete agreement with the shift-scale property of electromagnetic radiation presented here and furthermore, it is the particular case of the more general shift-scale property. As suggested by Eq. (42), in the sharp-edge approximation the Fourier-transform of the radiation field $\tilde{E}_{x,y}(\lambda)$ depends on six parameters, namely, two reduced angles Θ_x and Θ_y, reduced wavelength Λ, distance to the observation screen D_0, bending field amplitude B_0 and straight section length L. The use of parameter

$$\tilde{\rho} = \frac{mc^2}{eB_0}$$

with dimension of length rather than B_0 is more convenient for analysis. Since the electron charge e, its mass m and the speed of light c are the fundamental constants, the parameter $\tilde{\rho}$ is equivalent to B_0. By virtue of the relation $eB_0\rho = mc^2\gamma$, where ρ is the radius of the bend in the field B_0, we get: $\rho = \gamma\tilde{\rho}$. Radius of the trajectory bend ρ is much more convenient in practical use, but we should start our analysis with the parameter $\tilde{\rho}$ because it depends on B_0 only, while ρ depends both on B_0 and γ.

So, we can say that $\tilde{E}_{x,y}(\lambda)$ depends on the set of parameters

$$\left(\Theta_x, \Theta_y, \Lambda, \tilde{\rho}, D_0, L \right)$$

It is clear that the alternative set of the following parameters:

$$\left(\Theta_x, \Theta_y, \frac{2\pi L}{\Lambda}, \sqrt[3]{\frac{\tilde{\rho}^2 \Lambda}{2\pi L^3}}, D_0, L \right)$$

is mathematically equivalent to the previous one. Using the relations $\Lambda = \lambda\gamma^2$ and $\rho = \gamma\tilde{\rho}$, we get:

$$\frac{2\pi L}{\Lambda} = \frac{2\pi L}{\lambda\gamma^2} = \phi$$

and

$$\sqrt[3]{\frac{\tilde{\rho}^2 \Lambda}{2\pi L^3}} = \sqrt[3]{\frac{\rho^2 \lambda}{2\pi L^3}} = \delta \; ,$$

thus deriving the parameters φ and δ obtained by (Geloni et al., 2009) for the property of similarity.

5. Conclusion

The results of the theoretical analysis presented here show that electromagnetic radiation, generated by relativistic charged particles in external magnetic fields, offers shift-scale invariance properties which are analytically best expressed by the Eq. (42). It is significant to note that all previous analyses were based on assumptions which were very general in nature and the great bulk of insertion devices match these assumptions without any loss in generality. Let us list here the all constrains we have imposed.

A relativistic charged particle ($\gamma \gg 1$) has small transversal components of its reduced velocity: $\sqrt{\beta_x^2(z) + \beta_y^2(z)} \ll 1$. Electromagnetic radiation is observed at small angles: $\sqrt{n_x^2(z) + n_y^2(z)} \ll 1$. The external magnetic field is taken to be uniform in a transversal plane, and its longitudinal component is zero. This last requirement is the most essential one and there is no way of applying the results obtained here if the undulator focusing properties is to be included into consideration. In our study we did not consider any specific features of the external magnetic field such as its periodicity, etc. This means that any radiation, generated by relativistic charged particles in an external magnetic field under conditions just mentioned, is shift-scale invariant.

It is important that the Fourier-transform of the radiation field $\tilde{E}_{x,y}(\lambda)$ defined by Eq. (18) depends on the electron energy γ only implicitly, through the reduced angles $\Theta_{x,y}$ and reduced wavelength Λ, see Eq. 42. This fact is not evident at first glance because reduced energy γ is a dimensionless parameter and dimensional analysis cannot be applied here. In addition, the electron trajectory depends explicitly on its reduced energy γ while the dependence of the radiation field $E_{x,y}(\tau)$ on trajectory parameters is fairly intricate, see Eqs. (13), (20-24) (recall that the unit vector $\vec{n}(z)$ is trajectory-dependent).

The shift-scale property of electromagnetic radiation suggests the following elements in the algorithm for the simulation of radiation from an electron beam in a real-life experimental setup. With knowledge of external magnetic field (undulator field, fringe field at the ends of storage ring bending magnets and so on), it is possible to find four parameters $r_{x,y}(0)$ and $\beta_{x,y}(0)$, such that the electron with a mean energy is moving along the equilibrium trajectory. In particular, these parameters are zero if the correction fields are included in the undulator magnetic field map. For this single electron, we can compute the spectral-spatial distribution of radiation for a number of wavelengths λ and at different observation points x_o and y_o. With knowledge of this radiation distribution, the effects of electron beam emittance, its energy spread and finite width of the spectral device may be included in the simulation via numerical convolution. To do this would require no more than a three-dimensional integration.

For undulators with large number of periods $N \gg 1$, further simplifications can be made on the following qualitative grounds. The spectral width of the i -th harmonic is equal to

$$\Delta \lambda = \lambda / (iN),$$

which yields the corresponding range

$$\Delta \Lambda = \Lambda / (iN)$$

From this, it follows that variation in γ through the range

$$\Delta \gamma = \gamma / (2iN)$$

can radically alter the distribution of undulator radiation.

The cone of the i -th harmonic has the angular size

$$\theta_i = \frac{1}{\gamma} \sqrt{\frac{1 + 0.5K^2}{2iN}}$$

This means that the variables $\Theta_{x,y}$ change of the order of $\gamma \theta_i$ which will alter the undulator radiation distribution. On the other hand, $\Theta_{x,y}$ vary directly with the energy $\Delta \Theta_{x,y} = \Delta \gamma \cdot \Theta_{x,y}$. We will consider the radiation not far from the undulator axis $\theta_{x,y} = \delta_{x,y} \cdot \theta_i$, where $\delta_{x,y}$ are of the order of 1. The following range for γ variation is thus

$$\Delta \gamma = \gamma / \delta_{x,y},$$

which is considerably more than

$$\gamma / (2iN)$$

This means that γ variation affects the spectral-angular distribution of undulator radiation much more through the resulting change in variable Λ rather than through $\Theta_{x,y}$. If this is so, when averaging the radiation distribution over the energy spread, we only need to consider the γ variation for the variable Λ and use the mean value of γ for the variables $\Theta_{x,y}$. As a result this averaging process can be reduced simply to the corresponding integration over the radiation wavelength λ.

The shift-scale invariance of radiation distributions can be effectively employed in the raytracing computer codes such as RAY (Erko et al., 2008). For more sophisticated analyses, based on wavefront propagation simulation (Erko et al., 2008), it has been necessary to amplify the results presented here by the proper studying of the term $(2\pi c / \lambda) \tau(0)$ in the phase (24). This term is z -independent, has no effect on the radiation intensity and thus it was eliminated from our analysis here. At the same time it depends on the observation point and therefore plays an important role in wavefront propagation calculations. Such kind of theoretical analysis has been performed for the particular case of standard synchrotron radiation (Smolyakov, 1998). Some results of numerical simulations can be found in (Chubar et al., 1999), while the general case of radiation has yet to be analyzed.

6. References

Alferov, D.F.; Bashmakov, Yu.A. & Bessonov, E.G. (1976). Undulator radiation. *Proceedings (Trudy) of the P.N. Lebedev Physics Institute,* Vol.80, (1976), pp. 97-123, ISBN 0-306-10932-8, Consultants Bureau, New York

Anacker, D.C.; Hale, W. & Erskine, J.L. (1989). An algorithm for computer simulation of undulators and wigglers. *Nuclear Instruments and Methods in Physics Research A,* Vol.284, Nos.2-3, (December 1989), pp. 514-522, ISSN 0168-9002

Bahrdt, J; Scheer, M. & Wustefeld, G. (2006). Tracking simulations and dynamic multipole shimming for helical undulators. *Proceedings of the Tenth European Particle Accelerator Conference (EPAC'06),* pp. 3562-3564, ISBN 92-9083-278-9, Edinburgh, UK, June 26-30, 2006

Bessonov, E.G. (1981). On a class of electromagnetic waves. *Soviet Physics – JETP,* Vol.53, No.3, (March 1981), pp. 433-436, ISSN 0038-5646

Bosch, R.A. (1997). Long-wavelength radiation along a straight-section axis in an electron storage ring. *Nuclear Instruments and Methods in Physics Research A,* Vol.386, Nos.2-3, (February 1997), pp. 525-530, ISSN 0168-9002

Bossart, R.; Bosser, J.; Burnod, L.; Coisson, R.; D'Amico, E.; Hofmann, A. & Mann, J. (1979). Observation of visible synchrotron radiation emitted by a high-energy proton beam at the edge of a magnetic field. *Nuclear Instruments and Methods,* Vol.164, No.2, (August 1979), pp. 375-380, ISSN 0029-554X

Bossart, R.; Bosser, J.; Burnod, L.; D'Amico, E.; Ferioli, G.; Mann, J. & Meot, F. (1981). Proton beam profile measurements with synchrotron light. *Nuclear Instruments and Methods,* Vol.184, Nos.2-3, (June 1981), pp. 349-357, ISSN 0029-554X

Boyanov, B.I.; Bunker, G.; Lee J. M. & Morrison T. I. (1994). Numerical modeling of tapered undulator. *Nuclear Instruments and Methods in Physics Research A,* Vol.339, No.3, (February 1994), pp. 596-603, ISSN 0168-9002

Chapman, K.; Lai, B.; Cerrina, F. & Viccaro, J. (1989). Modelling of undulator sources. *Nuclear Instruments and Methods in Physics Research A,* Vol.283, No.1, (October 1989), pp. 88-99, ISSN 0168-9002

Chubar, O.V. & Elleaume, P. (1998). Accurate and efficient computation of synchrotron radiation in the near field region. *Proceedings of the Sixth European Particle Accelerator Conference (EPAC'98),* pp. 1177-1179, ISBN 0 7503 0579 7, Stockholm, Sweden, June 22-26, 1998

Chubar, O.V.; Elleaume, P. & Snigirev A. (1999). Phase analysis and focusing of synchrotron radiation. *Nuclear Instruments and Methods in Physics Research A,* Vol. 435, No.3, (October 1999), pp. 495-508, ISSN 0168-9002

Chubar, O.V. & Smolyakov, N.V. (1993). VUV range edge radiation in electron storage rings. *Journal of Optics (Paris),* Vol.24, No.3, (May 1993), pp. 117-121 ISSN 0150-536X

Coisson, R. (1979). Angular-spectral distribution and polarization of synchrotron radiation from a « short » magnet. *Physical Review A – General Physics*, Vol.20, No.2, (August 1979), pp. 524-528, ISSN 0556-2791

Dattoli, G.; Giannessi L. & Voikov G. (1994). Analytical and numerical computation of undulator brightness: a comparison. *Computers & Mathematics with Applications*, Vol.27, No.6, (March 1994), pp. 63-78, ISSN 0898-1221

Dattoli, G. & Voykov, G. (1993). Spectral properties of two-harmonic undulator radiation. *Physical Review E*, Vol.48, No.4, (October 1993), pp. 3030-3039, ISSN 1539-3755

Dejus, R. J. (1994). Computer simulations of the wiggler spectrum. *Nuclear Instruments and Methods in Physics Research A*, Vol.347, Nos.1-3, (August 1994), pp. 56-60, ISSN 0168-9002

Dejus, R. J. & Luccio, A. (1994). Program UR: General purpose code for synchrotron radiation calculations. *Nuclear Instruments and Methods in Physics Research A*, Vol.347, Nos.1-3, (August 1994), pp. 61-66, ISSN 0168-9002

Elleaume, P. & Marechal, X. (1991). Polarization characteristics of various ID's. *Internal Report ESRF-SR/ID-91-47*, (February 1991)

Elleaume, P. & Marechal, X. (1997). B2E: A software to compute synchrotron radiation from magnetic field data. Version 1.0. *Internal Report ESRF-SR/ID-91-54*, (March 1997)

Erko, A.; Idir, M.; Krist, T. & Michette, A.G. (2008). *Modern Developments in X-Ray and Neutron Optics, Springer Series in Optical Sciences*, Vol. 137, Springer, ISBN 978-3-540-74560-0, Springer Berlin Heidelberg New York

Geisler, A.; Ridder M. & Schmidt, T. (1994). SpontLight: A package for synchrotron radiation calculations. *DELTA Internal Report 94-7*, (April 1994), University of Dortmund, Institute for Acceleratorphysics, 44221 Dortmund, Germany

Geloni, G.; Kocharyan, V.; Saldin, E.; Schneidmiller, E. & Yurkov, M. (2009). Theory of edge radiation. Part I: Foundations and basic applications. *Nuclear Instruments and Methods in Physics Research A*, Vol.605, No.3, (July 2009), pp. 409-429, ISSN 0168-9002

Hirai, Y.; Luccio, A. & Yu, L. (1984). Study of the radiation from an undulator: Near fielf formation. *Journal of Applied Physics*, Vol.55, No.1, (January 1984), pp. 25-32, ISSN 0021-8979

Jacobsen, C. & Rarback, H. (1986). Prediction on the performance of the soft X-ray undulator. *Proceedings of SPIE*, Vol. 582, (January 1986), pp. 201-212, ISBN 0-89252-617-3

Kincaid, B.M. (1977). A short-period helical wiggler as an improved source of synchrotron radiation. *Journal of Applied Physics*, Vol.48, No.7, (July 1977), pp. 2684-2691, ISSN 0021-8979

Leubner, C. & Ritsch, H. (1986a). Accurate emission spectra from planar strong field wigglers with arbitrary field variation. *Nuclear Instruments and Methods in Physics Research A*, Vol.246, Nos.1-3, (May 1986), pp. 45-49, ISSN 0168-9002

Leubner, C. & Ritsch, H. (1986b). A note on the uniform asymptotic expansion of integrals with coalescing endpoint and saddle points. *Journal of Physics A: Mathematical and General*, Vol.19, No.3, (February 1986), pp. 329-335, ISSN 0305-4470

Lumpkin, A.; Yang, B.; Chung, Y.; Dejus, R.; Voykov G. & Dattoli G. (1995). Preliminary calculations on the determination of APS particle-beam parameters based on undulator radiation. *Proceedings of the Sixteenth Particle Accelerator Conference (PAC'95)*, pp. 2598-2600, ISBN 0780329376, Dallas, Texas, US, May 1-5, 1995

Nikitin, M.M.; Medvedev, A.F. & Moiseev, M.B. (1981). Synchrotron radiation from ends of straight-linear interval. *IEEE Transactions on Nuclear Science*, Vol.NS-28, No.3, (June 1981), pp. 3130-3132, ISSN 0018-9499

Nikitin, M.M.; Medvedev, A.F.; Moiseev, M.B. & Epp, V.Ya. (1980). Interference of synchrotron radiation. *Soviet Physics – JETP*, Vol.52, No.3, (September 1980), pp. 388-394, ISSN 0038-5646

Nikitin, M.M. & Epp, V.Ya. (1976). Effect of beam parameters on undulator radiation. *Soviet Physics: Technical Physics*, Vol.21, No.11, (November 1976), pp.1404-1407, ISSN 0038-5662

Rarback, H.; Jacobsen, C.; Kirz, J. & McNulty, I. (1988). The performance of the NSLS mini-undulator. *Nuclear Instruments and Methods in Physics Research A*, Vol.266, Nos.1-3, (April 1988), pp. 96-105, ISSN 0168-9002

Roy, P.; Guidi Cestelli, M.; Nucara A.; Marcouille, O.; Calvani, P.; Giura, P.; Paolone, A.; Mathis, Y.-L. & Gerschel A. (2000). Spectral distribution of infrared synchrotron radiation by an insertion device and its edges: A comparison between experimental and simulated spectra. *Physical Review Letters*, Vol.84, No.3 (January 2000), pp. 483-486, ISSN 1079-7114

Sanchez del Rio, M. & Dejus, R.J. (1998). XOP: recent developments. *Proceedings of SPIE*, Vol. 3448, (December 1998), pp. 340-345, ISSN 0277-786X

Shirasawa, K.; Smolyakov, N.V.; Hiraya, A. & Muneyoshi, T. (2003). Edge radiation as IR-VUV source. *Nuclear Instruments and Methods in Physics Research B*, Vol.199, (January 2003), pp. 526-530, ISSN 0168-583X

Smolyakov, N.V. (1985). Interference accompanying radiation from protons in the fringe field of dipole magnets in a synchrotron. *Soviet Physics: Technical Physics*, Vol.30, No.3, (March 1985), pp.291-295, ISSN 0038-5662

Smolyakov, N.V. (1986). Electromagnetic radiation from protons in the edge fields of synchrotron dipole magnets. *Soviet Physics: Technical Physics*, Vol.31, No.7, (July 1986), pp.741-744, ISSN 0038-5662

Smolyakov, N.V. (1998). Wave-optical properties of synchrotron radiation. *Nuclear Instruments and Methods in Physics Research A*, Vol. 405, Nos.2-3, (March 1998), pp. 235-238, ISSN 0168-9002

Smolyakov, N.V. (2001). Shift-scale invariance based computer code for wiggler radiation simulation. *Nuclear Instruments and Methods in Physics Research A*, Vol. 467-468, No.1, (July 2001), pp. 210-212, ISSN 0168-9002

Smolyakov, N.V. & Hiraya, A. (2005). Study of edge radiation at HiSOR storage ring. *Nuclear Instruments and Methods in Physics Research A*, Vol.543, No.1, (May 2005), pp. 51-54, ISSN 0168-9002

Steinmuller-Nethl, D.; Steinmuller, D. & Leubner, C. (1989). Spectral and angular distribution of radiation from advanced wigglers with sharply peaked magnetic fields. *Nuclear Instruments and Methods in Physics Research A*, Vol.285, Nos.1-2, (December 1989), pp. 307-312, ISSN 0168-9002

Tanaka, T. & Kitamura, H. (1995). Figure-8 undulator as an insertion device with linear polarization and low on-axis power density. *Nuclear Instruments and Methods in Physics Research A*, Vol.364, No.2, (October 1995), pp. 368-373, ISSN 0168-9002

Tanaka, T. & Kitamura, H. (2001). SPECTRA: a synchrotron radiation calculating code. *Journal of Synchrotron Radiation*, Vol.8, No.6 (November 2001), pp. 1221-1228, ISSN 0909-0495

Tatchyn, R.; Cox, A. D. & Qadri S. (1986). Undulator spectra: Computer simulations and modeling. *Proceedings of SPIE*, Vol. 582, (January 1986), pp. 47-65, ISBN 0-89252-617-3

Walker, R.P. (1988). Near field effects in off-axis undulator radiation. *Nuclear Instruments and Methods in Physics Research A*, Vol.267, Nos.2-3, (May 1988), pp. 537-546, ISSN 0168-9002

Walker, R.P. (1989). Calculation of undulator radiation spectral and angular distributions. *Review of Scientific Instruments*, Vol.60, No.7, (July 1989), pp. 1816-1819, ISSN 0034-6748

Walker, R.P. (1993). Interference effects in undulator and wiggler radiation sources. *Nuclear Instruments and Methods in Physics Research A*, Vol.335, Nos.1-2, (October 1993), pp. 328-337, ISSN 0168-9002

Walker, R.P. & Diviacco, B. (1992). URGENT – A computer program for calculating undulator radiation spectral, angular, polarization, and power density properties. *Review of Scientific Instruments*, Vol.63, No.1, (January 1992), pp. 392-395, ISSN 0034-6748

Wang, C. (1993). Monte Carlo calculation of multi-electron effects on synchrotron radiation. *Proceedings of SPIE*, Vol. 2013, (November 1993), pp. 126-137, ISBN 9780819412621

Wang, C. & Jin, Y. (1992) Undulator properties calculated with RADID. *Proceedings of the International Conference on Synchrotron Radiation Sources*, pp. 264-269, Indore, India, February 3-6, 1992

Wang, C.; Schlueter, R.; Hoyer E. & Heimann, P. (1994). Design of the advanced light source elliptical wiggler. *Nuclear Instruments and Methods in Physics Research A*, Vol.347, Nos.1-3, (August 1994), pp. 67-72, ISSN 0168-9002

Wang, C. & Xian, D. (1990). Radid: A software for insertion device radiation calculation. *Nuclear Instruments and Methods in Physics Research A*, Vol.288, Nos.2-3, (March 1990), pp. 649-658, ISSN 0168-9002

Wang, C. & Xiao, Y. (1992). On algorithms for undulator radiation calculation. *Proceedings of the International Conference on Synchrotron Radiation Sources*, pp. 270-277, Indore, India, February 3-6, 1992

Wang, Ch.; Molter, K.; Bahrdt, J.; Gaupp, A.; Peatman, W.B.; Ulm, G. & Wende, B. (1992). Calculation of undulator radiation from measured magnetic fields and comparison with measured spectra. *Proceedings of the Third European Particle Accelerator Conference (EPAC'92)*, Vol.2, pp. 928-930, ISBN 2-86332-115-3, Berlin, Germany, March 24-28, 1992

Yagi, K.; Yuri, M.; Sugiyama, S. & Onuki, H. (1995). Computer simulation study of undulator radiation. *Review of Scientific Instruments*, Vol.66, No.2, (February 1995), pp. 1993-1995, ISSN 0034-6748

Yamamoto, S. & Kitamura, H. (1987). Generation of quasi-circularly polarized undulator radiation with higher harmonics. *Japanese Journal of Applied Physics*, Vol.26, No.10, (October 1987), pp. L1613-L1615, ISSN 0021-4922

8

Photoemission from Metal Nanoparticles

Igor Protsenko and Alexander Uskov
Lebedev Physical Institute, Plasmonics ltd.
Russia

1. Introduction

It is well-known that oscillations of electron density of metal nanoparticles have resonant frequency in the visible or in the near IR spectral region: this is the localized plasmon resonance (LPR). New and rapidly developed branch of modern physics called "Nanoplasmonics", is dedicated to investigations of optical and electro-physical phenomena related with the excitation of LPR and others similar resonances in metal nanoparticles and nanostructures. Appearance of LPR is caused by the charge on the surface of nanoparticles (Brongersma & Kik, 2007; Klimov, 2009; Maier, 2007; Novotny & Hecht, 2006): the surface of the nanoparticle builds "potential well" for oscillations of the electron density of the metal. For small particles (which typical size is less than or about 40 nm) the frequency of LPR depends weakly on the size of nanoparticles and strongly on their shape, metal and external environment. LPR can be excited by the external electromagnetic field (EMF). Energy of LPR is stored in oscillations of the electron density of nanoparticles and in the EMF induced by oscillations of the electron density. When LPR is excited, the density of energy of EMF inside the nanoparticle and near it, on the distance of the order or less than the wavelength of EMF, is approximately Q times greater (Wang, 2006) than the energy density of the external FMF. Here Q is the quality factor of LPR, Q depends on losses due to absorption of EMF by the metal of nanoparticle and on the radiation of EMF from the nanoparticle to the environment. Usually Q≤10 in experiments (Brongersma & Kik, 2007; Hövel et al., 1993; Klimov, 2009; Maier, 2007; Novotny & Hecht, 2006; Schuller et al, 2010), though some theoretical estimations predict the maximum value of Q about several tens (Khlebtsov, 2008). With the excitation of LPR nanoparticle behaves as a "nano-cavity" for EMF. However in a difference from the usual cavity, as Fabri-Perot cavity, a *near-field* presents in the "nano-cavity". Near field is bounded with charges, near field zone is located on the distance $\leq \lambda$, where λ is the wavelength of EMF. Resonant (or „plasmonic" (Brongersma & Kik, 2007; Khlebtsov, 2008; Klimov, 2009; Maier, 2007; Novotny & Hecht, 2006)) properties of metal nanoparticles and the concentration of EMF near and inside particles are reasons for prediction and experimental observation a number of new phenomena (Brongersma & Kik, 2007; Khlebtsov, 2008; Klimov, 2009; Kneipp et al., 2006; Maier, 2007; Novotny & Hecht, 2006) as, for example, giant Relay scattering (Kneipp et al., 2006). New optoelectronic devices with plasmonic nanoparticles were designed and created, as sensors (Brongersma & Kik, 2007; Homola, 1999; Klimov, 2009; Kneipp et al., 2006; Maier, 2007; Novotny & Hecht, 2006), nano-scaled lasers (Bergman & Stockman, 2003; Noginov et al. 2009; Oulton et al., 2009; Protsenko et al., 2005), high efficient solar cells (Atwater & Polman, 2009; Catchpole & Polman, 2008; Pors, 2011).

Nanoparticles in optical devices can be considered as „nano-antennas" (Brongersma & Kik, 2007; Greffet, 2005; Klimov, 2009; Maier, 2007; Mühlschlegel et al., 2005; Novotny & Hecht, 2006). It is important and interesting to investigate also electrophysical properties of contacts of plasmonic nanoparticles with environment (for example, a surface) surrounding (supporting) nanoparticles. Transport of carriers (electrons or holes) or transport of the energy from the nanoparticle to the environment and back may cause noticerable influence to properties of the device with nanoparticles. Such influence may be positive or negative up to complete destruction of LPR and related phenomena. For example the effect of increase of photo-luminescence due to addition of metal nano-particles in luminescing media (as solutions of dye moleculas) is well-known. However if the distance between the nano-particle and the luminescing obgect is small (less than few nm) the luminescence can be fully dumped due to non-radiated recombination (Brongersma & Kik, 2007; Klimov, 2003, 2009; Maier, 2007; Novotny & Hecht, 2006).

Various constructions of "plasmonic" solar cells have been suggested. Both optical and electro-physical properties of metal nanoparticles are used in such solar cells, for example, for commutation (electrical connection) of cascades of the solar cell through metal nanoparticles using also „plasmonic" concentration of the light arround nanoparticles (Rand et al., 2006). Solar cell efficiency can be increased due to injection of carriers, photo-induced in metal nanoparticles, to the semiconductor substrate, while LPR is excited in nanoparticles (Westphalen et al., 2006). However the transport of carriers through the contacts of metal nanoparticles with the substrate are less studied than optical properties of metal nanoparticles. It is related with complexity of problems arisen in theoretical and experimental studies of electro-physical properties of nano-scaled objects. Investigation of the carrier transport and other electrophysical properties of junctions between substrates and metal nanoparticles must be carried out together with investigation of optical properties of nanoparticles. This is necessary for modeling plasmonic optoelectronic devices as solar cells (Monestier et al., 2007), photodetectors, nano-scaled LEDs and nano-lasers. This work gives an example of such „joint" study of optical and electro-physical phenomena in plasmonics.

An example of phenomena appeared at the interface between the nanoparticle and the environment is a photoemission from the nanoparticle studied here theoretically. The photoemission from the nanoparticle may be quite different from the photoemission from "large" (respectively to the wavelength of EMF) metal structures (as, for example, continues metal films widely used in photodetectors). One can note three major differences. *First*, the EMF inside the nanoparticle and near it is enhanced at the excitation of LPR. *Second*, the ratio of the area of surface of the nanoparticle to the volume of the nanoparticle is greater than for large "bulk" metal structures. This is important because of the "surface" photo-effect is more important for the photoemission than the "bulk" photo-effect (Brodsky, 1973). The "surface" photo-effect is occurred when the photon is absorbed at the collision of the electron with the surface, while the "bulk" photo-effect takes place when the photon is absorbed inside the metal at the collision of the electron with the lattice (i.e. with the phonon) or with another electron. Usually the "bulk" photo-effect is not taken into consideration at calculations of photo-emission as, for example, in the case of photo-emission from metal films (Brodsky, 1973). *Third*, in order to provide the surface photoemission the electric field must be non-parallel (perpendicular, at best) to the surface of the metal. This condition can be easily satisfied for nanoparticles than for continues metal films. Thus for metal nano-particles one can expect to obtain larger number of photo-electrons per unit of mass that it is obtained for bulk metal structures as metal films in well-known photo-detectors (She, 1981; Soole & Schumacher, 1991). Increase of the efficiency of photo-detectors by metal nano-particles helps, in particular, to increase the sensitivity of photo-detectors in the near and far IR spectral regions, which is important practical problem (Piotrowski et al., 1990; Yu et al., 2006).

For the study of photoemission from nano-particles and, in particular, for determination of conditions of the maximum of photoemission one has to know the cross-section of the photoemission from the nanoparticle, which is the main subject of calculations presented below. We present results for the case of the surface photo-emission from metal nano-particle at the excitation of LPR and show, as an example, that the photo-emission from metal nano-particles into p-doped Si is much more efficient that the photo-emission from the bulk metal film. It is worth to note that micro- and nano-structures on surfaces of metal photo-detectors is widely used for enhancement of the photo-emission, in particular, in photo-cathodes based on $A^{III}B^V$ semiconductors appeared at the beginning of 70-th years. It was shown in several publications (see, for example, the bibliography in (Schelev, 2000)) that the photo-emitting metal film of such photo-cathodes has dispersed structure. It is also well-known that all such photo-cathodes have larger photo-current respectively to photo-cathodes with photo-emission from flat metal layers. However LPR can be hardly excited in "dispersed" metal structures on surfaces of such photo-cathodes. These structures have good electric contact with surfaces, but LPR can be excited in metal nanoparticles electrically isolated from the environment. Thus the increase of the photo-emission due to the "concentration" of EMF at nano-structured surfaces of well-known photo-cathodes is hardly possible. Larger photoemission current of well-known photo-cathodes with dispersed surfaces was obtained due to lager surface, available for the photoemission. If LPR can be excited in nano-structures at surfaces of photo-cathodes, it will lead to more increase in the photoemission; this is why it is interesting and important to study photo-cathodes with LPR excited in nanoparticles. It was already suggested to use plasmonic properties of nano-particles for increase of the efficiency of photo-detectors (Hetterich, 2007), however without consideration of photoemission from nano-particles. Increase of the efficiency of photo-emission to vacuum from nano-particles has been observed experimentally (Nolle, 2004, 2005), it was suggested to use this effect for increase the efficiency of photo-detectors.

Here we carry out the analysis following by the well-known theory of photoemission (Brodsky, 1973). In particular we will find general expression for the probability amplitude and calculate the cross-section of the photoemission from metal nano-particle. Calculation shows the two order of magnitude increase of the photoemission current from gold nanoparticles to p-doped Si in comparison with the photocurrent from continues film of Au. We generalize the result of (Brodsky, 1973) for the probability of photoemission by taking into account changes (jumps) of EMF and electron mass at the surface of the nanoparticle. Taking into consideration these surface phenomena we see that the cross-section of photo-emission is changed (increased) several times with the respect to the case, when these "jumps" on the surface are neglected. Careful consideration of surface phenomena is important particularly for the photoemission from nanoparticles because of their large surface to volume ratio.

The probability and the cross-section of the photoemission found here let us to calculate the photocurrent from the layer of metal nanoparticles. We use analytical approach following details of physics of the increase of the photoemission from nanoparticles at LPR. It is shown, for example, that the increase of the cross-section of the photoemission is related not only with "concentration" of EMF but also with changes of EMF and free electron mass on the surface of nano-particle. It turns to be that quantum-mechanical interference phenomena in the dynamics of electron passing through the interface became important at the photoemission. Analytical approach lets us to separate three contributions in the photoemission: (1) the increase of the probability of photoemission, (2) the increase of EMF due to the excitation of LPR and (3) the factor of the shape of the surface of the nanoparticle. In principle, one can control photo-emission by changing these three factors. For example, by changing the shape of the nanoparticle one can shift the maximum of the photocurrent spectrum. However detailed analysis of possibilities of controlling the photoemission is out of the scope of this paper. Analytical approach lets us to take into consideration many peculiarities of optical "plasmonic" phenomena, for example, dynamical depolarization and radiation dumping phenomena. Analytical results can be used as test models for verification of numerical results found for more complex cases of photoemission as, for example, non-dipole approach at the interaction of nanoparticles with EMF.

Here we discuss the photoemission to vacuum or to the homogeneous medium surrounding nanoparticle. However the same approach may be applied for the analysis of other more complicated cases of the carrier transport at presence of nanoparticles. For example the nanoparticle may catch the carrier from the environment; the photoemission may occur as the tunnelling through the potential barrier (Nolle, 2007) also when the nano-particle stays near the surface of the semiconductor and separated from it by the tunnel layer. The last structure is typical for problems related with the increase of the efficiency of solar cells by meal nanoparticles (Atwater & Polman, 2009; Catchpole & Polman, 2008; Pors, 2011).

General expression for the probability amplitude of the photoemission with taking into account changes in the EMF and the mass of electron at the nanoparticle-environment interface is derived in the next subsection. Explicit expression for the probability amplitude is found for the step potential at the metal-environment interface. This result is used for the calculation of the cross-section of photoemission in the following sub-section. Example of the photoemission from spherical gold nano-particles into p-doped Si is considered after that. Results are summarised and directions for future studies of the photoemission from nano-particles are discussed in conclusion.

In order to calculate the probability of the photoemission from the metal we, following (Brodsky, 1973), use standard quantum-mechanic perturbation theory (Landau & Lifshitz, 1997) where the perturbation is the interaction of the electron with classical electromagnetic field at the metal-environment interface. We neglect by the curvature of the surface of the metal with the respect to de Broglie wavelength of the electron, i.e. we consider metal-environment interface as a flat surface. The key point is the calculation of "unperturbed" wave functions of the electron in order to insert them into the ready expression for the probability amplitude found from the perturbation theory. We found wave functions analytically supposing a step potential barrier between the metal and the environment. This lets us to obtain final expression for the cross-section of the photoemission "in quadratures" i.e. as an explicit expression containing the integral. The electromagnetic field causing the photoemission appears as a multiplier in this expression. Then we calculate the electromagnetic field in the nanoparticle and in the vicinity of it using the approach of classical electrodynamics well-known in plasmonics (Klimov, 2009).

2. Probability amplitude of the photoemission accounting "jumps" on the interface

2.1 General expression for probability amplitude of photoemission

Expression for the probability amplitude $C_+(\infty)$ of the photoemission of the electron was found in (Brodsky, 1973) by the perturbation theory. Electron moves in the medium (in the metal) along axes z perpendicular to the interface with the external environment. The electron interacts with the EMF of frequency ω,

$$C_+(\infty) = \frac{|e|m}{\hbar\omega W_1}\int_{-\infty}^{\infty}dz\left(E_m\frac{d\Psi_0}{dz}\Psi_{1-} + \frac{1}{2}\Psi_{1-}\Psi_0\frac{dE_m}{dz}\right) \qquad (1)$$

Here e is the (negative) charge of the electron, m is the mass of the electron. The interface of the medium with the environment is described by the one dimentional potential barrier $V(z)$, see Fig.1. Effective mass of the electron is changed on the interface, so that $m = m(z)$;

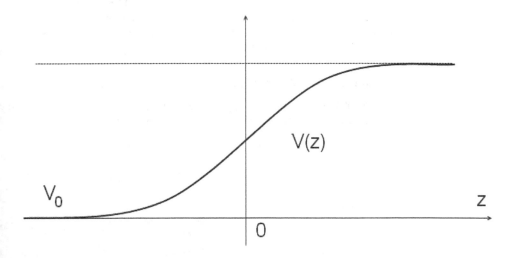

Fig. 1. Potential barrier where the electron moves.

Ψ_0, $\Psi_{1\pm}$ are wave functions of the electron in states with the energy E_i, $i = 0,1$ below and above the barrier respectively, $E_m = E / m(z)$, E is the amplitude of the component of the electric field polarized along axes z,

$$W(\Psi_{1-}\Psi_{1+}) = \Psi_{1-}\frac{d\Psi_{1+}}{dz} - \Psi_{1+}\frac{d\Psi_{1-}}{dz}$$

Final expression for the probability of the photoemission was obtained in (Brodsky, 1973) at the assumption that electric field E is constant along axes z. However this is not true at the interface where the normal component of E is changed due to the surface charge, so that one has to consider $E = E(z)$. Using the approach of (Brodsky, 1973) with $E = E(z)$ and $m = m(z)$, one can replace Eq. (1) by

$$C_+(\infty) = \frac{|e|m}{W_1(\hbar\omega)^2} \int_{-\infty}^{\infty} \frac{dz}{m}(c_V + c_E + c_m),$$ (2)

where c_V, c_E and c_m describes the photoemission taking into account jumps in the potential, in the electric field and in the effective mass of the electron in the interface, respectively,

$$c_V = -EV'\Psi_0\Psi_{1-}, \quad c_E = E'\left[\frac{\hbar^2}{2m}\Psi'_0\Psi''_{1-} + \left(E_0 - V + \frac{\hbar\omega}{2}\right)\Psi_0\Psi_{1-}\right],$$ (3)

$$c_m = -\frac{Em'}{m}\left(E_0 - V + \frac{\hbar\omega}{2}\right)\Psi_0\Psi_{1-}$$

Eq.(2) is integrated below for the case of step functions $V(z)$, $E(z)$ and $m(z)$ with the step at $z = 0$. For such functions $V' = V\delta(z)$, $E' = (E_+ - E_-)\delta(z)$, $m' = (m_0 - m)\delta(z)$, where $E_\pm$, are values of E to the right and to the left regions from z=0 respectively; m and m_0 are effective electron masses in the metal and outside it, respectively. Quantities $V(0) = V/2$, $E(0) = (E_+ + E_-)/2$, $m(0) = (m_0 + m)/2$, $\Psi' \equiv d\Psi/dz$.

2.2 Wave functions at the absence of perturbation

Let us find wave functions of the electron moving perpendicular to the interface between the metal and the environment, the interface is described by the step potential barrier, see Fig.2.

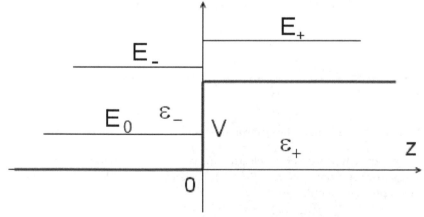

Fig. 2. Step potential barrier. The break in the electric field at z=0 is shown where the dielectric function ε_- of the metal is changed to the dielectric function ε_+ of the environment.

Electron of the charge e and the effective mass m at z<0 and m_0 at z>0 moves in the step potential

$$V(z) = \begin{cases} 0 & z < 0 \\ V/2 & z = 0 \\ V & z > 0 \end{cases} \tag{4}$$

Hamiltonian of the electron is $H = H_-$ at z<0 and $H = H_+$ at z>0,

$$H_- = -\frac{\hbar^2}{2m}\frac{d^2}{dz^2}, \quad H_+ = -\frac{\hbar^2}{2m_0}\frac{d^2}{dz^2} + V \tag{5}$$

We solve Schrödinger equation $i\hbar(\partial\bar{\Psi}_{0+}/\partial t) = H_+\bar{\Psi}_{0+}$ and find the wave function $\bar{\Psi}_{0+} = \Psi_0 \exp[-i(E_0/\hbar)t]$ of the state of the electron with total energy $E_0 < V$, this electron falls on the barrier and is reflected from it,

$$\Psi_0 = [\exp(ik_0 z) + A_0\exp(-ik_0 z)]_{z<0} + [B_0\exp(i\tilde{k}_0 z)]_{z>0}, \tag{6}$$

here wave numbers

$$k_0 = \frac{1}{\hbar}(2mE_0)^{1/2}, \quad \tilde{k}_0 = \frac{1}{\hbar}[2m_0(E_0 - V)]^{1/2} \tag{7}$$

Because $V > E_0$, the value $(E_0 - V)^{1/2} = i(V - E_0)^{1/2}$ is purely imaginary. The wave function Ψ_0 is normalized such that the coefficient in front of the term $\exp(ik_0 z)$ describing initial electron state at $z = -\infty$ is 1. Coefficients A_0 and B_0 in Eq.(6) are determined from conditions of regularity at z = 0

$$\Psi_0(z = -0) = \Psi_0(z = +0), \quad m^{-1}(\partial\Psi_0/\partial z)_{z=-0} = m_0^{-1}(\partial\Psi_0/\partial z)_{z=+0}, \tag{8}$$

Eqs.(8) are equivalent to $1 + A_0 = B_0$ and $1 - A_0 = \theta_0 B_0$, so that

$$A_0 = \frac{1-\theta_0}{1+\theta_0}, \quad B_0 = \frac{2}{1+\theta_0}, \quad \theta_0 = [(m/m_0)(1 - V/E_0)]^{1/2} \tag{9}$$

Similar way we find wave functions $\bar{\Psi}_{1\pm} = \Psi_{1\pm}\exp[-i(E_1/\hbar)t]$ of the electron state with the energy $E_1 > V$, where

$$\Psi_{1+} = [A_{1+}\exp(ik_1 z) + B_{1+}\exp(-ik_1 z)]_{z<0} + \exp(i\tilde{k}_1 z)_{z>0}, \tag{10}$$

$$\Psi_{1-} = [A_{1-}\exp(i\tilde{k}_1 z) + B_{1-}\exp(-i\tilde{k}_1 z)]_{z>0} + \exp(-ik_1 z)_{z<0}, \tag{11}$$

and real wave numbers

$$k_1 = \frac{1}{\hbar}(2mE_1)^{1/2}, \quad \tilde{k}_1 = \frac{1}{\hbar}[2m_0(E_1 - V)]^{1/2} \tag{12}$$

Coefficients $A_{1\pm}$ and $B_{1\pm}$ are determined from conditions of regularity of the wave function at $z = 0$, analogous to Eq. (8), which leads to equations $A_{1\pm} + B_{1\pm} = 1$, $A_{1+} - B_{1+} = \theta_1$, and $\theta_1(B_{1-} - A_{1-}) = 1$, where $\theta_1 = [(m / m_0)(1 - V / E_1)]^{1/2}$. One can find that

$$A_{1+} = (1 + \theta_1) / 2, \quad B_{1+} = (1 - \theta_1) / 2, \quad A_{1-} = (\theta_1 - 1) / 2\theta_1, \quad B_{1-} = (1 + \theta_1) / 2\theta_1 \qquad (13)$$

Wave functions Eqs. (10) and (11) make the fundamental set of solutions of Schrödinger equation with Hamiltonian (5).

2.3 Expression for probability amplitude of photoemission

We follow the procedure of (Brodsky, 1973) and consider the monochromatic EMF of frequency ω as a perturbation

$$\hat{U}_{pert}(z, d / dz)\cos(\omega t) = (1 / 2)\hat{U}_{pert}(z, d / dz)(e^{-i\omega t} + e^{i\omega t})$$

for the motion of the electron in Hamiltonian Eq. (5). If we apply operator $\hat{U}_{pert}$ to the wave function Ψ_i, $i = 0, 1\pm$ of the state of the electron we obtain

$$\hat{U}_{pert}\Psi_i = \frac{i\hbar e}{2c}\left[\frac{d}{dz}\left(\frac{A}{m}\Psi_i\right) + \frac{A}{m}\frac{d\Psi_i}{dz}\right] + e\varphi\Psi_i,$$

Where e is (negative) electron charge, c is the speed of light in vacuum, A is z-component of vector potential of EMF in the medium, φ is scalar potential of EMF. Calculations with wave functions (10) and (11) lead to expression

$$\left.\frac{W_1}{m}\right|_{z<0} = \left.\frac{W_1}{m_0}\right|_{z>0} = i\frac{k_1}{m}(1 + \theta_1) \equiv i\left(\frac{k_1}{m} + \frac{\tilde{k}_1}{m_0}\right) \qquad (14)$$

Taking $z=0$ and that Ψ'_0, Ψ'_{1-} have discontinuity so that $\Psi'_{0,1-}(0) = (1 / 2)[\Psi'_{0,1-}(-0) + \Psi'_{0,1-}(+0)]$ we find

$$\Psi_0\Psi_{1-} = \frac{2}{1 + \theta_0}, \quad \Psi'_0\Psi'_{1-} = \frac{2\tilde{k}_0 k_1 \bar{m}^2}{m_0 m(1 + \theta_0)}, \qquad (15)$$

Here and below $\bar{m} = (m_0 + m) / 2$, $\Delta m = m_0 - m$, $\bar{E} = (E_+ + E_-) / 2$, $\Delta E = E_+ - E_-$. Inserting Eqs.(14) and (15) into Eq.(2) and proceeding some calculations we find explicit expression for the probability amplitude of the photoemission

$$C_+(\infty) = \frac{2|e|m}{ik_1\bar{m}(\hbar\omega)^2(1 + \theta_0)(1 + \theta_1)}\left\{-V\bar{E} + \frac{\Delta E}{2}\left[E_1^{1/2} + (E_0 - V)^{1/2}\right]^2 - \frac{\bar{E}\Delta m}{2\bar{m}}(E_0 + E_1 - V)\right\} \qquad (16)$$

Now we express Eq.(16) through the variable $x = (\hbar k_0)^2 / (2mV) \equiv E_0 / V$. Taking into account that $k_1 = (\sqrt{2mV} / \hbar)(x + \hbar\omega / V)^{1/2}$, $\theta_0 = [(m / m_0)(1 / x - 1)]^{1/2}$, $\theta_1 = \{(m / m_0)[1 - (x + \hbar\omega / V)^{-1}]\}^{1/2}$ we find final expression for the probability of the photoemission

$$|C_+(\infty)|^2 = C_0 U(x) |K_{dis}(x)|^2, \tag{17}$$

where the dimensionless coefficient

$$C_0 = \frac{2|e|^2|E_-|^2 V}{m\hbar^2\omega^4}, \tag{18}$$

$$U(x) = \frac{4r_m^2}{(r_m + 1)^2} \times \frac{x}{[x + r_m(1-x)]\{(x + \hbar\omega / V)^{1/2} + [r_m(x + \hbar\omega / V - 1)]^{1/2}\}^2}, \tag{19}$$

$r_m = m / m_0$ and

$$K_{dis}(x) = \frac{1}{2}\left(1 + \frac{\varepsilon_-}{\varepsilon_+}\right)\left[1 + \frac{1 - r_m}{1 + r_m}\left(2x + \frac{\hbar\omega}{V} - 1\right)\right] + \frac{1}{2}\left(1 - \frac{\varepsilon_-}{\varepsilon_+}\right)\left[\left(x + \frac{\hbar\omega}{V}\right)^{1/2} + i(1-x)^{1/2}\right]^2 \tag{20}$$

Factor K_{dis} describes both the influence of the break of EMF and the break of the effective electron mass at the interface at z = 0; however Eq.(19) for U(x) also depends on the effective electron mass. In order to came back to the case of (Brodsky, 1973), where jumps of EMF and effective mass at the interface did not taken into account, one can set $r_m = 1$ and $\varepsilon_- = \varepsilon_+$ in Eqs., (19), (20) which leads to $K_{dis}(x) = 1$ and

$$U(x) = \frac{x}{\left[(x + \hbar\omega / V)^{1/2} + (x + \hbar\omega / V - 1)^{1/2}\right]^2}$$

Thus the probability of the photoemission of the electron in the case of the step potential in the interface and with account for the jumps of the electron effective mass and EMF in the interface is determined by Eq. (17), where C_0 is given by Eq. (18) and expressions for U(x) and $K_{dis}(x)$ are determined by Eqs. (19) and (20), respectively. These expressions will be used below for calculation of the cross-section of photoemission from the nano-particle.

3. Cross-section of photoemission from nano-particle for step potential.

3.1 Expression for cross-section of photoemission

Cross-section $\sigma_{ph\text{-}em}$ of photoemission from the nano-particle is, by definition,

$$\sigma_{ph\text{-}em} = \frac{J_{ph\text{-}em}}{I}, \tag{21}$$

where $J_{ph\text{-}em}$ is the total photocurrent from the nanoparticle in electrons per second, I is the intensity of the external monochromatic EMF causing photoemission in photons through cm^2 per second. Photocurrent from the nanoparticle is

$$J_{ph\text{-}em} = \int_{surface} j ds = \int_{surface} j(\theta, \varphi, r) r dr \sin\theta d\theta d\varphi, \tag{22}$$

where j is the photocurrent density in electrons through cm^2 per sec, the photocurrent is normal to the surface of the nanoparticle in the point of the surface determined by polar angle θ azimuth angle φ and by the distance r from zero of coordinate system to the surface of the nanoparticle, see Fig.3.

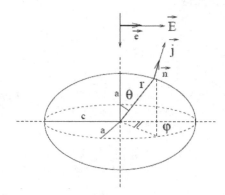

Fig. 3. The density j of the photocurrent in the point of the surface determined by angles θ and φ for spheroid nanoparticle excited by the external electric field of the amplitude E. The length of two semi-axes of the particle are the same, the length of the third semi-axes is c.

Now we came from the one-dimensional model of motion of electron to the three-dimensional model. Then in accordance with Eq. (2.30) of (Brodsky, 1973) we can write the photocurrent density dj of electrons with the energy in the interval $E_0 \div E_0 + dE_0$ as

$$dj = \frac{\hbar \tilde{k}_{1z}}{m} |C_+|^2 \, \Theta[k_{0z}^2 + (2m / \hbar^2)(\hbar\omega - V)]dn_0, \qquad (23)$$

where $\hbar \tilde{k}_{1z} / m$ is the speed of such electrons above the interface barrier, $dn_0 \equiv 2f_F(k_0)dk_{0x}dk_{0y}dk_{0z} / (2\pi)^3$ is the number of such electrons,

$$f_F(k_0) = [1 + \exp\{[(\hbar k_0)^2 / (2m) - \varepsilon_F] / k_B T\}]^{-1}$$

is Fermi distribution function, k_0 is the wave vector of the electron before absorption of the photon, $k_0^2 = k_{0x}^2 + k_{0y}^2 + k_{0z}^2$, $k_{0x,y}$ – components of the wave vector parallel to the surface of the nanoparticle, ε_F is Fermi energy of the metal of the nanoparticle k_B is Boltzmann constant, T is the temperature. Because of $2mE_0 / \hbar^2 = k_0^2$ then for the step potential

$$\tilde{k}_{1z} = \sqrt{(2m / \hbar^2)(E_0 + \hbar\omega - V) - (k_{0x}^2 + k_{0y}^2)} \equiv \sqrt{k_{0z}^2 + (2m / \hbar^2)(\hbar\omega - V)} \qquad (24)$$

is z-component of the wave vector of the electron above the barrier (it was denoted as $\tilde{k}_1$ in one-dimentional case), k_{0z} is the component of the wave vector of the electron inside the nanoparticle perpendicular to the interface, $|C_+|^2 \equiv |C_+(\infty)|^2$ is the probability of photoemission and Θ is theta-function. The density j of the photocurrent of electrons of all energies is

$$j = \frac{2}{(2\pi)^3} \int dk_{0x}dk_{0y}dk_{0z}f_F(k_0)\frac{\hbar \tilde{k}_{1z}}{m}|C_+|^2 \Theta[k_{0z}^2 + (2m/\hbar^2)(\hbar\omega - V)] \qquad (25)$$

Taking into account that only f_F depends on $k_{0x,y}$ in Eq.(25) one can take there the integral over $dk_{0x}dk_{0y}$ with replacement $k_{0x}^2 + k_{0y}^2 = \tilde{n}^2$ and using $\int_0^\infty dx(1 + e^x / b)^{-1} = \ln(1 + b)$, so that

$$\int dk_{0x}dk_{0y} f_F(\vec{k}_0) = \pi \int_0^\infty \frac{d\rho^2}{1 + \exp\{[\hbar^2(\rho^2 + k_{0z}^2)/(2m) - \varepsilon_F]/k_BT\}} =$$

$$\frac{2\pi m k_B T}{\hbar^2} \ln\{1 + e^{[\varepsilon_F - \hbar^2 k_{0z}^2/(2m)]/(k_BT)}\} \qquad (26)$$

Inserting Eqs. (23), (24), (26) into Eq. (25) and taking into account that according with Eqs. (17), (18) $|C_+|^2 \sim |C_-|^2$, we obtain the density of the photoemission current in some point of the surface of the nanoparticle

$$j = C_{emission} |E_-|^2, \qquad (27)$$

$$C_{emission} = \frac{|e|^2 k_B T V^2}{\pi^2 \hbar^5 \omega^4} \int_{0,1-\hbar\omega/V}^1 dx [1 + (\hbar\omega / V - 1)/x]^{1/2} \ln\left(1 + e^{\frac{\varepsilon_F - Vx}{k_BT}}\right) U(x) |K_{dis}(x)|^2,$$

where the low limit of the integration is 0 if $\hbar\omega > V$ and it is $1 - \hbar\omega / V$ if $\hbar\omega < V$; x<1 due to $E_0 < V$; $U(x)$ and $K_{dis}(x)$ are determined by Eqs. (19) and (20), respectively. If we neglect by the thermal excitation of electrons above Fermi surface, i.e. take in Eq. (27) the limit $T \to 0$ then, using $e^{(\varepsilon_F - Vx)/(k_BT)} \to \infty$, we write instead of Eq. (27)

$$C_{emission} = \frac{|e|^2}{\pi^2} \frac{V^3}{\hbar^5 \omega^4} \int_{0,1-\hbar\omega/V}^{\varepsilon_F/V} dx [1 + (\hbar\omega / V - 1)/x]^{1/2} (\varepsilon_F / V - x) U(x) |K_{dis}(x)|^2 \qquad (28)$$

It must be here that $\hbar\omega > V - \varepsilon_F$ and the low limit of integration is 0 if $\hbar\omega > V$. Thus the photocurrent from the nanoparticle is

$$J_{ph-em} = C_{emission} \int_{surface} |E_-|^2 ds, \qquad (29)$$

where the integral is taken over the surface of the nanoparticle and $C_{emission}$ does not depend on the point on the surface; normal component of the field $E_- = (\vec{E}_{int} \tilde{n})$, where $\vec{E}_{int}$ is the field inside the particle, $\tilde{n}$ is the unit vector normal to the surface. Components of EMF tangential to the surface have no influence to the photoemission. In principle, the motion of the electron along of the surface of the nanoparticle depends on tangential components of EMF and, therefore, has the influence to the distribution function of electrons. However the EMF at the photoemission is relatively weak, so that such influence is negligibly small with the respect, for example, to the heating of the particle at the absorption of EMF. EMF inside the nanoparticle is related with external EMF $\vec{E}$ incident to the nanoparticle by the relation $\vec{E}_{int} = \hat{F}(\vec{r})\vec{E}$, where $\hat{F}(\vec{r})$ is tensor. Spheroidal nanoparticles considered below have homogeneous EMF inside them so for such particles $\hat{F}$ is constant and does not depend on $\vec{r}$. For simplicity we suppose that $\vec{E}$ is parallel to one of the main axes of the spheroidal particle then $\vec{E}_{int} = F\vec{E}$, where F does not depend on $\vec{r}$. For non-spherical particles F depends on which main axes of the particle is parallel to $\vec{E}$. Thus

$$J_{ph-em} = C_{emission} \, |F|^2 \, K_{geometry} \, |E|^2, \tag{30}$$

where $K_{geometry} = \int_{surface}(\vec{n}\vec{e})$, $\vec{e}$ is unit vector parallel to the polarization of the external field, see Fig.3. Taking into account Eq. (21) and the intensity I of external EMF (in photons through cm² per sec), which is

$$I = \frac{1}{8\pi} \frac{cn_+ \, |E|^2}{\hbar\omega},$$

we find the cross-section of the photoemission

$$\sigma_{ph-em} = \frac{8\pi\hbar\omega}{cn_+} C_{emission} \, |F|^2 \, K_{geometry}, \tag{31}$$

where c is the speed of light in vacuum, $n_+ = \mathrm{Re}\sqrt{\varepsilon_+}$ is the refractive index of the medium outside the nanoparticle.

3.2 Parameters F and K$_{geometry}$

According with (Meier & Wokaun, 1983)

$$F = \frac{1}{1 + R_{dep} - iR_{rad}} \times \frac{\varepsilon_+}{\varepsilon_+ + (\varepsilon_- - \varepsilon_+)L}, \tag{32}$$

for spheroidal particles, where the second multiplier is the result of calculations in quazistatic approach,

$$L = \frac{r^2}{2} \int_0^\infty \frac{du}{(u+r^2)^2(u+1)^{1/2}},$$

aspect ratio $r = a/c$, a is the length of one of two equal semi-axes of ellipsoidal particle, c is the length of the third semi-axes, see Fig.3; the second multiplier in Eq. (32) takes into account dynamic depolarization (factor R_{dep}) and radiative losses (factor R_{rad}) (Bottcher, 1952)

$$R_{dep} = \frac{\varepsilon_- - \varepsilon_+}{\varepsilon_+ + (\varepsilon_- - \varepsilon_+)L}(A\varepsilon_+ y^2 + B\varepsilon_+^2 y^4), \quad R_{rad} = \frac{16\pi^3}{9} \frac{n_+^3}{r}\left(\frac{a}{\lambda}\right)^3 \frac{\varepsilon_- - \varepsilon_+}{\varepsilon_+ + (\varepsilon_- - \varepsilon_+)L}, \tag{33}$$

$$A = -0.4865L - 1.046L^2 + 0.848L^3, \quad B = 0.01909L + 0.1999L^2 + 0.6077L^3,$$

where $y = \pi a/\lambda$, λ is the wavelength of EMF in vacuum. Factor R_{dep} is characterized the nonhomogeneity of EMF inside the particle. Collisions of electrons with the surface of the nanoparticle lead to the deviation of the dielectric function ε_- of metal of nanoparticle from the dielectric function ε_{bulk} of the macroscopic (bulk) piece of the same metal. This difference can be taken into account according with (Brongersma, 2007)

$$\varepsilon_- = \varepsilon_{bulk} + \frac{\omega_{pl}^2}{\omega^2 + i\omega\gamma_0} - \frac{\omega_{pl}^2}{\omega^2 + i\omega(\gamma_0 + iAv_F/a)}, \tag{34}$$

where ω_{pl} and γ_0, are plasma frequency and the dumping increment, respectively, v_F is the rate of electrons near Fermi surface; A is constant of the order of 1 depending on the shape of the particle. According with (Meier, 1983)

$$K_{geometry} = \frac{\pi a^2}{r}\left[\frac{r}{1-r^2} + \frac{1-2r^2}{(1-r^2)^{3/2}}\arcsin(1-r^2)^{1/2}\right]$$ (35)

Thus the cross-section of the photoemission is determined by expression (31) where coefficient $C_{emission}$ is determined by Eq. (27), factors F and $K_{geometry}$ are given by Eqs. (32) and (35), respectively. One can compare the cross-section of the photoemission with cross-sections σ_{abs} and σ_{sc} of absorption and scattering of the light by nanoparticle (Meier & Wokaun, 1983)

$$\sigma_{abs} = \frac{8\pi^2 a^3 n_+}{3r\lambda}\text{Im}\alpha, \quad \sigma_{sc} = \frac{128\pi^5 a^6 n_+^4}{27r^2}|\alpha|^2, \quad \alpha = \frac{1}{1+R_{dep}-iR_{rad}}\times\frac{\varepsilon_- - \varepsilon_+}{\varepsilon_+ + (\varepsilon_- - \varepsilon_+)L}$$ (36)

Numerical estimations of the cross-sections of the photoemission from metal nanoparticles will be carried out in the next Subsection.

4. Photoemission from gold nanoparticles into Si

Let us calculate the cross-section of photoemission from the spherical gold nanoparticle into p-doped Si. We chose Si as the environment of the nanoparticle because of the work function χ_e for electron coming from Au to p-type Si is small $\chi_e = 0.34$ eV (Dutta, 2009). Because of Fermi energy for Au $\varepsilon_F = 5.1$ eV (Dutta, 2009), the height of the barrier in Au – p-Si interface is $V = \varepsilon_F + \chi_e = 5.44$ eV. The electron effective mass in Au and in Si is, respectively $m = 0.992m_l \approx m_l$ and $m_0 = 0.25m_l$ (Kittel, 1996), where m_l is the mass of free electron in vacuum; so that $r_m = 0.992/0.25 = 3.968$. The data for the dielectric function ε_{Au} of Au are taken from (Weber, 2002). It is convenient to write the dielectric function (34) of Au as a function of the wavelength λ of EMF in vacuum:

$$\varepsilon_-(\lambda) = \varepsilon_{Au}(\lambda) + \left(\frac{\lambda}{\lambda_p}\right)^2\left[\frac{1}{1+i\lambda/\lambda_f} - \frac{1}{1+(i\lambda/\lambda_f)(a_c/a+1)}\right],$$ (37)

we take $\lambda_p = 0.142$ μm and $\lambda_f = 55$ μm, they correspond to the best approximation

$$\varepsilon_{Au}(\lambda) = 12 + \left(\frac{\lambda}{\lambda_p}\right)^2\frac{1}{1+i\lambda/\lambda_f}$$ (38)

in the region of λ from 0.6 to 1.2 μm where LPR of spherical Au nanoparticle in Si is located; $a_c = Av_f\lambda_f/(2\pi c_0)$ is the parameter characterizing the collision of the electron with the surface of the nanoparticle, $A = 0.7$, $v_F = (2E_F|e|/m_0)^{1/2} \approx 1.3\times10^6$ m/sec; it is supposed that effective electron mass in Au is equal to free electron mass. Fig.4 shows Re and Im parts of $\varepsilon_{Au}(\lambda)$, its approximation according with Eq.(38) is quite close to $\varepsilon_{Au}(\lambda)$ and $\varepsilon_-(\lambda)$ found from Eq.(37) for a = 10 nm. One can see that $\text{Im}[\varepsilon_-(\lambda)]$ is noticeably greater than

$\mathrm{Im}\left[\varepsilon_{Au}(\lambda)\right]$, which points out to the necessity of taking into account collisions of electrons with the surface of the nanoparticle. We can use approximation (Adachi, 2002) for the dielectric function $\varepsilon_+(\lambda)$ of Si

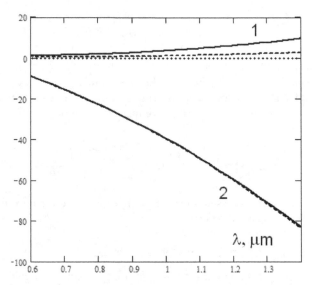

Fig. 4. Imaginary 1 and real 2 parts of the dielectric function of Au according with (Weber, 2003) are shown by dashed lines; according with Eq.(37) – by solid lines. Dotted horizontal line means 0.

$$\varepsilon_+(\lambda) = \varepsilon_\infty + \sum_{i=1}^{3} \frac{C_i}{1 - \left(\dfrac{1.242}{\lambda E_i}\right)^2 - i\dfrac{1.242}{\lambda E_i}\gamma_i} - F_1\,\chi_1^{-2}(\lambda)\ln\left[1 - \chi_1^2(\lambda)\right] - F_2\,\chi_2^{-2}(\lambda)\ln\frac{1 - \chi_1^2(\lambda)}{1 - \chi_2^2(\lambda)},$$

where

$$\chi_m(\lambda) = \left(\frac{1.242}{\lambda} + i\Gamma_m\right)\frac{1}{E_m} \quad \text{and} \quad \varepsilon_\infty = 0.2\,,\ C_1 = 0.77\,,\ C_2 = 2.96\,,\ C_3 = 0.3\,,\ F_1 = 5.22\,,\ F_2 = 4\,,$$

$$\gamma_1 = 0.05\,,\ \gamma_2 = 0.1\,,\ \gamma_3 = 0.1\,,\ E_1 = 3.38\,,\ E_2 = 4.27\,,\ E_3 = 5.3\,,\ \Gamma_1 = 0.08\,,\ \Gamma_2 = 0.1$$

Fig.5a shows cross-sections (36) of absorption and scaterring of Au nanoparticle in Si in πa^2 units. One can see that the absorption cross-section is greater than πa^2 more than one order of magnitude near $\lambda = \lambda_{LPR} = 0.857$ μm. Fig. 5b shows the factor $|F|^2$ appeared in Eq.(31), that is the factor of increase of intensity of EMF in the nanoparticle with the respect to the intensity of ENF outside the nanoparticle, F is determined by Eq.(32). At the excitation of LPR the intensity of EMF inside the nanoparticle is 150 times greater than the intensity of the external EMF outside the nanoparticle.

Cross-section σ_{ph-em} of the photoemission from spherical Au nanoparticle of radius $a = 10$ nm in Si is shown in Fig.6a, together with σ_{abs} and σ_{sc}, as function of the wavelength of the incident radiation, as it was found from the formula (31) with the use of Eqs. (27), (32) and (35).

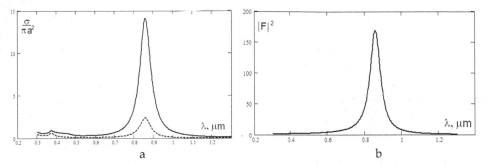

Fig. 5. (a) cross-section of absorption σ_{abs} (solid line) and scattering σ_{sc} (dashed line) of spherical Au nanoparticle in Si. (b) The factor of the enhancement of EMF in the nanoparticle.

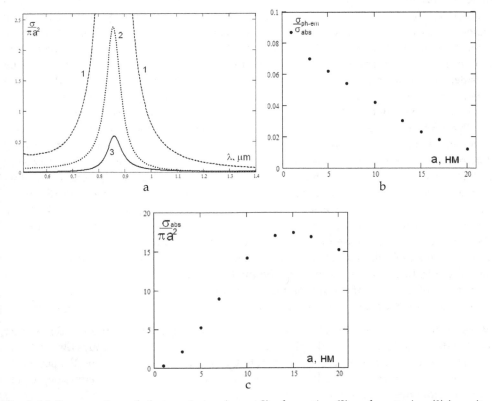

Fig. 6. (a) Cross-section of photoemission (curve 3), absorption (2) and scattering (1) in units of the geometrical cross-section πa^2 of spherical Au nanoparticle; (b) the ratio $\sigma_{ph\text{-}em} / \sigma_{abs}$ at the maximum of LPR for various radii of spherical nanoparticles; (c) cross-section of absorption of spherical Au nanoparticle versus its radius.

It can be seen from Fig.6a that $\sigma_{ph\text{-}em}$ reaches approximately the half of the geometrical cross-section πa^2 of the nanoparticle, which is about 4.2% of the maximum value of σ_{abs}. The ratio $\sigma_{ph\text{-}em} / \sigma_{abs}$ at the resonance, i.e. at $\lambda = \lambda_{LPR}$, is shown in Fig.6b, it characterizes relative part of the energy absorbed by nanoparticles and converted to the photocurrent. Though this relative part is not so large (about few percents), it is much bigger than for the case of continues Au film (see below). Fig.6b shows that the ratio $\sigma_{ph\text{-}em} / \sigma_{abs}$ is decreased almost linearly from 9% to 1% at the increase of the radius of nanoparticle from 1 to 20 nm. Thus the photoemission is relatively more effective for small particles. However the absorption cross-section itself is small at small radius a of the nanoparticle. This is because of the broadening of LPR due to collisions of electrons with the surface of the nanoparticle. Cross-section σ_{abs} reaches the maximum, for large a it goes down because of de-phasing and radiative losses, see Fig.6b. The optimum value of the radius of the nanoparticle can be estimated from calculations of the photoemission current made below.

Collective phenomena, as the interaction of particles with each other through EMF, may be quite important at the photoemission from the ensemble of nanoparticles. Detailed description of the influence of collective phenomena on the photoemission from nanoparticles is outside the score of present study; here we restrict ourselves only by some estimation. Noting that the number of photons absorbed per unit of time in the metal film can't exceed the number of photons falls per unit of time down to the surface of the film outside, one can write $\sigma_{abs} / (\pi a^2) < 1 / \eta$ where η is relative surface density of nanoparticles, i.e. σ_{abs} must decrease with the increase of η if $\sigma_{abs} > \pi a^2$. In practice σ_{abs} is decreased with η due to the broadening of LPR caused by collective phenomena. Quite possible that the broadening of LPR leads to the decrease of the factor F and the cross-section of photoemission for narrow spectrum of LPR, when λ is close to λ_{LPR}, however it does not mean that the photoemission from the ensemble of nanoparticles will be decreased with increase of η at broad spectrum of EMF as, for example, the solar spectrum. Note that narrow high-quality LPRs have been predicted and observed at certain conditions in ensembles of nanoparticles (Hicks, 2005), which means that the factor F may be quite big in encembles of nanoparticles even if σ_{abs} is not large. Thus one can't say *a-priori* that collective phenomena allways decrease the photoemission from nanoparticles. Detailed investigation of the influence of collective phenomena on the photoemission from nanoparticles will be carried out in future, here we compare the photoemission from the layer of nanoparticles with the photoemission from the continues metal film by using formulas obtained above without taking into account collective phenomana.

We estimate now the surface density of the photoemission current. We use Eq.(27) in order to estimate the density $j_{ph\text{-}em}$ of the photoemission current from thin continues film of Au; in Eq.(27) E_- is the component of EMF normal to the surface of the metal film. For simplicity we suppose that that $E_- = E$ this way we rather overestimate the photoemission current from the continues metal film.

Suppose that we have the layer of spherical Au nanoparticles in Si with relative surface density η; we estimate the surface density of the photocurrent if the layer is illuminated by solar radiation. Normalized solar spectrum is

$$w_s(\lambda) = \frac{\lambda^{-4} / 0.128}{\exp(2.616 / \lambda) - 1}$$

where λ is the wavelength in µm. The surface density $j_{ph\text{-}em}$ of the photocurrent normalized to the total intensity I of solar radiation is

$$\frac{j_{ph\text{-}em}}{I} = \frac{\eta}{\pi a^2} \int w_s(\lambda)\sigma_{ph\text{-}em}(\lambda)d\lambda \qquad (39)$$

Fig.7a shows the ratio of the surface density of photoemission versus the radius of the nanoparticle, - from the layer of spherical Au nanoparticles to the surface density of the photoemission current from continues Au layer for $\eta = 0.3$, i.e. when 30% of the surface is covered by Au nanoparticles. We take the spectral range of solar radiation from 0.32 to 2 nm, with about 80% of total energy of solar radiation.

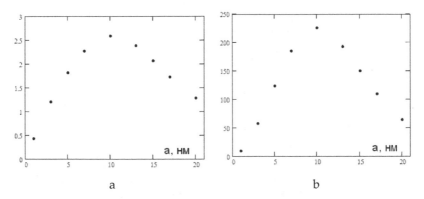

a   b

Fig. 7. (a) The ratio of the surface density of the photoemission current from the layer of nanoparticles with $\eta = 0.3$ to the surface density of the photoemission current from continues Au layer in Si for solar radiation for different radiuses of nanoparticles; (b) the same for monochromatic EMF at LPR.

Fig.7b shows the same quantity as Fig.7a but for the monochromatic EMF at the wavelength of LPR. One can see from Fig.7a that the photoemission current from the layer of nanoparticles is several times greater than from continues metal film for the case of solar radiation. There is optimal value of radius of nanoparticles when the photocurrent has maximum. For the monochromatic EMF near LPR the photocurrent from the layer of nanoparticles exceeds two orders of magnitudes the photocurrent from continues layer, see Fig.7b. Optimum radius of nanoparticles in this case is 10 nm. We consider here the "internal" efficiency of the photoemission from nanoparticles. The efficiency of collecting of the photocurrent into the external circuit is not considered. Estimations here do not take into account several factors as, for example, tunneling of photo-curriers through the potential barrier on the interface between the particle and the environment (Nolle, 2007); we can also note that the probability of the above barrier transition for the photo-carrier may be larger than the probability for the photo-induced carrier to leave potential well. Account for the potential barrier instead of the potential well can be done by using the approach presented above, it can be done analytically for step potential or numerically for more realistic smooth potentials also with an account for image forces. The photoemission from the volume of the nanoparticle may bring additional, but may be not so large, contribution to the photocurrent; it may be done following, for example, the approach of (Fowler, 1931).

In a difference with (Brodsky, 1973) here we take into account the change in the normal component of the electric field and in the electron effective mass at the interface between the metal and the environment. Let us estimate how important are these changes ("jumps") for the case of the photoemission from Au nanoparticle to p-doped Si. In order to came back to results of (Brodsky, 1973) (i.e. without "jumps") one can follow the procedure described after Eq. (20). Fig.8 show the cross-section of photoemission neglecting by the change of the effective mass of the electron, i.e. when $r_m = 1$, of neglecting by the change of EMF, when $\varepsilon_- / \varepsilon_+ = 1$, of when we neglect by both of them, when $r_m = \varepsilon_- / \varepsilon_+ = 1$. For comparison Fig.8 displays the cross-section of photoemission for $r_m \neq 1$ and $\varepsilon_- / \varepsilon_+ \neq 1$, that is the same as the curve 3 in Fig.6a.

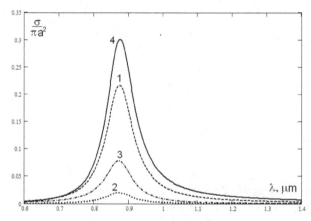

Fig. 8. Cross-sections of the photoemissions. We neglect by changes: of the effective mass of the electron at the interface (curve 1); EMF (2); the effective mass and EMF (3) that is the result of (Brodsky, 1973); all changes are taken into account (4).

According with Fig.8 account for the jump in EMF and in the effective mass considerably changes (increases) the value of the cross-section of photoemission. It is interesting to note that the maximum value of the cross-section is reached when all jumps are taken into account. The jump in the mass, without the jump in the EMF reduces photoemission (compare curves 2 and 3 in Fig.8), the jump in the EMF without the jump in the mass increases photoemission (compare curves 1 and 3 in Fig.8), but both of them lead to the maximum in photoemission (compare curves 4 with others in Fig.8). Such "non-additive" influence of jumps of EMF and electron mass to the cross-section of photoemission is the consequence of quantum-mechanical interference. Indeed, complex terms c_V, c_E and c_m describing, respectively, the photoemission with account for jumps in the potential, EMF and the electron mass at the interface appear as linear (additive) combination in Eq. (2) for the probability amplitude $C_+(\infty)$ of the photoemission. However the probability

$C_{emission} \sim |C_{+}(\infty)|^2$ appears in Eq. (31) for $\sigma_{ph\text{-}em}$, and noticerable quantum-mechanic interference between contributions from c_V, c_E and c_m is. This interference leads to incease of $\sigma_{ph\text{-}em}$, when all three terms are taking into account, but it leads to increase or to decrease of $\sigma_{ph\text{-}em}$ when there are only two or only one term. In practice it means that various interfaces may have quite different and unusual influence to the photoemission.

5. Conclusion

We calculated the probability amplitude and the cross-section of the photoemission from metal nanoparticles. It is shown that the cross-section of photoemission from metal nanoparticles is about the half of the geometrical cross-section of the nanoparticle, see Fig.6a; an example of the photoemission from Au spherical nanoparticles into p-doped Si is considered. It turns to be that if the surface density of nanoparticles in the layer is $\eta = 0.3$ then about 15% of all photons can be converted to photoelectrons at the excitation of the localized plasmon resonance (LPR), so that 15% is the internal quantum efficiency of photoemission at $\lambda = \lambda_{LPR}$. The photoemission current from the layer of nanoparticles is two orders of magnitude greater than from continues layer of Au at the monochromatic EMF exciting LPR. For the case of the broad spectrum of EMF (as solar spectrum) the photoemission from nanoparticles is several times more intensive that from continues metal layer, see Fig.7, there is the optimum radius of nanoparticles corresponding to the maximum of the photoemission current. Increase of the photoemission from nanoparticles with the respect to continues metal layer is occurred due to the increase of EMF inside and near nanoparticle at the excitation of EMF (see Fig.5) and due to relatively large surface of the nanoparticle, which surface is non-parallel to the polarization of the incident EMF.

We generalize the theory of (Brodsky, 1973) of the photoemission from metals by taking into account breaks in the EMF and in the electron effective mass in the interface of the metal and environment. It is shown that such breaks may considerably change the cross-section of photoemission from metals, as one can see from the example of photoemission from Au particles to p-doped Si considered here.

We do not take into account the volume photo-effect, which can increase the photocurrent even more; the potential in the metal-environment interface was approximated by rectangular potential well. However in reality there is a barrier in the interface. Account for the potential barrier in the interface and for the tunneling through the barrier will lead to more increase in the photoemission current. Calculations with more complicated potentials on the interface may be proceeded by direct generalization of the approach presented above. In the future one has to take into account collective phenomena as the interaction of nano-particles with each other through EMF, however it influences only the factor F describing the increase of EMF inside nanoparticle with the respect to EMF outside it. Approach of this work can be useful for description and study of recombination of carriers (also photo-induced carriers) on metal nanoparticles.

Results of this work can be used for creating of new high-sensitive photo detectors and photo converters of solar radiation into electric energy. There is important question about the minimum time of the photo effect related with the sensitivity of photodetectors (Schelev, 2000). It is possible, that the increase of the efficiency of the photoemission at the excitation of LPR lets to reduce the minimum time necessary to observe the photo effect. For example, if the femto-second laser pulse can excite LPR in the nanoparticle, then the photo detector with such nanoparticles can provide femtosecond time resolution even if the time of the photo-responce of the metal of the nanoparticle is larger than femtosecond. This way the question about shortest time of photo-response of the metal is replaced, therefore, by the question of the shortest time of the excitation of LPR.

6. References

Adachi, S., Mori, H., Ozaki, S. (2002). Model dielectric function for amorphous semiconductors. *Physical Review B,* Vol. 66, No.15, (2 October 2002), pp.(153201 [4 pages]), ISSN 1098-0121

Atwater, H. A., Polman, A. (2010). Plasmonics for improved photovoltaic devices. *Nature Materials,* Vol. 9, (March 2010), pp. (205-213), ISSN 1476-1122

Bergman, D. J., Stockman, M. I. (2003). Surface Plasmon Amplification by Stimulated Emission of Radiation: Quantum Generation of Coherent Surface Plasmons in Nanosystems. *Physical Review Letters,* Vol. 90, No. 20, (14 January 2003), pp. (027402 [4 pages]), ISSN 0031-9007

Bottcher, C. J. F. (1952). *Theory of Electric Polarization, Vol. 1,* Elsevier, Amsterdam

Brodsky, A.M. & Gurevich, Yu. Ya. (1973). *Theory of electron emission from metals,* Nauka, Moscow, Russia

Brongersma, M. L., Kik, P. G. (Eds.). (2007). *Surface Plasmon Nanophotonics,* Springer-Verlag, ISBN 978-140-2043-49-9, New York, NY, US

Catchpole, K. R., Polman, A. Plasmonic solar cells. (2008). *Optics Express,* Vol. 16, No. 26, (22 December 2008), pp. (21793-21800), ISSN 1094-4087

Dutta, A., Mazhari, B., & Visweswaran, G. S. (2009). Metal Semiconductor Contact. Schottky barrier height, In: *Semiconductor Devices (NPTEL Online – IIT Dehli 2011),* Available from:
http://nptel.iitm.ac.in/courses/Webcourse-contents/IIT-Delhi/Semiconductor%20Devices/index.htm

Fowler, R. H. (1931). The analysis of photoelectric sensitivity curves for clean metals at various temperatures. *Physical Review,* Vol.38, No.1, (July 1931), pp. (45-56), ISSN 1050-2947

Greffet, J.-J. (2005). Nanoantennas for Light Emission. *Science,* Vol. 308, No. 5728, (10 June 2005), pp. (1561-1563), ISSN 0036-8075

Hetterich, J., Bastian, G., Gippius, N.A., Tikhodeev, S.G., von Plessen, G., & Lemmer, U. (2007). Optimized Design of Plasmonic MSM Photodetector *IEEE Journal of Quantum Electronics,* Vol. 43, No. 10, (October 2007), pp. (855-859), ISSN 0018-9197

Hicks, E. M., Zou, S., Schatz, G.C., Spears, K.G., Van Duyne, R.P., Gunnarsson, L., Rindzevicius, T., Kasemo, B., & Käll, M. (2005). Controlling Plasmon Line Shapes through Diffractive Coupling in Linear Arrays of Cylindrical Nanoparticles Fabricated by Electron Beam Lithography *Nano Letters,* Vol. 5, No. 6, (18 May 2005), pp.(1065-1070), ISSN: 1530-6984

Homola, J., Yee S. S., & Gauglitz, G. (1999). Surface plasmon resonance sensors: review. *Sensors and Actuators B,* Vol. 54, No. 1-2, (25 January 1999), pp. (3-15), ISSN 0925-4005

Hövel, H., Fritz, S., Hilger, A., & Kreibig, U. (1993). Width of cluster plasmon resonances: Bulk dielectric functions and chemical interface damping. *Physical Review B*, Vol. 48, No. 24, (December 1993), pp. (18178-18188), ISSN 1098-0121

Khlebtsov, N. G. (2008). Optics and biophotonics of nanoparticles with a plasmon resonance. *Quantum electronics*, Vol. 38, No. 6, (June 2008), pp. (504–529), ISSN 1063-7818

Kittel, Ch. (1996). *Introduction to Solid State Physics*, Jon Wiley & Sons inc., London, New York, ISBN 978-047-1111-81-8

Klimov, V. V., (2003). Spontaneous atomic radiation in the presence of nanobodies. *Physics-Uspekhi*, Vol. 46, No. 10, (September 2003), pp. (979-984), ISSN 0042-1294

Klimov, V.V., (2009). *Nanoplasmonics*, Fizmatlit, ISBN 978-5-9221-1030-3, Moscow, Russia

Kneipp, K., Moskovits, M., & Kneipp, H. (Eds.). *Topics in Applied Physics Vol. 103*, Springer-Verlag, ISBN 978-3-540-33566-5 , Berlin, Heidelberg, New York

Landau, L. D., Lifshitz, L. M. (1997). *Quantum Mechanics: Vol 3*, Butterworth-Heinemann, ISBN 978-0-750-63539-4, International edition

Maier, S. A. (2007). *Plasmonics: Fundamentals and Applications*, Springer-Verlag New York Inc., ISBN 978-038-7331-50-8, New York, NY, US

Meier, M., Wokaun, A. (1983). Enhanced fields on large metal particles: dynamic depolarization. *Optics Letters*, Vol.8, No.11, (1 November 1983), pp. (581-583), ISSN 0146-9592

Monestier, F., Simona, J.-J., , Torchioa, P., Escoubasa, L., Florya, F., Baillyb, S., Bettigniesb, R., Guillerezb, S., & Defranouxc, C. (2007). Modeling the short-circuit current density of polymer solar cells based on P3HT:PCBM blend. *Solar Energy Materials and Solar Cells*, Vol. 91, No. 5, (6 March 2007), pp. (405-410), ISSN 0927-0248

Mühlschlegel, P., Eisler, H.-J., Martin, O. J. F., Hecht, B., & Pohl, D. W. (2005). Resonant Optical Antennas. *Science*, Vol. 308, No. 5728, (10 June 2005), pp. (1607-1609), ISSN 0036-8075

Noginov, M. A., Zhu, G., Belgrave, A.M., Bakker, R., Shalaev, V. M., Narimanov, E. E., Stout, S., Herz, E., Suteewong, T., & Wiesner U. (2009). Demonstration of a spaser-based nanolaser. *Nature*, Vol. 460, (27 August 2009), pp. (1110-1112), ISSN 0028-0836

Nollé, É. L., Shchelev, M. Ya. (2004). Photoelectron emission caused by surface plasmons in silver nanoparticles. (2004). *Technical Physics Letters*, Vol. 30, No.4, (April 2004), pp.(304-306), ISSN 1063-7850

Nollé, É. L., Shchelev, M. Ya. (2005). Photoelectron emission from granulated gold films activated by cesium and oxygen. *Technical Physics*, Vol. 50, No. 11, (November 2005), pp.(1528-1530), ISSN 1063-7842

Nollé, É. L. (2007). Tunneling photoeffect mechanism in metallic nanoparticles activated by cesium and oxygen. *Physics-Uspekhi*, Vol. 50, No. 10, (October 2007), pp.(1079–1083), ISSN 0036-8075

Novotny, L., Hecht, B. (2006). *Principles of Nano-optics*, Cambridge University Press, ISBN 978-052-1832-24-3, Cambridge, UK

Oulton, R. F., Sorger, V. J., Zentgraf, T., Ma, R-M., Gladden, C., Dai, L., Bartal, G., & Zhang, X. (2009). Plasmon lasers at deep subwavelength scale. *Nature*, Vol. 461, (1 October 2009), pp. (629-632), ISSN 0028-0836

Piotrowski, J., Galus, W., & Grudzi, M. (1990). Near room-temperature IR photo-detectors. *Infrared Phys.*. Vol. **31**, No. 1, (January 1990), pp. (1-48), ISSN 1350-4495

Pors, A., Uskov, A.V., Willatzen, M., & Protsenko, I. E. (2011). Control of the input efficiency of photons into solar cells with plasmonic nanoparticles. *Optics communications*, Vol. 284, No. 8, (15 April 2011), pp. (2226-2229), ISSN 0030-4018

Protsenko, I. E., Uskov, A. V., Zaimidoroga, O. A., Samoilov, V. N., & O'Reilly E. P. (2005). Dipole nanolaser. *Physical Review A*, Vol. 71, No. 6, (17 June 2005), pp. (063812 [7 pages]), ISSN 1050-2947

Rand B.P., Peumans, P. & Forrest, S.R. (2004). Long-range absorption enhancement in organic tandem thin-film solar cells containing silver nanoclusters. *Journal of Applied Physics*, Vol. 96, No.12, (15 December 2003), pp. (7519-7526), ISSN 0021-8979

Schelev, M. Ya. (2000) Femtosecond photoelectronics — past, present, and future. *Physics-Uspekhi*, Vol. 43, No.9, (September 2000), pp.(931–946), ISSN 0036-8075

Schuller, J., Barnard, E., Cai, W., Jun, Y., White J., & Brongersma M. (2010). Plasmonics for extreme light concentration and manipulation. *Nature Materials*, Vol. 9, No. 3, (March 2010), pp. (193–204), ISSN 1476-1122

Soole, J. B. D., Schumacher, H. (1991). InGaAs metal-semiconductor-metal photodetectors for long wavelength optical communications. *IEEE Journal of Quantum Electronics*, Vol. **27**, No.3, (March 1991), pp. (737 – 752), ISSN 0018-9197

Sze S.M., (1981). *Physics of semiconductor devices*, Wiley, ISBN 978-047-1056-61-4, New York, Chichester, Brisbane, Toronto, Singapore

Wang, F., Ron Shen, Y. (2006). General properties of local plasmons in metal nanostructures. *Physical Review Letters*, Vol. 97, No. 20, (November 2006), pp. (206806 [4 pages]), ISSN 0031-9007

Weber, M. J. (2002). *Handbook of optical materials*, CRC Press, Boca Raton, London, New York, Washington, D.C., ISBN 978-084-9335-12-9

Westphalen, M., Kreibig, U., Rostalski, J., Lüth, H., & Meissner, D. (2000). Metal cluster enhanced organic solar cells. *Sol.Energy Mater. Sol. Cells*, Vol. 61, No.1, (15 February 2000), pp. (97-105), ISSN 0021-8979

Yu, Z., Veronis, G., Fan, S., & Brongersma, M. (2006). Design of midinfrared photodetectors enhanced by surface plasmons on grating structures. *Applied Physics Letters,*Vol. 89, No.15, (9 October 2006), pp. (151116 3 pages), ISSN 0003-6951

9

Calculation and Measurement of Electromagnetic Fields

Hidajet Salkic[1], Amir Softic[1], Adnan Muharemovic[2],
Irfan Turkovic[3] and Mario Klaric[4]
[1]*PE Elektroprivreda BiH,*
[2]*Energoinvest d.d. Sarajevo,*
[3]*University of Sarajevo, Faculty of Electrical Engineering,*
[4]*Dalekovod d.d. Zagreb,*
[1,2,3]*Bosnia and Herzegovina*
[4]*Croatia*

1. Introduction

A man is exposed to electromagnetic fields in his environment. Electromagnetic fields always exist in nature – atmospheric static electric field, the Earth's magnetic static field, the fields of a wide range of frequencies due to the outbreaks in the atmosphere, etc. However, a man is today the most exposed to the artificial field, due to progress in technology and widespread use of electrical devices. Currents, induced by electric field of surface charges are the greatest if external electric field is parallel to the length of the body. Magnetic field induces a currents inside the body as well. Variable magnetic field acting by force on charged particles in the body and creates eddy currents according to Faraday's law. Such induced currents in the low frequency area can stimulate electrically excitable tissues, such as nerve and muscle fibers, through the mechanism of action potential triggering. The area of occupational exposure includes people who are exposed to electromagnetic fields in known circumstances during usual performing of work tasks in and around power facilities, but they are educated to take protective measures and they have all the tools and instructions provided. These people are aware of potential risks and take appropriate protective measures. The environment around the power facility falls within the area of increased sensitivity, which includes people of different ages and health conditions, including those particularly sensitive. In many cases people are not aware of exposure to electromagnetic fields and can not be expected to take protective measures to reduce exposure. Therefore, the restrictions for that area are stricter than those for area of occupational exposure. The most efficient way for reliable operation of the devices in high voltage substation is the calculation and measurement of low frequency electromagnetic fields in the substation, together with appropriate measurement procedure. Contemporary research of electromagnetic fields is based on the concept that complex theoretical research results in appropriate design solution, and developes almost exclusively as applied research. Generally, there are two directions; the first one based on calculation model development, and the second one based on models of objectivized physical measurements in hard conditions. In both cases, the goal is the same and can be summarized as follows: create the optimal variant of the electromagnetic fields calculation, in both the existing and in new substations. Research of the way of calculation, for low frequency electromagnetic fields area (Extra Low Frequency), in stationary regimes in distribution substations in urban areas, is performing in order to obtain the level of electric and magnetic field in the areas where primary power and secondary electronic equipment is located and where the temporarily or permanently stay people within one segment of the regulations of protection against electromagnetic fields. Pragmatically, it can be concluded that power equipment causes electromagnetic effects, and electronic equipment is subject to the activity of these influences. Routes of transmission of these influences, in such complex facility, are possible through conductors, inductive and capacitive links and radiation. The specific characteristics of electromagnetic field sources in power systems are: field intensity, frequency, waveform (content of harmonics), polarization, spatial and time variation of the field. The main sources of influence for this research are induced voltages, as consequence of variable electric, magnetic and electromagnetic fields. In the low frequency area (wavelength 6000 (km) at frequency 50 (Hz)) the irradiation is happening exclusively in closer zone, in which does not worth mutual perpendicularity of electric and magnetic field in the direction of wave propagation, constant ratio of amplitudes of the electric and magnetic field and dependence of the electric and magnetic field amplitudes of the distance from the source by the law $1/r$, and power density by the law $1/r^2$. Low frequency electric and magnetic fields can be observed separately, because at these frequencies the shifted currents are negligible. Mathematical models and numerical solving, as well as the method of experimental measurements of low frequency electric and magnetic fields of power facilities are presented in this research. Calculation and measurement of low frequency electric and magnetic fields, as well as their interconnection, are the main problems in electricity transmission and distribution, in terms of standardized electromagnetic compatibility and human exposure to non-ionizing electromagnetic radiation. For solving the electromagnetic influences with complex geometry in the low frequency area, the system of Maxwell's equations that fully describe the electromagnetic field is used. Maxwell's equations can be analytically solved only for a narrow class of one-dimensional problems of static and quasistatic fields. Each two-dimensional (2D) and three-dimensional (3D) geometric arrangement requires the application of numerical methods for solving of the field by using some of the famous software packages (for example MAXWELL 3D, EFC-400, FLUX 3D, MATLAB...) and other appropriate tools necessary for successful implementation of the research, and for which is necessary to make a detailed mathematical models of transformer stations with all geometrical and electrical parameters. For 3D distribution calculation of low frequency electric field the charge simulation method is applied (CSM-Charge Simulation Method) as well as the source element method (SEM-Source Element Method) that is usually considered as a special variant of the indirect boundary element method (IBEM-Indirect Boundary Element Method). For 3D distribution calculation of low frequency magnetic field, inside and around the transformer stations, a procedure based on the application of Biot-Savart's law for magnetic flux density of straight streamline of finite length and a principle of superposition is used. Based on the analysis of theoretical calculation, a detailed operational measurement program that includes all measurements in steady state with measurement location is shown. The measurement have to be conducted in accordance with the norm HRN IEC 61786-2001 - Measurement of low frequency magnetic and electric field with regard to exposure of human beinges-Spacial requirements for instrumants and quidance for measurements and the instructions given in the European recommendations ENV 50166 (People exposure to the electromagnetic radiation on low frequencies). For measurement methods under stationary conditions a modern measuring equipment (EFA–300 Field Analyzers) was used, which find application in researches and environmental studies for assessment of electric and magnetic fields in transmission and distribution power networks and facilities with belonging equipment. It is made to provide a sophisticated tool for precise researches of power low frequency influences for the engineers, experts in the field of health care, work protection and other profiles. Selection of the measurement points was made on the basis of the field intensity assessment, respectively the measurement is performed in places where the greatest intensity of electric and magnetic field was expected. The analysis of measurement results and their comparison with numerical calculations indicated the places in which the standards for people protection from electromagnetic fields are disturbed and gave suggestions for elimination of electromagnetic influences and their reduction to acceptable level. Confirmed satisfactory accuracy of the results obtained by calculations in comparison to the results of experimental measurements indicates the validity for introducing and developing a calculations for such practical problems related to design and reconstruction of substations, which is extremely important from an economic point of view, since, in this way, the demands for expensive experimental measurements and reparations are reduced. Presented mathematical models, calculation, measurement and visual 3D distribution of electric and magnetic field is a realistic assumption for research of interactions between electromagnetic fields and human bodies on the macroscopic and static level, with finding the certain optimization criteria in order to develop a new technology and process solutions and design methods. The research results are important from the scientific point of view but also they are important because of possibility for their practical application.

2. Background theory of models

Low frequency electromagnetic field around the substations is quasistatic. It has a conservative component of the electric field caused by charges and eddy component of the magnetic field caused by currents. The calculation of electric and magnetic fields at the points located far from the source (charges and currents) is obtained with thin-wired approximation and by representation of conductors with linear segments with current distribution calculation, and based on that, in the selected point of the space located in the air or in any ground layer the calculation of potentials is also obtained. The potentials and electromagnetic fields are firstly expressed in the form of components of the vector potentials, as a functions of the current in each segment of conductors network. The currents in the conductors segments are determined based on the voltage drop between a pair of network points, based on their own impedance. The ground influence on the conductor potential was taken into account by using a method of mirrors. The phasor of the electric potential at some point in the space is obtained by applying the superposition theorem, as a finite sum of potentials of elementary, time-varying charges on the surface of the conductor. The total value is represented as the integral of electrical potentials caused by charges density, which is located in the point given by a positional vector that is located on all thin-wired parts, including original conductors and their images, respecting land-air discontinuity, and given by the following integral equation:

$$\dot{\varphi} = \frac{1}{4\pi\varepsilon} \int_{l^{`}} \frac{\rho(r^{`})dl^{`}}{|r-r^{`}|} + \frac{1}{4\pi\varepsilon} \int_{l^{``}} \frac{\rho(r^{``})dl^{``}}{|r-r^{``}|} \tag{1}$$

It is necessary to discretize the equation by discretizing the field of the source with unknown distribution, for example, density of line charge by using appropriate combination of N linear, independent fundamental functions. Then, the discretization of conductor length on N segments and the discretization of observed points will be connected. We obtain the conductor division into segments of finite lengths and approximate the unknown distribution of the field with appropriate number of fundamental functions by following expression:

$$\dot{\rho} = \sum_{j=1}^{N} a_j' \rho_j' \text{ if it is } \dot{\rho} = \sum_{j=1}^{N} a_j'' \rho_j'' \tag{2}$$

where: ρ_j' is a fundamental function on segment j of the original conductor and ρ_j'' a fundamental function on segment j of the conductor in the mirror. Selected constant $a_j' = 1$ in the segment j is valid for that segment, while in other segments is 0. In that case, the potential equation can be represented as:

$$\dot{\varphi}(r) = \frac{1}{4\pi\varepsilon} \sum_{j=1}^{N} \int_{\Delta l_j'} \frac{a_j' \rho_j'(r')dl'}{|r-r'|} + \frac{1}{4\pi\varepsilon} \sum_{j=1}^{N} \int_{\Delta l_j''} \frac{a_j'' \rho_j''(r'')dl''}{|r-r''|} \tag{3}$$

In order to solve the expression, N observed points which are corresponding to energized conductors are selected in the space of known potential. It establishes a system of N equations with N unknown values, which is defined in the matrix form as:

$$[\varphi] = [M][\rho] \tag{4}$$

where the elements $M_{i,j}$ of matrix system [M] represent the potential of the observed point i, located on the conductor surface with current density ρ_j. Gauss-Seidel's method is used for solving this matrix equation. When the approximation of current density on the conductors is obtained, the vector-phasor of conservative component of the electric field intensity, at the observed point with position vector r, can be determined by using the equation:

$$\dot{E}(r) = \frac{1}{4\pi\varepsilon} \sum_{j=1}^{N} \int_{\Delta l_j'} \frac{a_j' \rho_j'(r)(r-r')dl'}{|r-r'|^3} + \frac{1}{4\pi\varepsilon} \sum_{j=1}^{N} \int_{\Delta l_j''} \frac{a_j'' \rho_j''(r)(r-r'')dl''}{|r-r''|^3} \tag{5}$$

In the 3D calculation, the vector of the electric field is in each point elliptically polarized, i.e. the peak of the vector E describes an ellipse in time. Each of three components have a different size and phase shift:

$$E_x(t) = E_{xmax}\cos(\omega t + \varphi)$$

$$E_y(t) = E_{ymax}\cos(\omega t + \varphi) \tag{6}$$

$$E_z(t) = E_{zmax}\cos(\omega t + \varphi)$$

The vector of the electric field is elliptically polarized and rotates in time. The effective value (RMS) of the absolute value of the electric field is used for the electric field presentation according to:

$$E_{ef} = \sqrt{\frac{1}{T} \int_0^T (E_x^2(t) + E_Y^2(t) + E_z^2(t))} \tag{7}$$

Generally, the size of the magnetic field or magnetic flux density can be decomposed into three components that are perpendicular to each other in the space. Each of these components is time-dependent:

$$\vec{B}(t) = \vec{B_x}(t) + \vec{B_y}(t) + \vec{B_z}(t) \tag{8}$$

The largest number of magnetic fields, around power facilities, are generated by basic harmonic with dominant frequency of 50 Hz, and have a negligible contribution of higher harmonics. The components are time-dependent by sinus dependence:

$$\vec{B}(t) = \sqrt{2}B_x \sin(\omega t + \varphi_x)\vec{i} + \sqrt{2}B_y \sin(\omega t + \varphi_y)\vec{j} + \sqrt{2}B_z \sin(\omega t + \varphi_z)\vec{k} \tag{9}$$

The effective value of magnetic flux density is expressed mathematically:

$$B_{ef} = \sqrt{\frac{1}{T}\int_0^T [B(t)]^2 dt} = \sqrt{B_x^2 + B_y^2 + B_z^2} \tag{10}$$

where are: t – time variable
                  T – period of time change
                  B_x, B_y, B_z – effective values of time-variable orthogonal components B

The absolute value of magnetic field intensity is determined by the equation:

$$|\vec{H}| = \frac{I}{2\pi r} \tag{11}$$

where are: I - electric current intensity through conductor
                  r - distance from the conductor

Three-phase AC system of 50 Hz generates elliptically polarized fields. The field vector rotates around a fixed ellipse whose radiuses of semi-axis represent the peak values. In the case of different frequencies existence, contribution to the field of individual segments are added together and integrated in time. The degree of polarization is defined by the ratio of the magnetic flux densities values between the, so-called, minor and major ellipse:

$$\frac{B_{min}}{B_{max}} \tag{12}$$

This axial ratio can take values between 0 (the magnetic field is linearly polarized) and 1 (the magnetic field is circularly polarized). For a particular value of frequency (50 Hz), the effective value of the magnetic induction can also be determined by the effective values of the magnetic flux density components along the two axis of the polarized magnetic field:

$$B_{ef} = \sqrt{B_{max}^2 + B_{min}^2} \tag{13}$$

Calculation of magnetic flux density distribution is performed based on the application of Biot-Savart's law for the induction of straight streamline of finite length and the law of superposition. Magnetic flux density at any point in the space can be calculated by superposition of contributions of each conductor in which current flows. The spatial position of conductor segments, their currents and phase angles represent the input sizes for the calculation of magnetic flux density at the desired points in the space. The direction of magnetic flux density vector is determined by the unit vector in cylindrical coordinate system related to the observed segment. As the position of segments is different in the space and so the directions of the induction vectors, it is necessary to decompose a vector of magnetic flux density to the components in the direction of each coordinate axis of the global system that is not tied to a particular segment. The direction of magnetic flux density vector is perpendicular to the boundary plane and defined as:

$$d\vec{B}(t) = \frac{\mu_0}{4\pi} \frac{d\vec{l} \times \vec{r}}{r^3} I(t) \tag{14}$$

Sizes dB and I are generally time-dependent and are transformed into complex values for easier calculation. Suppose that the i-th segment of length l is located at the origin of the coordinate system, parallel to the x-axis, its contribution to the field at point P (x, y, z) is:

$$\vec{B_i}(t) = \frac{\mu_0}{4\pi r} I_i(t) \left[\frac{L_i - x_p}{\sqrt{(L_i - x_p)^2 + r^2}} + \frac{x_p}{\sqrt{x_p^2 + r^2}} \right] \tag{15}$$

Vector components are:

$$B_{xi}(t) = 0$$

$$B_{yi}(t) = -\frac{z_p}{\sqrt{y_p^2 + z_p^2}} |\vec{B_i}(t)| \tag{16}$$

$$B_{zi}(t) = \frac{y_p}{\sqrt{y_p^2 + z_p^2}} |\vec{B_i}(t)|$$

This method divides each condustor into segments whose number is determined in advance. When working with computer programs for the electric and magnetic fields calculating, the number of segments is determined by the user. If the conductors have overhangs (for example transmission lines), the programs usually simulate this situation with equivalent parable. Sufficiently high accuracy is usually achieved by selecting 10 - 20 segments. For verification, the number of segments can be increased so the accuracy of that calculation is higher than in cases with 2, 3 or 4 segments. The difference between the actual situation and model depends on the division of conductors to a finite number of segments. When calculating the field intensity, the coordinates of the considered point are transforming to the local coordinate system of a given segment. The calculation gives the contribution of a given segment to the field that is transforming back to the original coordinate system. Vector sum of field contributions gives the total amount of magnetic flux density vector, which is caused by currents of N segments and obtained by adding the contributions of all segments:

$$B(t) = \sqrt{\left(\sum_{i=1}^{N} B_{x,i}(t)\right)^2 + \left(\sum_{i=1}^{N} B_{y,i}(t)\right)^2 + \left(\sum_{i=1}^{N} B_{z,i}(t)\right)^2} \tag{17}$$

where $B_{x,i}(t)$, $B_{y,i}(t)$ and $B_{z,i}(t)$ are the components of magnetic flux density of i-th segment.

For magnetic field presentation the effective value (RMS) of magnetic field density is used, according to:

$$B_{ef} = \sqrt{\frac{1}{T} \int_0^T \left(B_x^2(t) + B_y^2(t) + B_z^2(t)\right) dt} \tag{18}$$

3. Calculation of low frequency magnetic and electric fields of transformer station

The application of computer program EFC–400LF for calculation of the electric and magnetic fields is presented on the example of typical compact, concrete transformer station KBTS 10(20)/0.4 kV, 630 kVA. Transformer station equipment consists of:

- switchable power transformer, nominal ratio 10(20)/0,4 kV, rated power up to 630 (kVA), frequency 50 (Hz), Dyn5, shortcut voltage 4 %, voltage regulation ± 2 x 2,5 %,
- medium–voltage (MV) distribution switchgear, insulated by SF6 gas, completely shielded and protected from dangerous contact voltage, " Ring Main Unit " (RMU) type CCF 12/24 (kV), 400 (A), " SafeRing " with 3 fields, transformer and 2 conductive fields. Conductive fields are equipped with a three pole switch disconnector with the earthing switch, rated voltage of 24 (kV), rated current 400 (A), with auxiliary switch 2NO + 2 NC. Transformer field is equipped with a three pole switch disconnector 24 (kV), 200 (A), 16 (kA), fuses 24 (kV), 50 (A),
- low–voltage (LV) distribution switchgear, which consists of three fields, namely: load field and two distribution fields. Rated current is 1250 (A), shortcut withstand current 25 (kA), peak withstand current 52.5 (kA), the degree of protection is IP 21. In the load field the main three-pole switch disconnector is located, type OETL 1250, 1250 (A), 690 (V), " ABB ",
- the connecting line between the MV side of transformer and the field of MV switchgear, that is derived from three single–core cables, XLP insulated, of rated voltage 20 (kV), unit designation of the cable is 3x(XHE 49-A 1x50/16 mm2) or 3x(XHE 49-A 1x150/25 mm2), allowable current load is 210 (A). The space between clamps for fixing the cable is 600 (mm) maximum. Cables are at a distance of 1.00 (m), wrapped with plastic tapes and make a bundle,
- the connecting line between the LV side of transformer and LV distribution field that is derived from single–core cables, insulated with PVC which is resistant to temperatures up to 378.15 K/105 °C, rated voltage up to 1 (kV), unit designation of cable for phase conductors is 3x(2 x P/MT1 x 240mm2 1 kV) and for the neutral conductor 1x(P/MT1 x 240mm2 1 kV).

Since the main electrical equipment (MV and LV switchgears) is tested and certified in accordance with the applicable IEC standards (IEC439 for LV switchgears and IEC298 for MV switchgears), and considering that power transformer satisfies the requirements of standard IEC76, it can be concluded that listed technical parameters are consequently verified. Calculations of the magnetic and electric field were carried out by using a computer program EFC–400LF according to DIN VDE0848-1, which allows the simulation in 3D space. The corresponding MV 12 (kV) distribution switchgear is modeled with a maximum load current I'_m of 36.4 (A) at rated voltage of transformer secondary 0.4 (kV) and the maximum load current I'_m of 909 (A). The load value of 909 (A) rarely occurs, but the calculations are carried out for the worst case scenario, so on the basis of this case the other cases can be determined as cases that meet safety standards. It follows that the maximum current load of the LV side of transformer stations is evenly divided into four outputs, 227 (A) each. Significant sources of electric field in transformer station are MV and LV buses and MV outlets of power transformer, while the MV and LV circuit equipment is surrounded by grounded housings, shields or cable screens and is a negligible source of the electric field due to its complete obscurance. Calculation of the electric field was performed inside and outside the substation, with negligence of substation housing, due to the additional increase in safety with regard to the protection rules for electromagnetic fields. 2D and 3D views of the substation disposition in EFC–400LF are shown in Fig. 1, where the difference between reality and model depends of redistribution of conductors to a finite number of segments. Redistribution of substation conductors was performed on 635 segments, on resolution size $dx = dy = dz = 0.10$ (m). EFC–400LF program is able to solve a set of differential equations for the matrix with 16000 x 16000 elements (Methods: LU–decomposition or conjugate gradient). For given example calculation was done with a 261 x 261 matrix elements, which gives the values of electric and magnetic fields in the 68.121 points of the observed plane, surface 169 (m²), with the resolution size $dx = dy = dz = 0.05$ (m) and matrix of 261 x 101 elements, which gives values of electric and magnetic fields in the 26.361 points of the observed plane, surface 65 (m²), the resolution size $dx = dy = dz = 0.05$ (m). Visual view of the obtained results of magnetic flux density and electric field intensity was made in the computer program "Matlab", using "Runal.B" and "Runal.E" while subroutines "Crtajgraf.B" and "Crtajgraf.E" serve to open, load and display the results of calculations of magnetic flux density and electric field intensity.

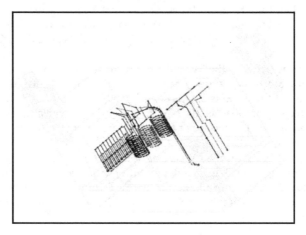

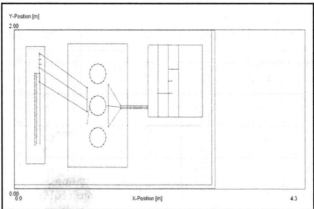

Fig. 1. 2D and 3D view of substation disposition in EFC–400.

The calculation of the electric and magnetic field was made:

i. In XY plane of transformer station at –5 (m) ≤ x ≤ 8 (m) and –5 (m) ≤ y ≤ 8 (m)
   - at a height z = 1.75 (m) from the ground. It is a plane with greatest values of electric and magnetic field outside the substation, in which a human head can be found,
ii. In XZ plane of transformer station at –5 (m) ≤ x ≤ 8 (m) and 0 ≤ z ≤ 5 (m)
   - for y = – 0.20 (m), accordingly 0.20 (m) of the longitudinal south side of substation,
   - for y = 2.10 (m), accordingly 0.20 (m) of the longitudinal north side of substation,
iii. In YZ plane of transformer station at –5 (m) ≤ y ≤ 8 (m) and 0 ≤ z ≤ 5 (m)
   - for x = – 0.20 (m), accordingly 0.20 (m) of the eastern side of substation,
   - for x = 3.10 (m), accordingly 0.20 (m) of the western side of substation.

3.1 Distribution calculation of electric and magnetic field in xy plane for z = 1.75 (m) (–5 (m) ≤ x ≤ 8 (m), –5 (m) ≤ y ≤ 8 (m))

The values of magnetic flux density and electric field intensity are observed in the areas I, II, III and IV of XY plane, at distances 0.20 (m), 1.00 (m), 1.50 (m) and 2.00 (m) from the walls of the substation, at height z = 1.75 (m) from the ground level, and shown in Fig. 2 and 4.

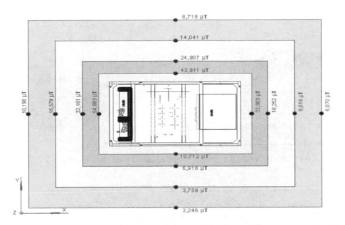

Fig. 2. The maximum magnetic flux density values calculated for areas I, II, III and IV at XY plane of observation (z = 1.75 m)

The maximum values of magnetic flux density in area I are in the range from 10.712 (μT) to 54.863 (μT), in area II from 6.918 (μT) to 32.161 (μT), in area III from 3.759 (μT) to 16.579 (μT), and in area IV from 2.246 (μT) to 10.198 (μT). Densities of magnetic flows inside the substation reache their maximum values at the intersection of XY plane with the primary transformer outlets, achieved by cable connections to buses of MV and LV switchgears, which, because of substation construction, can not be avoided, and are in the range from 0.05 (mT) to 6.40 (mT), while outside the housing they decreasing to the values from 100 (μT) to 50 (μT). Calculation results show that the value of magnetic flux density outside the substations do not exceed 54.863 (μT), in certain points of the area I, at a distance 0.20 (m) from the western cross side of the substation, at the level of transformer box. But, already at a distance from 0.50 (m) to 1.50 (m) from the substation the values decreasing to level from

32.161 (μT) to 2.246 (μT). Respecting the fact that magnetic flux density is proportional to the load force, and taking into account the usual loads of transformer station in normal operation of approximately 50 % of rated power, the maximum amount of magnetic flux density will not exceed the prescribed limit for the area of increased sensitivity. 2D and 3D distribution of magnetic flux density in the continuous distribution is given in Fig. 3 by using isolines. The maximum values of the electric field in area I are in the range from 0.052 (kV/m) to 0.352 (kV/m), in area II from 0.060 (kV/m) to 0.177 (kV/m), in area III from 0.023 (kV/m) to 0.081 (kV/m), and in area IV from 0.019 (kV/m) to 0.061 (kV/m). The maximum values of the electric field inside the substation are visible at the intersection of XY plane with a MV transformer outlets and cable connections of MV switchgear with the primary side of power transformer, and are in the range from 415.302 (kV/m) to 452.363 (kV/m), and with transformer box in the range from 2.194 (kV/m) to 16.912 (kV/m), while outside the housing they falling to the values between 1.00 (kV/m) and 0.50 (kV/m). 2D and 3D view of electric field distribution in the continuous distribution is shown in Fig. 5.

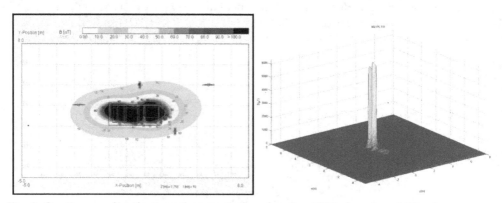

Fig. 3. Continuous distribution of magnetic flux density at XY plane (z = 1.75 m)

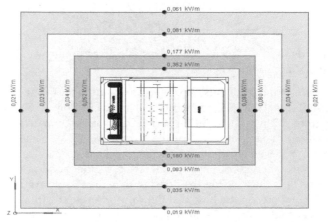

Fig. 4. The maximum electric field values calculated for areas I, II, III and IV at XY plane of observation (z = 1.75 m)

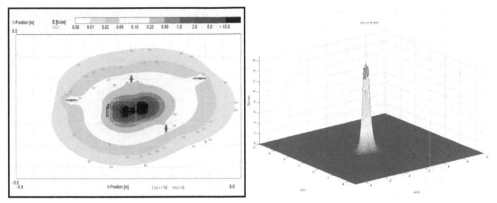

Fig. 5. Continuous distribution of electric field at XY plane (z = 1.75 m)

3.2 Distribution calculation of electric and magnetic field in xz plane for y = – 0.20 (m) (–5 (m) ≤ x ≤ 8 (m), 0 ≤ z ≤ 5 (m))

At a distance 0.20 (m) from the longitudinal south side of the substation (y = - 0.20 m), for observed XZ plane (–5 (m) ≤ x ≤ 8 (m), 0 ≤ z ≤ 5 (m)), the value of the magnetic flux density at z = 0.50 ÷ 1.75 (m) across the LV switchgear, is in the range from 14.051 (µT) to 10.686 (µT), while across the MV distribution switchgear the value is 8.111 (µT). The values of magnetic flux density across the power transformer are from 10.095 (µT) to 12.539 (µT) at z = 1.00 ÷ 1.50 (m). The highest values of the electric field are from 0.180 (kV/m) to 0.186 (kV/m) across the power transformer at z = 1.50 ÷ 1.75 (m). 2D and 3D view of magnetic flux density and electric field distribution in the continuous distribution is shown in Fig. 6 and 7.

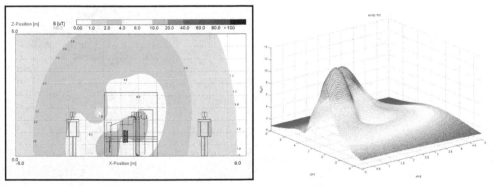

Fig. 6. Continuous distribution of magnetic flux density at XZ plane (y = - 0.20 m)

3.3 Distribution calculation of electric and magnetic field in xz plane for y = 2.10 (m) (–5 (m) ≤ x ≤ 8 (m), 0 ≤ z ≤ 5 (m))

At a distance 0.20 (m) from the longitudinal north side of the substation (y = 2.10 m), for observed XZ plane (–5 (m) ≤ x ≤ 8 (m), 0 ≤ z ≤ 5 (m)), the value of magnetic flux density is in the range from 101.102 (µT) to 145.202 (µT) at z = 1.00 (m) across the LV distribution switchgear, from 51.521 (µT) to 80.082 (µT) at z = 1.00 ÷ 1.75 (m) across the MV distribution switchgear, respectively from 35.197 (μT) to 74.145 (μT) across the power transformer. The highest values of the electric field are in the range from 0.500 (kV/m) to 0.795 (kV/m) at z = 1.00 ÷ 1.75 (m) across the power transformer and implemented MV and LV cable connections. 2D and 3D view of magnetic flux density and electric field distribution in the continuous distribution is shown in Fig. 8 and 9.

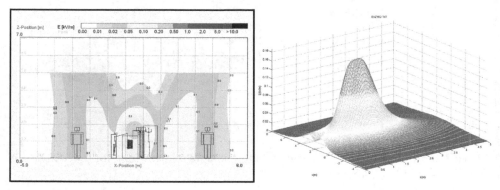

Fig. 7. Continuous distribution of electric field at XZ plane (y = – 0.20 m)

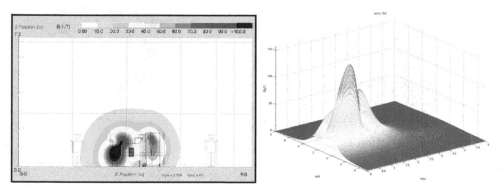

Fig. 8. Continuous distribution of magnetic flux density at XZ plane (y = 2.10 m)

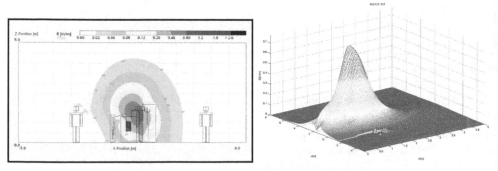

Fig. 9. Continuous distribution of electric field at XZ plane (y = 2.10 m)

3.4 Distribution calculation of electric and magnetic field in yz plane for x = – 0.20 (m) (–5 (m) ≤ y ≤ 8 (m), 0 ≤ z ≤ 5 (m))

At a distance 0.20 (m) of the western side of substation (x = – 0.20 m), for observed YZ plane (–5 (m) ≤ y ≤ 8 (m), 0 ≤ z ≤ 5 (m)), the value of magnetic flux density is in the range from 96.238 (µT) to 131.326 (µT), at z = 0.20 ÷ 0.50 (m) across the LV distribution switchgear, while at z = 1.00 ÷ 1.75 (m) the value decreases to 54.843 (µT). The highest values of the electric field are in the range from 0.049 (kV/m) to 0.050 (kV/m) at z = 1.75 (m) across the LV distribution switchgear, and implemented cable connections with LV power transformer outlets. 2D and 3D view of magnetic flux density and electric field distribution in the continuous distribution is shown in Fig. 10 and 11.

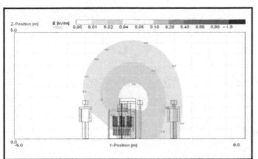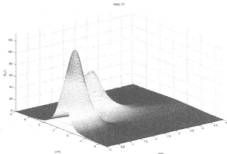

Fig. 10. Continuous distribution of magnetic flux density at YZ plane (x = – 0.20 m)

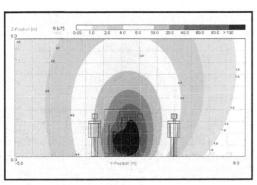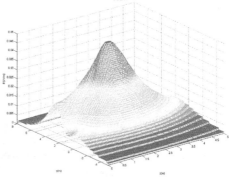

Fig. 11. Continuous distribution of electric field at YZ plane (x = – 0.20 m)

3.5 Distribution calculation of electric and magnetic field in yz plane for x = 3.10 (m) (–5 (m) ≤ y ≤ 8 (m), 0 ≤ z ≤ 5 (m))

At a distance 0.20 (m) from the western side of substation (x = 3.10 m), for observed YZ plane (–5 (m) ≤ y ≤ 8 (m), 0 ≤ z ≤ 5 (m)), the value of magnetic flux density is in the range from 40.194 (µT) to 68.846 (µT), at z = 0.20 ÷ 1.00 (m) across the MV distribution switchgear and connecting MV network cables, while at z = 1.00 ÷ 2.00 (m) toward the buses it falls to 27.954 (µT). The highest values of the electric field are in the range from 0.083 (kV/m) to

0.097(kV/m), at z = 1.00 ÷ 1.55 (m) across the MV switchgear and implemented cable connections with MV power transformer outlets. 2D and 3D view of magnetic flux density and electric field distribution in the continuous distribution is shown in Fig. 12 and 13.

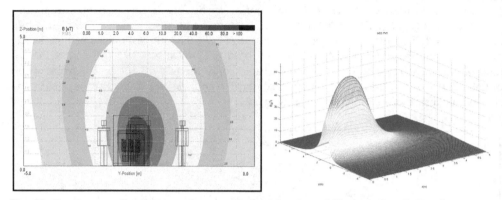

Fig. 12. Continuous distribution of magnetic flux density at YZ plane (x = 3.10 m)

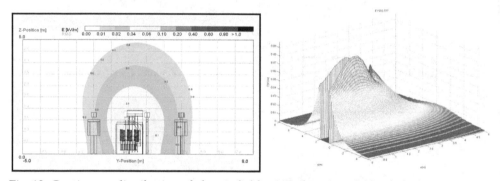

Fig. 13. Continuous distribution of electric field at YZ plane (x = 3.10 m)

In XY plane for z = 1.75 (m) and XZ plane for y = – 0.20 (m) calculated values of magnetic flux density satisfy limited values for area of occupational exposure (B_{max} = 500 (µT), E_{max} = 10 (kV/m)) and area of increased sensitivity (B_{max} = 100 (µT), E_{max} = 5 (kV/m)) in accordance with ICNIRP (1998) Internacional commission on non-ionizing radiation protection.

In XZ plane for y = 2.10 (m) and YZ plane for x = – 0.20 (m) calculated values of magnetic flux density do not satisfy limited values according to ICNIRP. Based on the above, the conclusion can be, that for reduction of magnetic field, it is necessary implementing of high–permeability protection in the form of shielding of LV switchgear with metal plates of steel and aluminum, thickness from 1 (mm) to 5 (mm). The values of the electric field are less than the maximum allowed according to ICNIRP.

In YZ plane for x = 3.10 (m) calculated values of magnetic flux density satisfy limited values for area of occupational exposure, and for most part of the observed area at a height z = 0.80 ÷ 1.00 (m) do not satisfy limited values for area of increased sensitivity, but it is also very unlikely that people will stay longer in this area. The values of the electric field are less than the maximum allowed according to ICNIRP.

4. Measurements of low frequency magnetic and electric fields of transformer station

Calculations of the values of low frequency electric and magnetic fields in electric power networks and facilities are usually limited by configurations, for which a fields sources can be quite simplified. Generally, when calculating, observing of all relevant emissions of individual supplements is carried out to be able to estimate their contribution to the resulting field. To calculate the electric field of electric power networks the voltage must be known, while magnetic fields are defined by currents. The phase voltage is generally constant, but the phase currents can vary within a wide area, depending of the load. Modern computer programs can calculate the distribution of the field of very complex power systems, and there is a need for confirmation of the results by measurements.When measuring the magnetic and electric fields it is necessary that the source of electromagnetic radiation and its environment in which measurements take place be precisely defined. The source of the field is each conductor flowed by current. Extremely dangerous are the sources which have winding conductors flowed by current (inductors, transformers). Shielding of such devices in ferromagnetic materials significantly reduces the field in their vicinity. The magnetic field can easily penetrate into buildings from external sources and therefore is considered as more dangerous than the electric field, which is usually attenuated by the first physical obstacle. Selection of points where the measurement will take place should be made based on assessment of the field intensity, or in the way that measuring have to be done in places where the largest values of electric and magnetic fields are expected. Measurement is necessary to be done in accordance with the regulations of the HRN IEC 61786-2001 – Measurement of low–frequency magnetic and electric field with regard to exposure of human beinges – Spacial requirements for instrumants and quidance for measurements and the instructions given in ENV 50166 European recommendations (People exposure to electromagnetic radiation on the low frequencies). For measuring the magnetic and electric fields intensity a digital measuring instrument EFA–300 is used, which finds application in researches and environmental studies for assessment of the electric and magnetic fields of electric power transmission and distribution networks and facilities with appropriate equipment and devices. It is designed to provide a sophisticated tool for precise studies of low–frequency energy impacts for engineers, experts in the field of health, safety and other profiles. The best choice to measure the fields that have one frequency component is the broadband mode. Broadband measurement in the range from 5 (Hz) to 32 (kHz) is performing by using the built–in isotropic probes. In the broadband mode, large display allows simultaneous viewing of measurement results and frequencies. There is a possibility of adding option, so–called "plug–in" which will extend the measurements possibilities. Smaller, "sniffer" probe, has a radius of 3 (cm) while a larger, more sensitive probe, has a surface of 100 (cm²). The user selects between the measurement of the effective or peak value in dynamic range from 1 (nT) to 31.6 (mT) for magnetic fields and from 1 (V/m) to 100 (kV/m) for electric fields.

The construction of the instrument can cause the error in measuring in a number of ways:

- measuring probe and a source of electromagnetic field can be capacitively coupled, which causes the increased values of the field on the instrument up to 100 times higher than the actual values. Such phenomenon can be caused by the parasitic capacitance between the mass of the instrument and the Earth,
- problem of instrument frequency band. It happens that instrument is sensitive outside of the nominal frequency area or even has a higher sensitivity. Then the signal, which is not expected, because it is outside the measurement range, creates the appearance of large values of the measuring field,
- dispersion phenomena that occurs on the surface of the irradiated object. Due to the reflection of the secondary radiation caused by the induced currents, on the uneven surfaces of the irradiated object field is deformed. The problem is particularly expressed in the electric field due to perturbations in the vicinity of any conductive object, including humans. Be sure to measure the electric field using dielectric tripods and probes holders. During measuring of the magnetic field the problem is less expressed, because the field perturbation occurs only in the vicinity of ferromagnetic materials, and the presence of people in the measuring field (metrologist) has no effect on his perturbation.

Uncertainty in measuring of the electric and magnetic field intensity with these instruments is complex measuring. Uncertainty consists of two components:

- uncertainty of calibration of the instrument (u_{cal}) - establishing a relationship between the values of parameters shown by the instrument and the corresponding values realized by standards. It is expressed by calibration certificate (certificate of calibration),
- uncertainty of instrument digital indicator resolution (u_{rez}) - caused by the resolution (the largest number of decimal places) of image of the instrument digital pointer.

When assessing the overall uncertainty, some partial uncertainties are taken for the range at which the measurement was made, and the total measurement uncertainty is expressed by the equation:

$$U = \sqrt{u_{cal}^2 + u_{rez}^2}$$

4.1 Measuring results of magnetic and electric field (KBTS 10(20)/0.4 kV, 630 kVA)

During the measurement preparations, based on project documentation according to which the subject substation is constructed (KBTS 10(20)/0.4 kV, 630 kVA), a total number of 36 measuring points was selected, outside the substation, where the maximum level of electric and magnetic field was expected, as well as 7 measurement points inside the substation. At distances 0.50 (m), 1.00 (m) and 1.50 (m), outside the walls of substation, and heights 1.75 (m), 1.50 (m) and 1.00 (m) above the ground level, corresponding to the head, chest and lower human extremities, a 108 measurement points were located. Inside the substation, a measurement points were chosen in the vicinity of MV and LV switchgears, transformer and transformer outlets, as well as in the vicinity of the implemented cable connections to the MV and LV buses (Fig. 14).

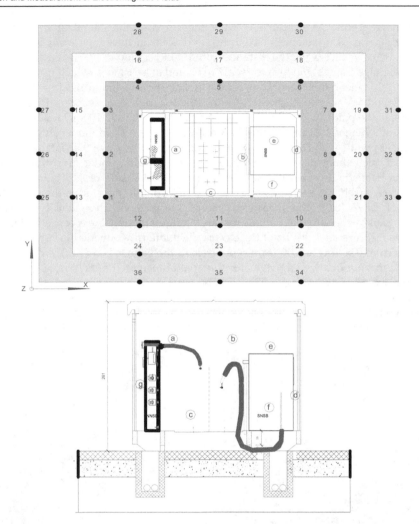

Fig. 14. Measurement points of electric and magnetic field outside and inside the substation

Measurement area was related to occupational exposure area and an area of increased sensitivity. The measurement was performed from 14:05 AM to 16:25 PM, at a temperature 20.9°C and relative humidity 28.4 % in the substation, and air temperature 22°C and relative humidity 27 % outside the building, with constant substation load. After locating the measurement points, measuring instrument EFA–300 Field Analyzers has been checked, climatic conditions were collected, and measurements and analysis of the magnetic and electric fields were performed in the substation that is connected to 50/60 Hz power transmission system and distribution network with devices that use such an energy (Fig. 15). The results of the measured values of magnetic flux density and electric field intensity at measurement points are related to the currently load of substation of 40 % of rated power, with measured current at LV side of 375 (A) and MV side load current of 15 (A). The highest values of magnetic flux density outside the substation at a heights 1.75 (m), 1.50 (m), and

1.00 (m) above ground level (Table 1) were measured at the lateral of the LV side of transformer station, at a distance of 0.50 (m), and were in the range from 57.699 (μT) to 24.892 (μT), while by increasing the distance from the substation to 1.00 (m) that values were falling to the range from 27.750 (μT) to 16.937 (μT), and at a distance of 1.50 (m) they droped to 8.378 (μT). Magnetic flux density measured inside the substation reaches its maximum values at LV and MV transformer outlets, implemented cable connections with MV and LV switchgears, in the point "a" 172.150 (μT) and in the point "b" 195.100 (μT), while in the LV switchgear, at a height 1.00 (m) above the ground level, the maximum value of magnetic flux density at point "g" is 119.185 (μT). Measurement results prove the statement obtained by numerical calculations of substation magnetic flux density, under maximum load, that the value of magnetic flux density outside the lateral of LV side of substation exceed the value of 57.699 (μT), but even at a distance of 1.50 (m) from the substation they are falling to the value of 2.897 (μT). The measured values of magnetic flux density outside the substations satisfy the limited values for area of occupational exposure according to ICNIRP. The measurement results of the electric field intensity outside the substation walls, at distances from 0.50 (m) to 2.00 (m) do not exceed the value of 0.176 (kV/m), and are far less than the maximum allowed values for the area of increased sensitivity and the area of occupational exposure according to ICNIRP. The measured values of electric field inside the substation are at point "a" 8.120 (kV/m), point "b" 10.155 (kV/m), and point "c" 6.550 (kV/m) but outside the equipment housing they are falling to the values from 0.583 (kV/m) to 0.087 (kV/m), and thus satisfy limited values for the area of professional exposure.

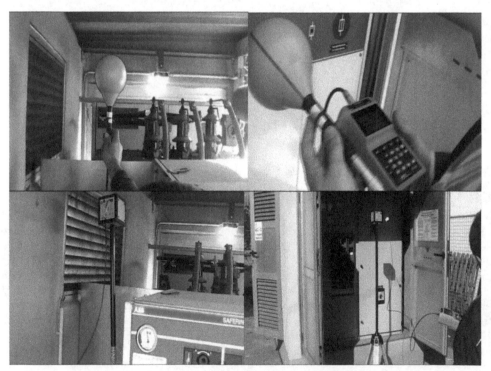

Fig. 15. The measurement of magnetic and electric fields in KBTS 10(20)/0.4 kV

4.2 Calculation results of magnetic and electric fields (KBTS 10(20)/0.4 kV, 630 kVA)

For a given substation loaded with 40 % of rated power, with measured current at LV side of 375 (A) and on the MV side of 15 (A), the numerical calculation of the magnetic field distribution was performed in the XY plane of the substation and at a heights 1.75 (m), 1.50 (m) and 1.00 (m) from the ground level. Calculation of the electric field distribution in the XY plane of the substation, at –5 (m) ≤ x ≤ 8 (m) and –5 (m) ≤ y ≤ 8 (m), at a height z = 1.75 (m) from the ground level is identical to the calculation of the electric field outside the walls and inside the substation at full load, because the electric field depends on the voltage which does not change its value at any substation load. Calculated values of magnetic flux density satisfy limited values for area of occupational exposure (B_{max} = 500 (µT), E_{max} = 10 (kV/m)) and area of increased sensitivity (B_{max} = 100 (µT), E_{max} = 5 (kV/m)) in accordance with ICNIRP (1998) Internacional commission on non-ionizing radiation protection. The obtained calculation results are presented in different variants of graphical formats that describe 2D and 3D continuous distribution of magnetic flux density (Fig. 16 to 18).

4.2.1 Calculation results of magnetic field distribution in xy plane for z = 1.75 (m) (–5(m) ≤ x ≤ 8(m) and –5 (m) ≤ y ≤ 8(m))

The values of magnetic flux density for z = 1.75 (m), at a distance of 0.50 (m) from the substation sites are in the range from 4.932 (µT) to 18.240 (µT), at a distance of 1.00 (m) are in the range from 3.125 (µT) to 14.355 (µT), and at a distance of 1.50 (m) are in the range from 3.041 (µT) to 10.634 (µT). Magnetic flux density inside the substation reaches its maximum values at the intersection of XY plane with the primary transformer outlets, implemented cable connections to MV and LV switchgears and is in the range from 0.150 (mT) to 0.366 (mT). Calculation results show that the values of magnetic flux density outside the substation do not exceed 22.433 (µT) in certain points at a distance of 0.20 (m) from the northern transversal side of the substation, in the level of transformer box. Already, at a distance from 0.50 (m) to 2.00 (m) from the substation they are decreasing to the value from 18.240 (µT) to 7.810 (µT).

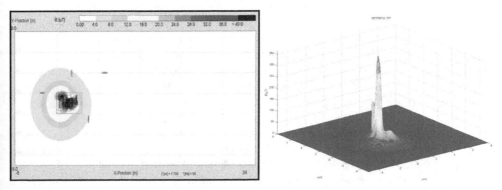

Fig. 16. Continuous distribution of magnetic flux density at XY plane (z = 1.75 m)

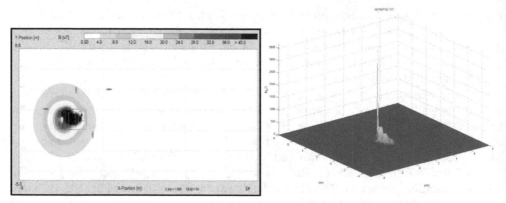

Fig. 17. Continuous distribution of magnetic flux density at XY plane (z = 1.50 m)

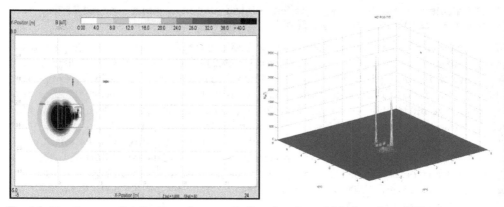

Fig. 18. Continuous distribution of magnetic flux density at XY plane (z = 1.00 m)

4.2.2 Calculation results of magnetic field distribution in xy plane for z = 1.50 (m) (−5 (m) ≤ x ≤ 8 (m) and −5 (m) ≤ y ≤ 8 (m))

The values of magnetic flux density for z = 1.50 (m), at a distance of 0.50 (m) from the substation sites, are in the range from 5.299 (µT) to 26.518 (µT), at a distance of 1.00 (m) are in the range from 4.032 (µT) to 18.876 (µT) and at a distance of 1.50 (m) are in the range from 3.169 (µT) to 12.672 (µT). Magnetic flux density inside the substation reaches its maximum value at the intersection of XY plane with the primary transformer outlets, implemented cable connections with MV switchgear and the block of MV buses, and is in the range from 1.067 (mT) to 3.671 (mT). Calculation results show that the value of magnetic flux density outside the substation does not exceed 33.461 (µT) in certain areas points, at a distance of 0.20 (m) from the western lateral side of the substation, in the level of LV switchgear. Already, at a distance from 0.50 (m) to 2.00 (m) from the substation it falls to the value from 26.518 (µT) to 8.706 (µT).

4.2.3 Calculation results of magnetic field distribution in xy plane for z = 1.00 (m) (–5 (m) ≤ x ≤ 8 (m) and –5 (m) ≤ y ≤ 8 (m))

The values of magnetic flux density for $z = 1.00$ (m), at a distance of 0.50 (m) from the substation sites are in the range from 5.742 (µT) to 60.964 (µT), at a distance of 1.00 (m) are in the range from 4.293 (µT) to 29.109 (µT) and at a distance of 1.50 (m) are in the range from 3.333 (µT) to 16.367 (µT). Magnetic flux density inside the substation reaches its maximum value at the intersection of XY plane with the primary transformer outlets, implemented cable connections with MV switchgear, where is in the range from 1.184 (mT) to 3.610 (mT), and at the LV cable outlets and LV busbar connections, where is in the range from 1.048 (mT) to 2.015 (mT), while outside of the equipment housing decreases to the value from 100 (µT) to 50 (µT).

4.3 Comparison of calculation results and measurements of low frequency magnetic and electric fields (KBTS 10 (20)/0.4 kV, 630 kVA)

If we want to make a comparison of calculation results and measurements we will note certain differences. By calculation a distribution of magnetic and electric fields is obtained in the XY, XZ, YZ planes of the substation, while measuring only give distribution of magnetic and electric fields in XY plane at a height 1.75 (m), 1.50 (m) and 1.00 (m) from the ground level. From above, follows the importance of calculation for determining the levels of emitted magnetic and electric fields, caused by the substation. It is also important to notice the importance of projection of diagram parts that connects the points at which the magnetic flux density or electric field intensity are approximated. In this way, this ensures that in any case, human bodies will not lead to a situation that they are exposed to the values of magnetic and electric fields that exceed the limits prescribed by the Regulations on Non-Ionising Radiation. In addition, it is interesting to observe where can be terminate the consideration of substation as a significant source of magnetic and electric fields, because this releases substation from the prescribed periodic measurements. From a comparison the problem that appears in the measurement of magnetic and electric fields of the substation can be noted. By measuring, the value of magnetic and electric field can be obtained only at a certain height. But, for a precise view, the field should be measured at a different heights, and only then a data that would be relevant to determine the nature and level of magnetic and electric fields will be obtained. In addition, there is a problem with number of suitable places at which measuring can be done. Number of places is limited because sometimes the field conditions greatly complicate the implementation of measurement. Beside, the measurement of magnetic field is performed under certain substations load, which changes according to the daily and annual load diagram, so that comes into play only measuring of the electric field that is mostly constant. In Tables 1 to 4, the measured and calculated values of electric and magnetic fields are shown, as well as the percentage measurement error in relation to the calculation.

Diagrams on Fig. 19 to 22 present calculated and measured values of magnetic flux density inside and outside the substation, as well as the errors of measured relative to the calculated values.

Mark of measurement place outside transformer station		Calculated B above ground of TS			Measured B above ground of TS			Error		
		z=1.75 m	z=1.50 m	z=1.00 m	z=1.75 m	z=1.50 m	z=1.00 m			
(m)		B (μT)			B (μT)			(%)		
1	x = 0.50	y = 0.00	15.183	33.291	20.395	14.304	31.450	19.280	5.79	5.53	5.47
2	x = 0.50	y = 1.05	16.853	60.772	26.486	15.720	57.699	25.176	6.72	5.06	4.95
3	x = 0.50	y = 2.10	15.533	26.433	19.131	14.570	24.892	18.186	6.20	5.83	4.94
4	x = 0.00	y = 2.60	13.219	17.187	14.952	12.376	15.996	13.950	6.38	6.93	6.70
5	x = 1.45	y = 2.60	9.989	14.108	11.822	9.263	13.139	11.050	7.27	6.87	6.53
6	x = 2.90	y = 2.60	5.236	6.392	5.731	4.897	6.056	5.436	6.47	5.26	5.15
7	x = 3.40	y = 2.10	4.932	5.742	5.299	4.613	5.391	4.990	6.47	6.11	5.83
8	x = 3.40	y = 1.05	6.001	6.720	6.370	5.607	6.305	5.980	6.57	6.18	6.12
9	x = 3.40	y = 0.00	5.275	5.747	5.513	4.954	5.386	5.168	6.09	6.28	6.26
10	x = 2.90	y = -0.50	5.845	6.437	6.122	5.492	6.120	5.824	6.04	4.92	4.87
11	x = 1.45	y = -0.50	10.297	14.981	12.287	9.644	14.174	11.555	6.34	5.39	5.96
12	x = 0.00	y = -0.50	13.700	25.061	17.217	12.801	23.758	16.356	6.56	5.20	5.00
13	x = 1.00	y = 0.00	12.073	20.467	14.891	11.464	19.247	14.128	5.04	5.96	5.12
14	x = 1.00	y = 1.05	14.355	29.025	18.863	13.644	27.750	18.052	4.95	4.39	4.30
15	x = 1.00	y = 2.10	11.755	17.939	13.932	11.246	16.935	13.248	4.33	5.60	4.91
16	x = 0.00	y = 3.10	9.067	11.102	9.971	8.514	10.490	9.365	6.10	5.51	6.08
17	x = 1.45	y = 3.10	7.494	9.525	8.379	6.942	8.804	7.790	7.37	7.57	7.03
18	x = 2.90	y = 3.10	4.361	5.171	4.517	4.087	4.865	4.242	6.28	5.92	6.09
19	x = 3.90	y = 2.10	3.825	4.293	4.032	3.565	4.029	3.775	6.80	6.15	6.37
20	x = 3.90	y = 1.05	4.373	4.903	4.581	4.147	4.615	4.315	5.17	5.87	5.81
21	x = 3.90	y = 0.00	4.031	4.355	4.181	3.822	4.085	3.951	5.18	6.20	5.50
22	x = 2.90	y = -1.00	4.867	5.316	5.073	4.516	4.922	4.705	7.21	7.41	7.25
23	x = 1.45	y = -1.00	8.347	10.481	9.328	7.886	9.828	8.827	5.52	6.23	5.37
24	x = 0.00	y = -1.00	9.985	14.406	11.568	9.296	13.655	10.920	6.90	5.21	5.60
25	x = 1.50	y = 0.00	9.186	13.155	10.645	8.731	12.381	10.185	4.95	5.88	4.32
26	x = 1.50	y = 1.05	10.630	16.328	12.661	9.980	15.350	11.910	6.11	5.99	5.93
27	x = 1.50	y = 2.10	8.863	11.974	10.059	8.378	11.274	9.460	5.47	5.85	5.95
28	x = 0.00	y = 3.60	6.540	7.642	7.036	6.125	7.143	6.628	6.35	6.53	5.80
29	x = 1.45	y = 3.60	5.690	6.781	6.169	5.277	6.286	5.725	7.26	7.30	7.20
30	x = 2.90	y = 3.60	3.667	4.221	3.899	3.426	3.942	3.650	6.57	6.61	6.39
31	x = 4.40	y = 2.10	3.041	3.333	3.169	2.895	3.148	3.011	4.80	5.55	4.99
32	x = 4.40	y = 1.05	3.352	3.625	3.480	3.158	3.397	3.276	5.79	6.29	5.86
33	x = 4.40	y = 0.00	3.170	3.373	3.268	3.028	3.189	3.095	4.48	5.46	5.29
34	x = 2.90	y = -1.50	4.690	4.396	4.220	4.386	4.090	4.005	6.48	6.96	5.09
35	x = 1.45	y = -1.50	6.372	7.517	6.838	6.013	7.072	6.493	5.63	5.92	5.05
36	x = 0.00	y = -1.50	7.305	9.278	8.070	6.802	8.652	7.515	6.89	6.75	6.88

Table 1. Comparison of measured and calculated values of magnetic flux density outside TS

Mark of measurement place inside transformer station			Measured value	Calculated value	Error
(m)			B (µT)		(%)
A	x = 1.30	y = 1.10	z = 1.75	172.150	183.498	6.18
B	x = 1.70	y = 1.10	z = 1.75	195.100	211.018	7.54
C	x = 1.45	y = 1.80	z = 0.30	5.876	6.385	7.97
D	x = 2.80	y = 1.00	z = 1.00	17.152	18.401	6.79
E	x = 2.50	y = 1.05	z = 2.15	5.858	6.355	7.82
F	x = 2.50	y = 1.00	z = 1.50	16.155	17.304	6.64
G	x = 0.15	y = 1.05	z = 1.00	119.185	129.367	7.87

Table 2. Comparison of measured and calculated values of magnetic flux density inside TS

Mark of measurement place outside transformer station		Calculated value	Measured value	Error
(m)		z=1.75m		
		E (kV/m)		(%)
1	x = 0.50	y = 0.00	0.015	0.014	6.67
2	x = 0.50	y = 1.05	0.028	0.030	-7.14
3	x = 0.50	y = 2.10	0.033	0.035	-6.06
4	x = 0.00	y = 2.60	0.046	0.043	6.52
5	x = 1.45	y = 2.60	0.176	0.164	6.82
6	x = 2.90	y = 2.60	0.057	0.055	3.51
7	x = 3.40	y = 2.10	0.046	0.043	6.52
8	x = 3.40	y = 1.05	0.057	0.060	-5.26
9	x = 3.40	y = 0.00	0.039	0.040	-2.56
10	x = 2.90	y = -0.50	0.037	0.035	5.41
11	x = 1.45	y = -0.50	0.083	0.085	-2.41
12	x = 0.00	y = -0.50	0.020	0.019	5.00
13	x = 1.00	y = 0.00	0.010	0.010	0.00
14	x = 1.00	y = 1.05	0.017	0.016	5.88
15	x = 1.00	y = 2.10	0.022	0.020	9.09
16	x = 0.00	y = 3.10	0.031	0.028	9.68
17	x = 1.45	y = 3.10	0.079	0.080	-1.27
18	x = 2.90	y = 3.10	0.038	0.035	7.89
19	x = 3.90	y = 2.10	0.030	0.028	6.67
20	x = 3.90	y = 1.05	0.033	0.030	9.09
21	x = 3.90	y = 0.00	0.025	0.025	0.00
22	x = 2.90	y = -1.00	0.022	0.020	9.09
23	x = 1.45	y = -1.00	0.033	0.030	9.09
24	x = 0.00	y = -1.00	0.011	0.010	9.09
25	x = 1.50	y = 0.00	0.010	0.011	-10.00
26	x = 1.50	y = 1.05	0.011	0.010	9.09
27	x = 1.50	y = 2.10	0.001	0.001	0.00
28	x = 0.00	y = 3.60	0.021	0.020	4.76
29	x = 1.45	y = 3.60	0.050	0.047	6.00

30	x = 2.90	y = 3.60	0.024	0.025	-4.17
31	x = 4.40	y = 2.10	0.020	0.020	0.00
32	x = 4.40	y = 1.05	0.021	0.020	4.76
33	x = 4.40	y = 0.00	0.015	0.014	6.67
34	x = 2.90	y = -1.50	0.014	0.015	-7.14
35	x = 1.45	y = -1.50	0.018	0.020	-11.11
36	x = 0.00	y = -1.50	0.009	0.010	-11.11
37	x = 1.45	y = -2.00	0.011	0.010	9.09
38	x = 2.00	y = 1.05	0.008	0.008	0.00
39	x = 1.45	y = 4.10	0.023	0.021	8.70
40	x = 4.90	y = 1.05	0.014	0.015	-7.14

Table 3. Comparison of measured and calculated values of electric field intensity outside TS

Mark of measurement place inside transformer station			Measured value	Calculated value	Error
(m)			E (kV/m)		(%)
A	x = 1.30	y = 1.10	z = 1.75	8.120	8.757	7.27
B	x = 1.70	y = 1.10	z = 1.75	10.155	10.957	7.32
C	x = 1.45	y = 1.80	z = 0.30	6.550	7.105	7.81
D	x = 2.80	y = 1.00	z = 1.00	0.001	0.001	0.00
E	x = 2.50	y = 1.05	z = 2.15	0.001	0.001	0.00
F	x = 2.50	y = 1.00	z = 1.50	0.001	0.001	0.00
G	x = 0.15	y = 1.05	z = 1.00	0.085	0.090	5.56

Table 4. Comparison of measured and calculated values of electric field intensity inside TS

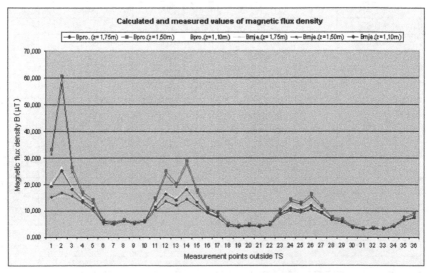

Fig. 19. Diagram of calculated and measured values of magnetic flux density outside the substation

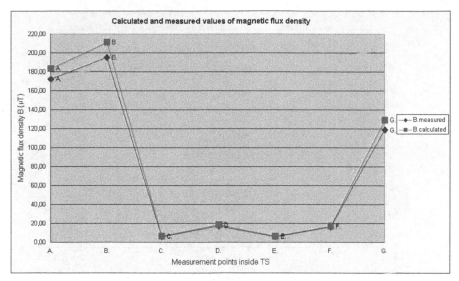

Fig. 20. Diagram of calculated and measured values of magnetic flux density inside the substation

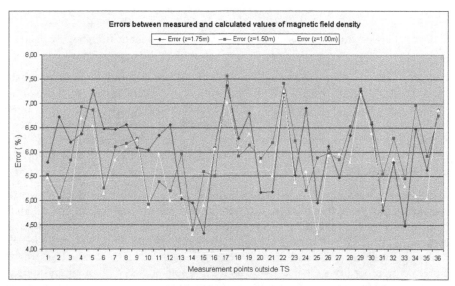

Fig. 21. Errors between measured and calculated values of magnetic flux density outside the substation

From the diagrams it is evident that the values of magnetic flux density obtained by calculation appropriately follow changes in the measured values. The presented calculation gives the percentage error between measured and calculated values for some measuring points, and which ranges from 4.32 % to 7.25 % outside, or from 6.18 % to 7.97 % inside the substation.

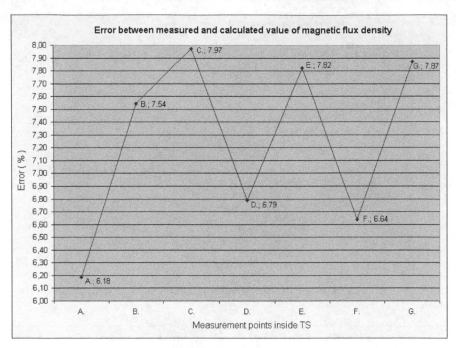

Fig. 22. Errors between measured and calculated values of magnetic flux density inside the substation

Calculated and measured values of electric field intensity inside and outside the substation are presented at a diagrams on Fig. 23 to 26, as well as the errors of measured in relation to the calculated values. From a diagrams it is evident that the value of electric field intensity obtained by calculation appropriately follow changes in the measured values. The presented calculation gives the percentage error between measured and calculated values for some measuring points outside the substation, which ranges from -11.11 % to 9.68 % and 7.81 % inside the substation.

The calculation results give a satisfactory coincidence with the results of experimental measurements, indicating the validity of implementing and developing such a calculations for practical purposes related to the design and reconstruction of existing substations. From the economic point of view, it is possible to achieve significant savings because an expensive experimental measurements and repairs can be reduced. For evaluation the field distribution both procedures are necessary, as they supplemented each other and thus allow a safe assessment of fields sizes. Interestingly, the maximum value of magnetic flux density is calculated and measured at a height z = 1.00 (m), the area of human hips, so this height is imposed as a referece regarding the allowable sizes of exposure to the non–ionizing radiation of electromagnetic fields. When measuring the values of the magnetic field density around the substation, some differences in values measured at different transformer station walls were detected. The reason for this is the way of placement of the main sources of magnetic field inside the substation which are the MV and LV transformer outlets and LV distribution outlets.

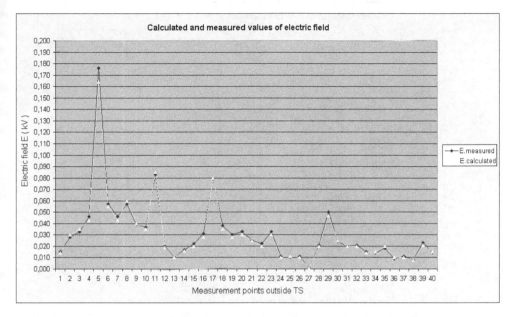

Fig. 23. Diagram of calculated and measured values of electric field intensity outside the substation

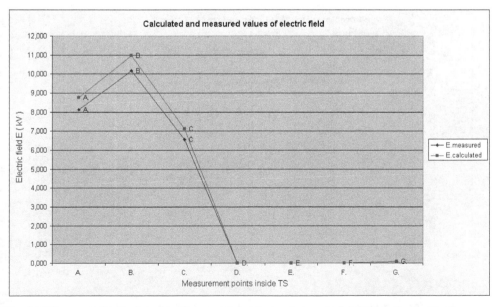

Fig. 24. Diagram of calculated and measured values of electric field intensity inside the substation

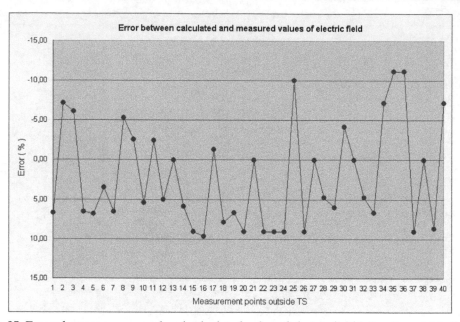

Fig. 25. Errors between measured and calculated value of electric field intensity outside the substation

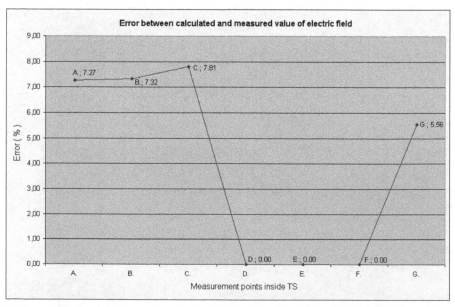

Fig. 26. Errors between measured and calculated value of electric field intensity outside and inside transformer station

5. Conclusion

Calculation and measurement of LV electric and magnetic fields and their interconnectivity are the main problems in transmission and distribution of power energy in terms of standardized electromagnetic compatibility and human exposure to non-ionizing electromagnetic radiation. Solving of these problems is reduced to solving nonlinear differential equations by modeling and application of numerical methods and experimental measurements of low-frequency electric and magnetic fields. The research presents the application of mathematical models and charge simulation method (CSM) for calculating of low-frequency electric field distribution, while the calculation of magnetic flux density distribution inside and around the substation, which indicates the level of low-frequency magnetic field, is performed by the procedure based on the application of Biot-Savart's law for the induction of straight stream line of finite length. The original scientific contribution of this research is determining the 3D distribution of low-frequency electric and magnetic fields, their interaction under conditions of complex geometry and standardized substation electromagnetic compatibility (EMC) in the area of biological effects of electromagnetic fields. Obtained 3D models represent very complex functional dependence of electric and magnetic fields distribution, as the basis for the objectified physical measurements to develop an optimal variants for solving electromagnetic compatibility (EMC) in existing and new power facilities. Satisfactory accuracy of the results obtained by calculations with the results of experimental measurements of EFA-300 Field Analyzers instrument is confirmed, which indicate validity of implementation and development of such calculations for the executable solutions of transformer stations. From an economic point of view this method of calculation could reduce the need for expensive experimental measurements and repairs of power plants, providing confirmation that complicated theoretical researches resulting in appropriate executable solutions. Presented mathematical models, calculation, measurement and visual 3D distribution of electric and magnetic fields represent a realistic assumption for the study of interactions between electromagnetic fields and human bodies at the macroscopic and static level, with finding of certain optimization criterias, in order to create a new technological, process solutions and design methods. The research results are important both from the scientific point of view and from the standpoint of possibilities for practical applications.

6. References

[1] I.Kapetanovic, V. Madzarevic, A. Muharemovic, H.Salkic, *"Exposure to Low Frequency Magnetic Fields of a Transformer Station"*, IJESSE -International Journal of Electrical Systems Science and Engineering, pp. 120-128, ISSN 2070-3953, Volume 1, Number 2, 2008.

[2] V.Madzarevic, A.Nuhanovic, A.Muharemovic, H.Salkic, *"Numerical calculation of magnetic dissipation in power transformers,* "Boundary Elements XXVII: Incorporating Electrical Engineering and Electromagnetics", WIT Transactions on Modelling and Simulation p.p. 673-683, ISNN 1743-355X, Vol 39, 2005.

[3] H.Salkic, V.Madzarevic, I.Kapetanovic, A.Muharemovic, *"Numerical calculation of magnetic dissipation and forces on coil in power transformers"*, CIRED 2005, 6-9 June 2005, Turin, Italy

[4] H.Salkic, V.Madzarevic, E.Hukic, *"Calculation and Measuring of Low-frequency Electric Field Distribution of 10(20)/0,4 kV, 630 kVA Transformer Station"*, 43rd international universities power engineering conference (UPEC2008) University of Padova, and the Department of Industrial Engineering, University of Cassino September 1-4, 2008. Padova, Italy

[5] H.Salkic, V.Madzarevic, A. Muharemovic, E.Hukic, "Numerical Solving and Experimental Measuring of Low frequency Electromagnetic Fields in Aspect of Exposure to Non- ionizing Electromagnetic Radiation", The 14th International Symposium on Energy, Informatics and Cybernetics: ISAS 2008. WMSCI 2008, ISBN-10: 1-934272-32-9 (Vol II), may 2008. Orlando, Florida, USA

[6] D. Poljak, *"Human exposure to non-ionizing radiation"*, Zagreb: Kigen ltd., 2006.

[7] V. Madzarevic, A. Nuhanovic, A. Muharemovic, H. Salkic, *"Numerical calculation of magnetic dissipation and power tranformers"*, Boundary Elements XXVII: Incorporating Electrical Engineering and Electromagnetics, WIT Transactions on Modelling and Simulation, pp.673-683. ISNN 1743-355X, Vol. 39, 2005.

[8] V. Madzarevic, H. Salkic, N. Mehinovic, E. Hukic, *"Calculation And Measuring Of Low-frequency Magnetic Field Of 10(20)/0,4 kV Transformer Station"*, XVIII International conference on Electrical Machines Vilamoura, Portugal 6-9 semptember 2008.

[9] EFC-400 Simulation Software- Narda Safety Test Solutions and Partner FGEU Wandel&GoltermannGmbH&Co,.Elektronische Meûtechnik Postfach 1262, 72795 Eningen, Allemagne

[10] S. Kraljevic, D. Poljak, V. Doric, *A simplified method for the assessment of ELF magnetic fields from three-phase power lines*, Boundary Elements XXVII Southampon, UK, Boston, USA: WIT Press, Computational Mechanics Inc, 2005.

[11] D. Poljak, *Electromagnetic Modelling of Wire Antenna Structures*, WIT Press, Southampton-Boston 2002.

[12] S. H. Myung, B. Y. Lee, J. K. Park, *Three Dimensional Electric Field Analysis of Substation Using Nonuniform Optimal Charge Simulation*, 9 International Symposium on High Voltage Engineering, Graz Austria August 1995.

[13] B.Y. Lee, J.K. Park, S. H. Myung, S. W. Min, E. S. Kim, *An Effective Modelling Method to Analyze Electric Field around Transmission Lines and Substations Using Generalized Finite Line CHarge*, IEEE Trans. On Power Delivery, Vol. 12, No. 3, pp. 1143-1150, July 1997.

[14] D. Poljak, C.A Brebbia, *Boundary Elements for Electrical Engineers*, WIT Press, Southampton-Boston, 2005.

[15] Z.Haznadar , S.Zeljko , *Electromagnetic Field, Waves and Numerical Methods*, Amsterdam, Berlin, Oxford, Tokyo, Washington, D.C. : IOS Press, 2000 (monography).

[16] Numerical Calculations of Induced Currents: Electric and Magnetic fields, Sources of EMFs, National Grid EMF, 2006.

10

Beneficial Effects of Electromagnetic Radiation in Cancer

I. Verginadis[1], A. Velalopoulou[1], I. Karagounis[1], Y. Simos[1],
D. Peschos[1], S. Karkabounas[1] and A. Evangelou[1]
Laboratory of Physiology, Faculty of Medicine,
University of Ioannina, Ioannina,
Greece

1. Introduction

In the last three decades, a large number of studies have been arisen, which dealt with the effects of electromagnetic fields (EMFs) in biological systems (Aaron & Ciombor, 1993; Tao & Henderson, 1999; Tofani et al., 2002b; Walker et al., 2007). The EMFs have been used in very important technological applications that concern in diagnosis (e.g. MRI, X-rays, CT). A part of scientific community turned its interest to the application of EMFs in the treatment of various pathological conditions, mainly in experimental level, such as osteoporosis, the bone fractures, muscle regeneration, diabetes, arthritis and neurological disorders (Barker et al., 1984; Bassett et al., 1974a; Dortch & Johnson, 2006; Fischer et al., 2005; László et al., 2011; Otter et al., 1998; Tabrah et al., 1990; Wang et al., 2010). The last decades EMFs are gradually being used in the research field of one of the most deadly disease known to man, cancer.

Cancer constitutes one of the most serious causes of death worldwide and according to World Health Organization (WHO), it accounted for 7.6 million deaths (around 13% of all deaths) in 2008 (World Health Organization [WHO], 2011). Deaths from cancer are projected to continue rising to over 11 million in 2030.

The modern methods of cancer treatment include: chemotherapy, radiation therapy, surgery, immunotherapy, monoclonal antibody therapy etc. Clinicians select the suitable treatment for the patient, examining, apart from the general situation of his health, the type of cancer, the location and grade of the tumour as well as the stage of the disease. Certain types of cancer, due to their complexity, require combination of treatments. The patients, however, are called to face the side effects, which often accompany the therapeutic methods, such as fatigue, nausea and vomiting, loss of appetite, pain, hair loss, nerve and muscle effects, metastasis and many others. The final aim of the scientists is to increase the effectiveness of the existing treatments, eliminate the side effects and to improve as much as possible the quality of life of patient. Data provided by several studies support the possible development of new alternative forms of treatment, which in combination with the already existing ones, could contribute in the achievement of this aim.

Several epidemiological studies have implicated the EMFs with the induction of mutations, leukaemia and neurological and cardiovascular disorders (Ahlbom et al., 2001). Although there are indications of the adverse effects of EMFs, the authors' opinion is that this is the one side of the coin. Therefore, the other side of the coin is the study of possible anticancer effects of EMFs; this constitutes a challenge.

There is data supporting the opinion that the use of EMFs has effects in the cell proliferation and in malignant tumours in animals (Tofani et al., 2001; Yamaguchi et al., 2006). It has also been reported that EMFs could act synergistically with chemotherapeutic agents (Gray et al, 2000; Ruiz Gómez et al., 2002), and reverse the resistance of cancer cells in chemotherapy (Hirata et al., 2001; Janigro et al., 2006). Certain clinical studies have shown that the application of EMFs in cancer patients, does not present side effects or toxicity (Barbault et al., 2009; Roncheto et al., 2004). Existed data, also indicate that they prolong the survival time of patients and inhibit the disease progression (Barbault et al., 2009; Kirson et al., 2007). Consequently, EMFs can be used as a low-cost, safe and adjuvant treatment of the existing anticancer therapy.

2. *In vitro* studies

Various *in vitro* techniques have been employed to examine the effects of EMFs on cancer as well as normal cells. These studies investigated whether exposure of cells to EMFs results in modulation of cell growth and also the type of cell death (apoptosis or necrosis). Data regarding the mechanism of action are scarce and will be discussed in more detail in a later paragraph (Section 7).

According to Chen Y.C. et al., exposure of HeLa (human cervical cancer) and PC-12 (rat pheochromocytoma) cells, for 72h continuously, to ELF-EMF of 1.2±0.1 mT, at 60 Hz, results in a significant decrease in cell proliferation, at about 18.4% and 12.9%, respectively (Chen Y.C. et al., 2010). The same effect on HeLa cells was also seen, after exposure to PEMF of 0.18T, at 0.8Hz, for 16h. A decrease, at about 15%, in cell proliferation, was observed 24h later (Tuffet et al., 1993). The application of ELF-EMF (at 50Hz) of different intensities and durations on PC-12 cells, results not only in a slight transient decrease of the proliferation rate but also in morpholigical differentiation (Morabito et al., 2010).

The ability of human colon adenocarcinoma cells to proliferate was tested, when ELF-EMF of 1.5mT peak and 1Hz or 25Hz, for 15min or 360min, was applied. Results revealed a significant decrease in cell growth, in cells exposed to 1Hz for 360min (Ruiz Gómez et al., 1999). The same research team studied the exposure of HCA-2/1cch (human colon adenocarcinoma) cells to 25 Hz, 1.5 mT, for 2h and 45 min. In presence of dexamethasone, a decrease of 55.84±7.35% of the relative cell number occurred (Ruiz Gómez et al., 2000).

The effects of static magnetic fields were also studied on cell proliferation. A 64h exposure under a 7T uniform static magnetic field leads to reduction of viable cell number by 19.04±7.32%, 22.06±6.19%, and 40.68±8.31% in three human cancer cell lines: the HTB 63 (melanoma), HTB 77 IP3 (ovarian carcinoma), and CCL 86 (lymphoma; Raji cells) cell line, respectively (Raylman et al., 1996). Furthermore, the static magnetic fields seem not to affect the proliferation of normal cell lines according to Tofani et al., 2001. Two transformed cell lines, WiDr (human colon adenocarcinoma) and MCF-7 (human breast adenocarcinoma), and the untransformed cell line, MRC-5 (embryonal lung fibroblast), were exposed to 3mT static MF, modulated in amplitude with 3mT ELF-MF, at 50 Hz, with a superimposition of

ELF magnetic field, for 20min. Both WiDr and MCF-7 cells, showed morphological evidence of increased apoptosis, while MRC-5 cells remained intact and did not show any increase in apoptosis (Tofani et al., 2001).

Apoptosis was determined as the cause of cell death after exposure to EMF according to a study by Hisamitsu et al. and another study by Simkó et al. In the first one, HL-60 and ML-1 (Human Myeloblastic Leukemia) cells undergo apoptosis (detected through ladder type-DNA fragmentation), after exposure at 50Hz, 45mT ELF-EMF for time periods of 1 and 2.5h. The same study revealed that normal human peripheral blood leukocytes did not undergo DNA fragmentation, when exposed to ELF-EMF under the same exposure conditions (Hisamitsu et al., 1997). Simkó et al. studied the effects of ELF-EMF in the SCL II (human squamous cell carcinoma) cells and AFC (human amniotic fluid) under different field intensities and durations. It was observed that when a 50Hz, 0.8-1.0mT EMF was applied, for 48h and 72h of continuous exposure, a significant increase in the frequency of micronucleus (MN) formation and induction of apoptosis in SCL II, occurred. Moreover, exposure of AFC cells did not reveal any significant differences, compared to control, at different EMF intensities and various exposure periods (Simkó et al., 1998).

SCOV3 (human ovarian carcinoma) cells undergo apoptosis, when exposed to a pulsed electric field of 10kV/cm, 100ns, 1 Hz, for 5min (Yao et al., 2008). The authors proposed that apoptosis induction was due to the increase of the intracellular concentration of Ca^{2+}. High resolution ^{1}H-NMR spectroscopy also revealed an apoptosis like behavior, when K562 (human muelogenous leukaemia) cells were exposed to ELF 50 Hz sinusoidal magnetic field of 1mT or 5mT, for 2h (Santini et al., 2005). Moreover, continuous exposure of SH-SY5Y (human neuroblastoma) cells to a 900MHz radiofrequency radiation (SAR: 1W/kg), for 24h, leads to significant reduction in the viability of neuroblastoma cells (Buttiglione et al., 2007), whereas exposure of human epidermoid cancer KB cells at a 1.95MHz non-thermal electromagnetic field (SAR 3.6±0.2 mW/g) induced a time-dependent apoptosis, which reached 45% after 3h of exposure (Caraglia et al., 2005).

In order to detect whether the potency of anticancer drugs (vincristine (VCR), mitomycin C (MMC), and cisplatin) was enhanced in the presence of a pulsed electromagnetic fields (PEMF) of 1.5mT, at 1 and 25Hz, Ruiz Gómez et al. used HCA-2/1cch (a multidrug resistant human colon adenocarcinoma [HCA]) cells, as a cancer model. Two different modes of exposure were implemented: (a) exposure to drug and PMF for 1h simultaneously and (b) drug exposure for 1 h, and then exposure to PEMF for the next 2 days (2 h/day). The results showed that vincristine's cytotoxicity was increased at 1Hz PEMF, while that of mitomycin C and cisplatin was significantly increased at 25Hz PEMF (Ruiz Gómez et al., 2002).

In another study, the experimental data obtained by Miyagi et al. indicated that when murine osteosarcoma cells, resistant to doxorubicin, were treated with DOX in the presence of 10 x 10^{-3} mT PEMF at 10Hz, the inhibition growth rate was significantly higher compared to both non-exposed resistant cells and those non-treated with doxorubicin (Miyagi et al., 2000). Similarly, the application of PEMFs (at 10Hz and intensity of 4G) increased doxorubicin (DOX) binding ability to nuclear DNA and inhibited cell growth of MOS/ADR1 (P-gp positive multidrug resistant murine osteosarcoma) cells. Also, data indicated that this type of field reversed the DOX resistance of the MOS/ADR1 cells (Hirata et al., 2001).

K562 (human muelogenous leukaemia) and U937 (histiocytic lymphoma) cells were tested for their viability under different modes of exposure. When cancer cells exposed to a 50Hz sinusoidal ELF-PEMF, it was observed that the electromagnetic field induced both apoptosis and necrosis. Furthermore, application of ELF-PEMF in the presence of the chemotherapeutic agent actinomycin-C (ACM) resulted in strong enhancement of the cytotoxic effect of ACM in cancer cells. (Traitcheva et al., 2003).

In a study of Chen et al., K562 cells were treated with cis-platin (DDP) under the exposure in a static magnetic field (SMF) of 8.8mT for 12h. It was found that the cytotoxic effect of DDP was enhanced in the presence of SMF, as well as the DNA breakage was increased. Also, as atomic force microscopy revealed, the cell surface ultrastructure was modified (Chen W.F. et al., 2010). In a similar study, K-562 cell line was also used in order to investigate the potential synergistic effects between adriamycin (ADM) and exposure to a moderate-intensity static magnetic field (SMF) of 8.8mT, for 12h. The cytotoxic effect of ADM was enhanced in the presence of SMF, through the significant inhibition of the metabolic activity (cell proliferation) of these cancer cells (Hao et al., 2011).

Additionally, BEL-7402 (human hematoma cell line) cells were treated with a variety of X-ray irradiation doses (0, 2, 4, 6, 8 and 10 Gy) combined to 100 Hz EMF (sine wave with a mean intensity of 0.7mT), for two or six exposure times (duration of each exposure session was 30mins, with 12h intervals). Two periods of EMF exposure combined with X-ray irradiation, at a dose of 2 Gy, increased the apoptosis rates of BEL-7402 cells, compared to those subjected to X-ray irradiation alone. Furthermore, six periods of EMF exposure caused higher apoptosis rates than two periods did. Thus, repetitive EMF exposure periods may cause accumulation of apoptotic effects in BEL-7402 cells (Jian et al., 2009).

3. *In vivo* studies

Several scientific teams have turned their interest to the effects of EMFs, in apoptosis, angiogenesis and tumour growth, blood flow and platelet adherence as well as in the transcription of genes that is related with the appearance of cancer, *in vivo* models.

In more detail, Syrian Golden hamsters bearing A-Mel-3 melanomas were exposed to SMFs with varying field strength (<600 mT) at different exposure times (1 min to 3h). Short time of exposure, at a magnetic flux density of 150mT, presented a significant reduction of red blood cell velocity and segmental blood flow in tumour microvessels. An extended exposure to SMFs (up to 3h), resulted in comparable reductions (Strieth et al., 2008). One year later the same scientific team used Syrian Golden Hamsters bearing syngenic A-Mel-3 melanomas, which were exposed to a SMF of 586mT for three hours. A deceleration of tumour growth was observed whereas angiogenesis was attenuated (Strelczyk et al., 2009).

Wang et al. used moderate-intensity and spatial gradient static magnetic fields (GSMF) (0.2–0.4 T, 2.09 T/m, 1–11 days) on two *in vivo* models, a chick chorioallantoic membrane (CAM) and a matrigel plug, in order to investigate their potential effects on angiogenesis. The *in vivo* findings revealed that GSMF caused reduction of vascular numbers and contents of hemoglobin, and inhibited vascularization (Wang et al., 2009).

In two independent experiments, exposure of nude mice bearing a subcutaneous human colon adenocarcinoma (WiDr), in static magnetic fields of intensity of 5.5 mT, daily for 70 min (first experiment) and for four consecutive weeks (second experiment), respectively, resulted in a significant increase of survival time as well as a significant inhibition of tumour growth. In addition, a reduction of cell proliferation and an increase of apoptosis in tumours of treated animals were observed. These findings were accompanied by the evidence of reduction of the expressed p53 (Tofani et al., 2002b).

Antitumour and immunomodulatory effects of pulsed magnetic fields have also been investigated in a study, which utilized the following conditions: pulse width = 238 μs, peak magnetic field = 0.25 T (at the center of the coil), frequency = 25 pulses/s, 1000 pulses/sample/day and magnetically induced eddy currents in B16-BL6 melanoma model mice = 0.79–1.54 A/m². Exposure of mice in pulsed magnetic fields lasted 16 days. The experimental data showed anticancer and immunomodulatory properties of pulsed magnetic stimulation such as decrease of tumour growth and elevated production of tumour necrosis factor (TNF-a) in mouse spleens (Yamaguchi et al., 2006). In addition, extremely low-frequency pulsed-gradient magnetic field (with the maximum intensity of 0.6–2.0 T, gradient of 10–100 T/m, pulse width of 20–200 ms and frequency of 0.16–1.34 Hz) presented antitumour and antiangiogenic properties, in exposed Kunming mice bearing murine tumour (Zhang et al., 2002). Analogous effects have also been determined in several other studies (de Seze et al., 2000; Williams et al., 2001).

While it is rendered known, henceforth, that low level electromagnetic fields present very interesting effects in physiology of cancer, particular attention has begun to be given in ELF-EMFs. Jimenez-Garcia et al. used male Fischer-344 rats, which were subjected to the modified resistant hepatocyte model and exposed to 4.5 mT - 120 Hz ELF-EMF. The results showed a decrease of more than 50% of the number and the area of γ- glutamyl transpeptidase-positive preneoplastic lesions, glutathione S-transferase placental expression, as well as a significant decrease of proliferating cell nuclear antigen, Ki-67, and cyclin D1 expression. These findings showed inhibition of preneoplastic lesions, through antiproliferative activity of ELF-EMF (Jimenez-Garcia et al., 2010).

Several studies come to prove the anticancer activity of certain electric fields. In one of them, low intensity, intermediate-frequency (100–300 kHz), alternating electric fields were used in *in vivo* treatment of tumours in C57BL/6 and BALB/c mice (B16F1 and CT-26 syngeneic tumour models, respectively) and induced significant slowing of tumour growth and extensive destruction of tumour cells within 3–6 days (Kirson et al., 2004). These findings have been extended by another study, in which additional animal tumour models (intradermal B16F1 melanoma and intracranial F-98 glioma) were used (Kirson et al., 2007). Mi et al. utilized 48 BALB/c mice, which were inoculated with U_{14} cervical cancer cells and then subjected in steep pulsed electric field (SPEF). The presented data indicated irreversible destruction of integrality of tumour cell, retardation of tumour growth and prolongation of survival time (Mi et al., 2004).

Certain studies have utilized chemotherapeutic agents for the study of synergistic phenomena between chemotherapy and electromagnetic fields. Female B6C3F1 mice, with transplanted mammary adenocarcinoma, were exposed to static magnetic or electric fields and presented significantly greater tumour regression compared to that of mice treated only with 10 mg/kg of adriamycin (Gray et al., 2000). Potential anticancer activity of magnetic field has also been investigated through similar study, in which nude mice, bearing a subcutaneous human breast tumour (MDA-MB-435), were exposed for 70 min daily, for six consecutive weeks, to modulated MF (static with a superimposition of extremely low-frequency fields at 50 Hz), of total intensity of 5.5 mT. The anticancer activity of MF was compared to that of cyclophosphamide. The inhibition on spread and growth of lung metastases caused by MF was greater than that caused by cyclophosphamide. It is worth to mention that no toxic or abnormal effects were observed (Tofani et al., 2002a). Moreover, cisplatin, one of classic anticancer drugs, when used in combination with ELF-MF exposure, extended the survival time of immunocompetent mice bearing murine Lewis Lung carcinomas (LLCs) compared with that of mice treated only with cisplatin (Tofani et al., 2003).

4. Clinical studies

As described previously in sections 2 and 3, a large number of *in vitro* and *in vivo* studies, support the anticancer effects of EMFs. On the other hand, there is only a small number of data concerning the application of EMFs in clinical studies, which deal with cancer management.

Salvatore J, has designed and completed a Phase I clinical study, using a combination of static magnetic field (SMF) and antineoplastic chemotherapy, in patients with advanced malignancy (lung cancer, non-Hodgkin's Lymphoma, and colon/rectum cancer). The aim of this study was to establish the safety and toxicity of this combination and not the efficacy of treatment. Data were collected from 10 patients, by estimating the white blood cell and platelet count. There were no differences in the previous markers in control and participants during the treatment plan. Results from this work suggest that the combination is safe without increasing the severity of chemotherapy toxicity, and set the bases for the Phase II and III clinical trials. In these trials, the efficacy of a SMF as an anti-neoplastic agent either alone or in combination with chemotherapy, is going to be determined (Salvatore et al., 2003).

In another study of Barbault et al., it is proposed that a combination of tumour-specific frequencies may have a therapeutic effect. A total of 1524 frequencies, ranging from 0.1 to 114 kHz, were identified from 163 cancer patients, while a compassionate treatment was offered to 28 patients with advanced cancer (breast, ovarian pancreas, colon, prostate, sarcoma and other types of cancer). The patients received a total of 278.4 months of experimental treatment and the median treatment duration was 4.1 months per patient. None of the patients, who received experimental therapy, reported any side effects of significance. Two of the patients presented a complete and partial response to the treatment and four patients presented stable disease. A woman, with breast cancer, showed a complete disappearance of some lesions, according to PET-CT (Positron emission tomography - computed tomography), and significant improvement of the overall condition. Thus, the tumour-specific frequencies provide an effective and well tolerated treatment which may present antitumour properties in end-stage patients (Barbault et al., 2009).

Eleven patients with mean age of 60 years and with stage IV, locally advanced or metastatic disease (adenocarcinoma, duct carcinoma, squamous cell carcinoma and other types), were enrolled in a human pilot study conducted by Ronchetto et al. Patients were exposed for 5 days/week, over 4 weeks, according to two different static magnetic fields schedules: 20 min daily (4 patients) and 70 min daily (7 patients). Results showed that MF-exposed patients present mild or no side effects. Furthermore, this pilot study supports the evidence that human exposure to MF with specific physical characteristics is associated with a favourable safety profile and good tolerability (Ronchetto et al., 2004).

Kirson et al. has reported, that exposure of cancer cell lines and tumour bearing animals to low-intensity, intermediate-frequency (100–300 kHz), alternating electric fields, revealed significant anticancer effects *in vitro* and *in vivo* (Kirson et al., 2004). Based on these findings, he proceeded in a pilot clinical trial, including 10 patients with recurrent glioblastoma, using the above described fields. No serious adverse events were observed in all patients, after >70 months of cumulative treatment. The median time of disease progression and median overall survival were more than double than the reported medians of historical control patients. The authors concluded that this type of fields can be used as a safe and effective treatment for cancer patients (Kirson et al., 2007).

In 2008, Salzberg et al. designed a prospective, pilot study to investigate the safety and efficacy of low-intensity, intermediate-frequency electric fields in 6 patients (heavily pre-treated with several lines of therapy) with metastatic solid tumours, while no additional standard treatment option was available to them. A device was used to emit the frequencies 100–200 kHz, at a field intensity of 0.7 V/cm. A patient presented 51% reduction in tumour size, after 4 weeks of fields' treatment. Also, an arrest of tumour growth was seen in three patients, during treatment. Despite the small number of patients, this study revealed that this type of electric fields presented lack of toxicity and significant efficacy in patients' treatment (Salzberg et al., 2008).

In a recent phase I/II clinical study, 41 patients with advanced hepatocellular carcinoma (HCC), were subjected to very low levels of electromagnetic fields modulated at HCC-specific frequencies (410.2Hz-20365.3Hz). Patients were being administered with three-daily 60min outpatient treatments, till the disease progression or death. During treatment no NCI grade 2, 3 or 4 toxicities (grades based on National Cancer Institute Common Terminology Criteria (CTC) for adverse events), were observed, while most of the patients reported complete disappearance or decrease of pain shortly after treatment initiation. Four patients presented a partial response to the treatment, while 16 patients (39%) had a stable disease for more than 12 weeks. This type of EMFs provided a safe and well tolerated treatment, as well as evidence of antitumour effects in HCC-patients (Costa, 2011).

A brief description of the beneficial effects per EMF category is presented in Figure 1.

5. Studies of our research team

5.1 Studies in resonant frequencies

After several experimental studies, Benveniste concludes that "molecular signal could be mimicked by electromagnetic signals" (Benveniste, 2004). This means that an interaction between specific electromagnetic frequencies and molecules may exist, through the phenomenon of resonance. Our studies are based on this phenomenon, using only resonant frequencies derived from the initial target.

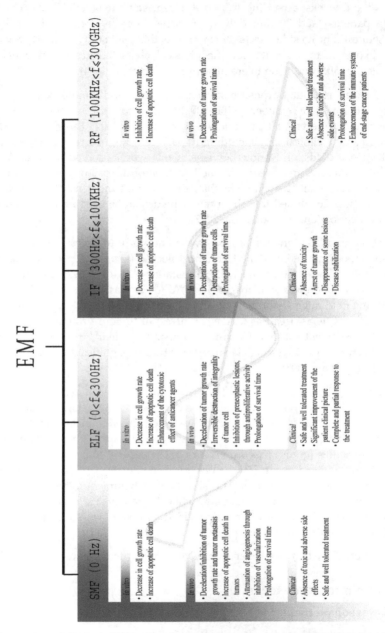

Fig. 1. Illustration of the beneficial effects per EMF category according to the level of the scientific research. Electromagnetic fields have been categorized according to Scientific Committee on Emerging and Newly Identified Health Risks (SCENIHR). Abbreviations: EMF, electromagnetic fields; SMF, static magnetic field; ELF, extremely low frequency; IF, intermediate frequency; RF, radiofrequency; f, frequency

The investigation of the effects of electromagnetic radiation against various cancer cell lines as well as bearing tumour rat models has been subsumed in our research interests. A series of an *in vitro*, *in vivo* experiments and a clinical study have been published, dealing with the antitumour and immunomodulatory effects of resonant low intensity or intermediate or radiofrequency fields.

In 2006, our research group published an *in vitro* study, using low intensity radiofrequency electromagnetic field, causing no thermal effects, against leiomyosarcoma cells (LMS) and smooth muscle cells (SMC). A total of 492 frequencies, from 10 to 120 kHz, were used for the exposure of both cell lines to resonant EMFs. These frequencies were generated by a device, with the following characteristics: the intensity for electric field was 1.1 to 1.11 ± 0.01 V/m, for the magnetic field 0.0027 to 0.0029 ± 0.00005 A/m and the power density of the electromagnetic field was 3.22 mWatt/m^2 approximately. During the EMF or sham exposure, all cell cultures were placed in a Faraday cage, at room temperature, in order to exclude the external electromagnetic field interactions. Both cell lines were exposed to the 492 resonant frequencies, for 45 min, for two consecutive sessions (at 72 h and 120h from zero time). After the first session of exposure (at 72h), LMS cell growth was significantly decreased, more than 98% ($P< 0.0001$ compared to the control cells). The remaining cells (2%) were cultured and re-exposed (at 120h-second session). Cells presented a remarkable resistance to EMFs, induced only a 20% decrease in proliferation, after the second session of exposure. Then the exposed LMS (from the second session) were preserved in liquid nitrogen for a short time period. After that, LMS cells were defrosted, cultured and exposed once again, at the same EMF pattern for two more sessions as described previously, in order to estimate the apoptosis and cell cycle arrest. Exposed cells presented a significant increase in apoptosis, compared to the control group (sham exposed). On the other hand, SMC did not present any significant alteration on their cell growth when exposed to EMFs. Data revealed that the specific electromagnetic spectrum of LMS cells causes cell death by apoptosis. Another point that has to be stated is that after repeated exposures, the phenotype of LMS cells was altered, and so, the initial electromagnetic resonance fingerprint should be reconsidered (Karkabounas et al., 2006).

Our *in vivo* study presents the effects of a resonant low intensity electromagnetic field, causing no thermal effects, on Wistar rats. LMS cells were exposed for two sessions as described previously according to the protocol of the *in vitro* experiment. Female Wistar rats were inoculated with, exposed (one group) and non-exposed (three groups) LMS cells to EMF. Animals belonging to experimental Group-II (EG-II) were inoculated with cells exposed to EMF and were not further exposed to irradiation. Animals which were inoculated with cells non-exposed to EMF, were randomly separated into three groups: The control Group-CG, in which animals were sham exposed, the experimental control Group-ECG, in which animals were exposed to a non-resonant EMF radiation pattern, for 5 h per day till the death of all animals and experimental Group-I, EG-I, in which animals were exposed to the resonant EMF radiation for 5 h per day, for a maximum of 60 days. Animals of both EG-II and EG-I demonstrated a significant prolongation of the survival time and a decreased tumour growth rate, in comparison to the animals of CG. Furthermore, the survival time of EG-I animals was found to be significantly longer compared to that of EG-II animals. Concluding, results revealed that exposure of tumour bearing Wistar rats to resonant electromagnetic frequencies caused significant prolongation of the survival time and decrease of tumour growth rate (Avdikos et al., 2007).

The aim of our recent study, published in 2011, was to investigate whether the coherent electromagnetic fields were able to enhance the immune system in end-stage cancer patients. Fifteen end-stage patients (5 male and 10 female) were recruited for this study while a complete medical history was received. No female patient was pregnant. All of them had completed their chemotherapy radiation, and/or adjuvant antioxidant treatment, at least 4 weeks, before participation in the study, while none of them received any medications. All patients were tested for the type of malignancy, by histology and CT (Computed Tomography) or MRI (Magnetic Resonance Imaging). Data from blood biochemistry, haematological analysis and tumour markers were also included in the study. All patients were exposed at radiofrequencies ranging from 600 kHz–729 kHz, for 8 h/day, 6 days/week for 4 weeks. The population of NK cells and cytotoxicity of NK T-lymphocytes versus K562 cancer cell line were estimated by flow cytometry, before and after exposure. Results revealed that no side effects were recorded in patients, while data from biochemical and haematological analysis remained stable. The populations of NK cells and NKT lympocytes and the NK cytotoxicity (at ratio of 12.5:1) against K562 cells were significantly increased in all exposed-patients ($p < 0.001$). In conclusion, increase in number and cytotoxicity of NK cells seems to be critical for the prolongation of the survival time and quality of life of end-stage cancer patients (Evangelou et al., 2011).

5.2 ^{1}H-NMR-retrieved resonant frequencies from biological and chemical molecules

Every molecule emits specific frequencies ("fingerprint") providing a distinct electromagnetic spectrum. Based on this concept and on results of our previous *in vitro* and *in vivo* studies, in 2008, we have started a new set of experiments, in order to investigate the biological effects of specific molecules' resonant frequencies emission (RFRs-not to be confused with the abbreviation of radiofrequencies [RFs]), obtained from their ^{1}H-NMR spectrum analysis. We hypothesized that the emission of these RFRs could produce the same or similar effects with the molecules themselves, in cellular and animal systems. Molecules' RFRs were obtained by transforming the chemical shifts (in ppm) of their NMR spectra, using the equation given by Keeler (Keeler, 2005). The resultant set of frequencies constitutes, to our opinion, the above mentioned "fingerprint".

Two different experiments have been conducted, using a chemically-synthesized compound (SnMNA) with anticancer properties (Verginadis et al., 2011a), as well as morphine. In the first experiment, a set of RFRs (26 frequencies) was obtained from the ^{1}H-NMR spectrum analysis of the SnMNA complex. Leiomyosarcoma (LMS), human breast adenocarcinoma cells (MCF-7) and normal human fetal lung fibroblast (MRC-5), were exposed to SnMNA-RFRs, for 5h/day for two consecutive days. MTT assay was used for estimation of cell growth proliferation. Significant cell death ($p < 0.01$) was observed in the two cancer cell lines, whereas there was no cytotoxic effect against MRC-5 cells. Additionally, tumour bearing Wistar rats, were exposed to the same SnMNA-RFRs, for 5h/day, till the first animal death. Experimental findings revealed significant prolongation of the mean survival time ($p < 0.05$) and reduction of the mean tumour growth rate ($p < 0.05$) of the exposed-animals, compared to the non-exposed or exposed ones to randomly selected non resonant frequencies (non-RFRs which possess the same energy to the corresponding RFRs) (control groups) (Evangelou et al., 2008; Verginadis et al., 2008). In the second experimental procedure, the analgesic effect of morphine-RFRs (45 frequencies obtained from ^{1}H-NMR spectrum analysis) were evaluated using the hot-plate and the tail-flick test (analgesia tests). Healthy Wistar rats were exposed to morphine-RFRs for 5h and measurement of latency times were taken, after 1 and 5 hours of exposure, by both analgesia tests. Exposed to morphine-RFRs animals, presented significant increase in the analgesia (p<0.05) compared to those exposed to non-RFRs (Verginadis et al., 2011b). Preliminary results showed that, when animals were treated with naloxone (a μ-opioid receptor competitive antagonist of morphine) and after being exposed to morphine-RFRs, did not present any analgesia. The latter indicates that the morphine-RFRs analgesic effect is probably exerted through direct or indirect activation of the μ-opioid receptors.

6. Reproducibility

According to bibliography, there are scientific teams arguing about the reproducibility and variance of experimental findings, because of not clearly described protocols or not accurate application of them. Malyapa et al. (Malyapa et al., 1998) tried to replicate the work of Lai and Singh (Lai & Singh 1995) without any success, because there were significant differences in comet assay analyses. Two different research groups (Lacy-Hulbert et al., 1995; Saffer & Thurston, 1995) tried to replicate the work of Goodman (Goodman & Henderson, 1991; Goodman et al., 1992) with controversial results. Jin et al. (Jin et al., 1997), published the possible parameters, not being considered, which were responsible for the inability of these two groups of investigators to replicate Goodman's work: different HL60 cell populations, mRNA extraction procedure, the stability/variability of internal standards and sham exposure set-up.

There is a number of specific EMF-exposure parameters which have to be outlined in publications, such as used frequencies, duration and pattern of exposure (continuous and/or intermittent), pulse shape (pulsed or sinusoidal fields), intensity and depth of penetration. Also, the researchers should mention the specific intensity field at the target site and not at the surface or close to the generating device (Markov, 1994). Thus, in order to achieve reproducibility of the biological results, analytical experimental protocols and a complete report of the exposure conditions must be described.

7. Mechanism of action

The clarification of mechanisms of action of EMFs, on the cellular systems, has been proved difficult work for the scientists so far. Several models of interaction of cells with chemical phenomena, caused by EMFs, have been discussed, depending on the physical parameters of their emission. The complexity of problem increases, when different cellular types and different radiation "windows" are used, something that leads to different cellular responses. The diversity of these responses is referred to changes of free charges on cellular membrane, alterations of membrane-related proteins and enzymes, as well as to the activation of signal transduction pathways.

Little have been done for the determination of relationship between EMFs and their potential anticancer activity. Most studies that have dealt with this, lead to conclusions, which are limited in their findings. They do not propose any more general mechanisms, but even if they do so, there is no correlation between them.

As proposed by Chen Y et al., when cancer cells were exposed to ELF-EMF, their bioactivity was disturbed, resulting in an abnormal cell signal transduction process, which is possibly responsible for the inhibition mechanisms of cell growth. This can be assumed by the theoretical calculation, of the tangential ionic motion (such as K^+, Na^+, Ca^{2+} and Cl^-) in living cells, governed by the exposure in the time-variant MF and induced EF of the associated ELF-EMF. In theory, this calculation suggests that the oscillating motion of ions in the vicinity of the cell membrane with a net tangential displacement, could exert a significant electromagnetic force acting on the voltage sensors in the voltage-gated channels, screening ionic flux into or out of the cell membrane, which results in the failure of the signal transduction process and the inhibition of the cell growth (Chen Y.C. et al., 2010).

According to the study of Hisamitsu et al., HL-60 and ML-1 leukemic cells underwent ladder-type nucleosomal DNA fragmentation, when exposed to ELF-EMF of 50Hz and 45mT, for 1 and 2.5h. Several *in vitro* studies have associated the magnetic field exposure (50Hz, 22mT, for 1h) with increased intracellular Ca^{2+} concentration (Lindström et al., 1993; Walleczek & Budinger, 1992), which in turn affects endonuclease activity. Based upon reported data, Hisamitsu supports that the ladder type nucleosomal DNA fragmentation was caused by enhanced endonuclease activity, because of the elevated intracellular Ca^{2+} (Hisamitsu et al., 1997).

ELF-EMFs (4.5mT - 120Hz) have been proved to decrease the levels of expressed PCNA, involved in DNA replication and in the RAD6-dependent DNA repair pathway, of Ki-67, associated with DNA replication and of cyclin D, which participates in cell cycle progression. According to these results and to the authors' opinion, ELF-EMF influence the continuity of cell cycle and DNA synthesis of liver cancer cells, possibly via Ca^{2+} flow regulation or via radical chemistry interactions, under the reported conditions of radiation (Jimenez-Garcia et al., 2010).

Moreover, cell adhesion molecules (CAMs), are proteins located on the cell surface, involved in cell adhesion, the process of binding with other cells or with the extracellular matrix. A 50 Hz magnetic field (with a magnetic flux density of 0.5 mT) caused significant changes in cell growth, fibronectin and CD44 expression in MG-63, a human osteosarcoma cell line. In fact, there was a decrease in fibronectin receptor expression, whereas an increase in hyaluronan receptor expression was seen. CAMs are involved in cancer cell functions, such as proliferation and metastasis. Integrins and CD44 participate in the above processes, as members of CAMs. The adhesion of cells, via integrins regulates cellular shape, motility and cell cycle. Moreover, the levels of expressed CD44 influence cell–cell interactions, cell adhesion and migration. According to Rudzki and Jothy (Rudzki & Jothy, 1997) and proposed mechanism of action by Santini, it is indicated that MF influence these molecules' expression, responsible for the transmission of vital signals, in the growth and metastasis of cancer (Santini et al., 2003).

Furthermore, repetitive magnetic stimulation has been shown antitumour and immunomodulatory properties, since tumour necrosis factor (TNF-a) production was increased in mouse spleens, after exposure of B16-BL6 melanoma model mice to pulsed magnetic field. Yamaguchi et al. explanation and bibliography (Ashkenazi, 2002; Aggarwal, 2003), correlates TNF-a, which initiates the TNFR1–TRADD–FADD–Caspase-8–Caspase-3 apoptotic pathway, with the antitumour effects shown after pulsed magnetic field stimulation (Yamaguchi et al., 2006).

The most difficult part of explanation of EMFs' mechanism of action, regarding to their antineoplastic properties, is the connection of provided energy by EMFs in the cellular system and primary response of cancer cells, with some of "classical" signal transduction pathways contributing to cellular death. Various models of explanation of interaction between MFs and Ca^{2+} have been proposed, from physics viewpoint such as ion parametric resonance model by Lednev (Lednev, 1991) and ion interference model by Binhi (Binhi, 1997), but they do not cover the knowledge gap about the biochemical interaction site for MFs, in the cellular system. Gartzke and Lange supported that the ion-conducting actin filament bundle within microvilli, could play the role of the cellular target system for MF (Gartzke & Lange, 2002). The analytic description of the above mechanistic models does not fall into the aims of this paragraph.

8. Conclusion and recommendation for further work

As the applications of EMFs have influenced a lot of aspects of everyday life, life sciences and medicine were also meant to be influenced. At the present, however, the nature of their action on the cellular systems remains enigmatic, and particularly, in presence of such a complex disease, as is cancer. A continuously increasing number of studies come to prove the anticancer activity of EMF emission, but in a specific "window" of action, and explaining only certain mechanistic parts of it. A lot of pieces are still missing from the puzzle, because of the many parameters, such as the type of information being transmitted, the conditions of emission, the frequencies, the doses, the type of experimental cancers as well as the genetic material of cells, on which cellular response depends. Our research aims to outflank these "windows" of action, using resonant frequencies derived from the ^{1}H-NMR spectrum of biological and chemical molecules ("fingerprint") and to determine the molecular pathways which are triggered.

Based on our results so far, we have indications to support our hypothesis. Several experiments are in progress, investigating the possible signaling pathways triggered by interactions between RFRs emission and cancer cellular targets. Briefly, we are going to study the effects of RFRs, derived from anticancer agents with determined mechanism of action, on various cancer cell lines. Modulation of gene expression and specific signaling pathways activation, by RFRs, will provide us significant information about their mechanism of action and will support the idea that electromagnetic signals could imitate the molecular signal through the resonance phenomenon.

A new research horizon lies ahead of us, as the potential clinical application of EMFs could be proved an innovative, alternative or/and adjuvant therapeutic approach, in cancer treatment.

9. Acknowledgments

The Authors gratefully acknowledge the assistance of Dr. Michaela Filiou and Mr. Dimitris Palitskaris for reviewing and editing the manuscript.

10. References

Aaron, R. K., & Ciombor, D. M. (1993). Therapeutic effects of electromagnetic fields in stimulation of connective tissue repair. *Journal of cellular biochemistry*, Vol. 52, No. 1, (May 1993), pp. 42-46, ISSN 1097-4644

Aggarwal, B. B. (2003). Signalling pathways of the TNF superfamily: A double-edged sword. *Nature Reviews. Immunology*, Vol. 3, No. 9, (September 2003), pp. 745–756, ISSN 1474-1741

Ahlbom, I. C., Cardis, E., Green, A., Linet, M., Savitz, D., Swerdlow, A. (2001). Review of the epidemiologic literature on EMF and health. *Environ Health Perspect*, Vol. 109, Suppl 6, (December 2001), pp. 911-933. ISSN 1552-9924

Ashkenazi, A. (2002). Targeting death and decoy receptors of the tumour-necrosis factor superfamily. *Nature Reviews. Cancer*, Vol. 2, No. 6, (June 2002), pp. 420–430, ISSN 1474-1768

Avdikos, A., Karkabounas, S., Metsios, A., Kostoula, O., Havelas, K., Binolis, J., Verginadis, I., Hatziaivazis, G., Simos, I., & Evangelou, A. (2007). Anticancer effects on leiomyosarcoma-bearing Wistar rats after electromagnetic radiation of resonant radiofrequencies. *Hell J Nucl Med*, Vol. 10, No. 2, (May-August 2007) pp. 95-101, ISSN 1790-5427

Barbault, A., Costa, F. P., Bottger, B., Munden, R. F., Bomholt, F., Kuster, N., & Pasche, B. (2009). Amplitude-modulated electromagnetic fields for the treatment of cancer: discovery of tumour-specific frequencies and assessment of a novel therapeutic approach. *J Exp Clin Cancer Res*, Vol. 28, No. 1, (April 2009), pp. 51, ISSN 1756-9966

Barker, A. T., Dixon, R. A., Sharrard, W. J., & Sutcliffe, M. L. (1984). Pulsed magnetic field therapy for tibial non-union. Interim results of a double-blind trial. *Lancet*, Vol. 1, No. 8384, (May 1984), pp. 994-996, ISSN 1474-547X

Bassett, C. A., Pawluk, R. J., & Pilla, A. A. (1974a). Augmentation of bone repair by inductively coupled electromagnetic fields. *Science*, Vol. 184, No. 136, (May 1974), pp. 575-577, ISSN 1095-9203

Benveniste, J. (2004). A fundamental basis for the effects of EMFs in biology and medicine. The interface between matter and function, In: *Bioelectromagnetic Medicine*, Rosch, P. J. & Markov, M. S., pp. 207-211, Taylor & Francis, ISBN 0-8247-4700-3, , Boca Raton, FL, USA

Binhi V. N. (1997). Interference ion quantum states within a protein explains weak magnetic field's effects in biosystems. *Electromagnetobiology*, Vol. 16, No. 3, (January 1997), pp. 203–214

Buttiglione, M., Roca, L., Montemurno, E., Vitiello, F., Capozzi, V., & Cibell, G. (2007). Radiofrequency radiation (900 MHz) induces egr-1 gene expression and affects cell-cycle control in human neuroblastoma cells. *Journal of cellular physiology*, Vol. 213, No. 3, (December 2007), pp. 759-767, ISSN 1097-4652

Caraglia, M., Marra, M., Mancinelli, F., D' Ambrosio, G., Massa, R., Giordano, A., Budillon, A., Abbruzzese, A., & Bismuto, E. (2005). Electromagnetic fields at mobile phone frequency induce apoptosis and inactivation of the multi-chaperone complex in human epidermoid cancer cells. *Journal of cellular physiology*, Vol. 204, No. 2, (August 2005), pp. 539–548, ISSN 1097-4652

Chen, W. F., Qi, H., Sun, R. G., Liu, Y., Zhang, K., & Liu, J. Q. (2010). Static magnetic fields enhanced the potency of cisplatin on K562 cells. *Cancer Biother Radiopharm*, Vol. 25, No. 4, (August 2010), pp. 401-408, ISSN 1557-8852

Chen, Y. C., Chen, C. C., Tu, W., Cheng, Y. T., & Tseng, F. G. (2010). Design and fabrication of a microplatform for the proximity effect study of localized ELF-EMF on the growth of *in vitro* HeLa and PC-12 cells. *J. Micromech. Microeng*; Vol. 20, No. 12, ISSN 0960-1317

Costa, F.P., de Oliveira, A.C., Meirelles, R., Machado, M.C., Zanesco, T., Surjan, R., Chammas, M.C., de Souza Rocha, M., Morgan, D., Cantor, A., Zimmerman, J., Brezovich, I., Kuster, N., Barbault, A., & Pasche, B. (2011). Treatment of advanced hepatocellular carcinoma with very low levels of amplitude-modulated electromagnetic fields. *Br J Cancer*, Vol. 105, No. 5, (August 2011), pp. 640-648. ISSN 1532-1827

de Seze, R., Tuffet, S., Moreau, J. M., & Veyret, B. (2000). Effects of 100 mT time varying magnetic fields on the growth of tumours in mice. *Bioelectromagnetics*, Vol. 21, No. 2, (February 2000), pp. 107-111, ISSN 1521-186X

Dortch, A. B., & Johnson, M. T. (2006). Characterization of pulsed magnetic field therapy in a rat model for rheumatoid arthritis. *Biomed Sci Instrum*, Vol. 42, (2006), pp. 302-307, ISSN 0067-8856

Evangelou, A., Verginadis, I., Avdikos, A., Simos, I., Havelas, K., Zouridakis, A., & Karkabounas, S. (2008). Comparison of the cytotoxic effects of a Sn-mercaptonicotinic acid complex to the coherent electromagnetic resonant radiofrequency spectra of the same complex, on a rat leiomyosarcoma cell line, *Proceedings of 10th International Symposium on Metal Ions in Biology and Medicine*, ISBN 978-2-7420-0714-1, Corsica, France, May 2008

Evangelou, A., Toliopoulos, I., Giotis, C., Metsios, A., Verginadis, I., Simos, Y., Havelas, K., Hadziavazis, G., & Karkabounas, S. (2011). Functionality of natural killer cells from end-stage cancer patients exposed to resonant electromagnetic fields. *Electromagn Biol Med*, Vol. 30, No. 1, (March 2011) pp. 46-56, ISSN 1536-8386

Fischer, G., Pelka, R. B., & Barovic, J. (2005). Adjuvant treatment of knee osteoarthritis with weak pulsing magnetic fields. Results of a placebo-controlled trial prospective clinical trial. *Z Orthop Ihre Grenzgeb*, Vol. 143, No. 5, (September-October 2005), pp. 544-550, ISSN 0044-3220

Gartzke, J., & Lange, K. (2002). Cellular target of weak magnetic fields: ionic conduction along actin filaments of microvilli. *American journal of physiology. Cell physiology*, Vol. 283, No. 5, (November 2002), pp. C1333-C1346, ISSN 1522-1563

Goodman, R., & Henderson, A. S. (1991). Transcription and translation in cells exposed to extremely low frequency electromagnetic fields. *Bioelectrochem Bioenerg*, Vol. 25, No. 3, (June 1991), pp. 335–355, ISSN 0302-4598

Goodman, R., Bumann, J., Wei, L. X., Shirley-Henderson, A. (1992). Exposure of human cells to electromagnetic fields: effect of time and field strength on transcript levels. *Electromagn Biol Med*, Vol. 11, No. 1 (January 1992), pp. 19–28, ISSN 1536-8386

Gray, J. R., Frith, C. H., & Parker, J. D. (2000). *In vivo* enhancement of chemotherapy with static electric or magnetic fields. *Bioelectromagnetics* Vol. 21, No. 8, (December 2000), pp. 575-583, ISSN 1521-186X

Hao, Q., Wenfang, C., Xia, A., Qiang, W., Ying, L., Kun, Z., & Runguang, S. (2011). Effects of a moderate-intensity static magnetic field and adriamycin on K562 cells. *Bioelectromagnetics*, Vol. 32, No. 3 (April 2011), pp. 191-199, ISSN 1521-186X

Hirata, M., Kuzuzaki, K., Takeshita, H., Hashiguchi, S., Hirasawa, Y., & Ashihara, & T. (2001). Drug resistance modification using pulsing electromagnetic field stimulation for multidrug resistant mouse osteosarcoma cell line. *Anticancer Research*, Vol. 21, No. 1A, (January-February 2001), pp. 317-320, ISSN 1791-7530

Hisamitsu, T., Narita, K., Kashara, T., Seto, A., Yu, Y., & Asano, K. (1997). Induction of apoptosis in human leukemic cells by magnetic fields. *Japanese Journal of Physiology*, Vol. 47, No. 3, (June 1997), pp. 307-310, ISSN 1881-1396

Janigro, D., Perju, C., Fazio, V., Hallene, K., Dini, G., Agarwal, M.K, & Cucullo, L. (2006). Alternating current electrical stimulation enhanced chemotherapy: a novel strategy to bypass multidrug resistance in tumour cells. *BMC Cancer*, Vol. 17, No. 6, (March 2006), pp. 72, ISSN 1471-2407

Jian, W., Wei, Z., Zhiqiang, C., & Zheng, F. (2009). X-Ray-induced apoptosis of BEL-7402 cell line enhanced by extremely low frequency electromagnetic field *in vitro*. *Bioelectromagnetics*, Vol. 30, No. 2, (January 2009), pp. 163-165, ISSN 1521-186X

Jimenez-Garcia, M. N., Arellanes-Robledo, J., Aparicio Bautista, D. I., Rodriguez- Segura, M. A., Villa-Trevino, S., & Godina-Nava, J. J. (2010). Anti-proliferative effect of an extremely low frequency electromagnetic field on preneoplastic lesions formation in the rat liver. *BMC Cancer*, Vol. 24, No. 10, (April 2010), pp. 159, ISSN 1471-2407

Jin, M., Lin, H., Han, L., Opler, M., Maurer, S., Blank, M., & Goodman, R. (1997). Biological and technical variables in myc expression in HL60 cells exposed to 60 Hz electromagnetic fields. *Bioelectrochem Bioenerg*, Vol. 44, No. 1, (1997), pp. 111–120, ISSN 0302-4598

Karkabounas, S., Havelas, K., Kostoula, O. K., Vezyraki, P., Avdikos, A., Binolis, J., Hatziavazis, G., Metsios, A., Verginadis, I., & Evangelou, A. (2006). Effects of low intensity static electromagnetic radiofrequency fields on leiomyosarcoma and smooth muscle cell lines. *Hell J Nucl Med*, Vol. 9, No. 3, (September-December 2006), pp. 167-172, ISSN 1790-5427

Keeler, J. (2005). NMR and energy levels, In: *Understanding NMR spectroscopy*, Keeler, J., pp. (1-19), John Willey and Sons Ltd, ISBN 978-0-470-01786-9, West Sussex, England

Kirson, E. D., Gurvich, Z., Schneiderman, R., Dekel, E., Itzhaki, A., Wasserman, Y., Schatzberger, R., & Palti, Y. (2004). Disruption of cancer cell replication by alternating electric fields. *Cancer Research*, Vol. 64, No. 9, (May 2004), pp. 3288–3295, ISSN 1538-7445

Kirson, E. D., Dbaly, V., Tovarys, F., Vymazal, J., Soustiel, J. F., Itzhaki, A., Mordechovich, D., Steinberg-Shapira, S., Gurvich, Z., Schneiderman, R., Wasserman, Y., Salzberg, M., Ryffel, B., Goldsher, D., Dekel, E., & Palti, Y. (2007). Alternating electric fields arrest cell proliferation in animal tumour models and human brain tumour, *Proceedings of the National Academy of Sciences of the United States of America*, Vol. 104 , No. 24,(June 2007), pp. 10152-10157, ISSN 00278424

Lacy-Hulbert, A., Wilkins, R. C., Hesketh, T. R., & Metcalfe, J. C. (1995). No effect of 60 Hz electromagnetic fields on MYC or beta-actin expression in human leukemic cells. *Radiat Res*, Vol. 144, No. 1, (October 1995), pp. 9-17, ISSN 1938-5404

Lai, H., & Singh, N. P. (1995). Acute low-intensity microwave exposure increases DNA single-strand breaks in rat brain cells. *Bioelectromagnetics*, Vol. 16, No. 3, (1995), pp. 207-210, ISSN 1521-186X

László, J. F., Szilvási, J., Fényi, A., Szalai, A., Gyires, K., & Pórszász, R. (2011). Daily exposure to inhomogeneous static magnetic field significantly reduces blood glucose level in diabetic mice. *Int J Radiat Biol*, Vol. 87, No.1, (January 2011), pp. 36-45, ISSN 1362-3095

Lednev, V. V. (1991). Possible mechanism for the influence of weak magnetic fields on biological systems. *Bioelectromagnetics*, Vol. 12, No. 2, (1991), pp. 71-75, ISSN 1521-186X

Lindström, E., Lindström, P., Berglund, A., Mild, K. H., & Lundgren, E. (1993). Intracellular calcium oscillations induced in a T-cell line by a weak 50 Hz magnetic field. *J Cell Physiol*, Vol. 156, No. 2, (August 1993), pp. 395-398, ISSN 1097-4652

Malyapa, R. S., Bi, C., Ahern, E. W., & Roti, J. L. (1998). Detection of DNA damage by the alkaline comet assay after exposure to low-dose gamma radiation. *Radiat Res*, Vol. 149, No. 4, (April 1998), pp. 396-400, ISSN 1938-5404

Markov, M. (1994). Biological effects of extremely low frequency magnetic fields, In: *Biomagnetic Stimulation*, Ueno, S., pp. (91-103), Plenum Press, ISBN 030644707X. New York, USA

Mi, Y., Sun, C., Yao, C., Xiong, L., Liao, R., Hu, Y., & Hu, L. (2004). Lethal and inhibitory effects of steep pulsed electric field on tumour-bearing balb/c mice, *Proceedings of the 26th Annual International Conference of the IEEE EMBS*, San Francisco, CA, USA, September 2004

Miyagi, N., Sato, K., Rong, Y., Yamamura, S., Katagiri, H., Kobayashi, K., & Iwata, H. (2000). Effects of PEMF on a murine osteosarcoma cell line: drug-resistant (p-glycoprotein-positive) and non-resistant cells. *Bioelectromagnetics*, Vol. 21, No. 2, (February 2000), pp. 112-121, ISSN 1521-186X

Morabito, C., Guarnieri, S., Fanò, G., & Mariggiò, M. A. (2010). Effects of acute and chronic low frequency electromagnetic field exposure on PC12 cells during neuronal differentiation. *Cell Physiol Biochem*, Vol. 24, No. 6, (October 2010), pp. 947-958, ISSN 1421-9778

Otter, M. W., McLeod, K.J., & Rubin, C.T. (1998). Effects of electromagnetic fields in experimental fracture repair. *Clin Orthop Relat Res*, Vol. 355, (October 1998), pp. S90-S104, ISSN 1528-1132

Raylman, R. R., Clavo, A. C., & Wahl, R. L. (1996). Exposure to strong static magnetic field slows the growth of human cancer cells *in vitro*. *Bioelectromagnetics*, Vol. 17, No. 5, (1996), pp. 358-363, ISSN 1521-186X

Ronchetto, F., Barone, D., Cintorino, M., Berardelli, M., Lissolo, S., Orlassino, R., Ossola, P., & Tofani, S. (2004). Extremely low frequency-modulated static magnetic fields to treat cancer: A pilot study on patients with advanced neoplasm to assess safety and acute toxicity. *Bioelectromagnetics*, Vol. 25, No. 8, (December 2004), pp. 563-71, ISSN 1521-186X

Rudzki, Z., & Jothy, S. (1997). CD44 and the adhesion of neoplastic cells. *Mol Pathol*, Vol. 50, No. 2, (April 1997), pp. 57-71, ISSN 1472-4154

Ruiz Gómez, M. J., Pastor Vega, J. M., de la Peña, L., Gil Carmona, L., & Martínez Morillo, M. (1999). Growth modification of human colon adenocarcinoma cells exposed to a low-frequency electromagnetic field. *Journal of physiology and biochemistry*, Vol. 55, No. 2, (June 1999), pp. 79-83. ISSN 1877-8755

Ruiz Gómez, M. J., de la Peña, L., Pastor, J. M., Martínez Morillo, M., & Gil, L. (2000). 25 Hz electromagnetic field exposure has no effect on cell cycle distribution and apoptosis in U-937 and HCA-2/1[cch] cells. *Bioelectrochemistry*, Vol. 53, No. 1, (January 2001), pp. 137–140, ISSN 1878-562X

Ruiz Gómez, M. J., De la Peña, L., Prieto-Barcia, M. I., Pastor, J. M., Gil, L., & Martínez-Morillo, M. (2002). Influence of 1 and 25 Hz, 1.5 mT magnetic fields on antitumour drug potency in a human adenocarcinoma cell line. *Bioelectromagnetics*, Vol. 23, No. 8 (December 2002), pp. 578-585, ISSN 1521-186X

Saffer, J. D., & Thurston, S. J. (1995). Short exposures to 60 Hz magnetic fields do not alter MYC expression in HL60 or Daudi cells. *Radiat Res*, Vol. 144, No. 1, (October 1995), pp. 18-25, ISSN 1938-5404

Salvatore, J. R., Harrington, J., & Kummet, T. (2003). Phase I clinical study of a static magnetic field combined with anti-neoplastic chemotherapy in the treatment of human malignancy: initial safety and toxicity data. *Bioelectromagnetics*, Vol. 24, No. 7 (October 2003), pp. 524-527, ISSN 1521-186X

Salzberg, M., Kirson, E., Palti, Y., & Rochlitz, C. (2008). A pilot study with very low-intensity, intermediate-frequency electric fields in patients with locally advanced and/or metastatic solid tumours. *Onkologie*, Vol. 31, No. 7, (July 2008), pp. 362-5, ISSN 1423-0240

Santini, M. T., Rainaldi, G., Ferrante, A., Indovina, P. L., Vecchia, P., & Donelli, G. (2003). Effects of a 50 Hz sinusoidal magnetic field on cell adhesion molecule expression in two human osteosarcoma cell lines (MG-63 and Saos-2). *Bioelectromagnetics*, Vol. 24, No. 5, (July 2003), pp. 327-338, ISSN 1521-186X

Santini, M. T., Ferrante, A., Romano, R., Rainaldi, G., Motta, A., Donelli, G., Vecchia, P., & Indovina, P. L. (2005). A 700 MHz 1H-NMR study reveals apoptosis-like behavior in human K562 erythroleukemic cells exposed to a 50 Hz sinusoidal magnetic field. *International journal of radiation biology*, Vol. 81, No. 2, (February 2005), pp. 97-113, ISSN 1362-3095

Simkó, M., Kriehuber, R., Weiss, D. G., & Luben, R. A. (1998). Effects of 50 Hz EMF exposure on micronucleus formation and apoptosis in transformed and nontransformed human cell lines. *Bioelectromagnetics*, Vol. 19, No. 2, (1998), pp. 85–91, ISSN 1521-186X

Strelczyk, D., Eichhorn, M. E., Luedemann, S., Brix, G., Dellian, M., Berghaus, A., & Strieth, S. (2009). Static magnetic fields impair angiogenesis and growth of solid tumours *in vivo*. *Cancer Biology & Therapy*, Vol. 8, No. 18, (September 2009), pp. 1756-1762, ISSN 1555-8576

Strieth, S., Strelczyk, D., Eichhorn, M. E., Dellian, M., Luedemann, S., Griebel, J., Bellemann, M., Berghaus, A., & Brix, G. (2008). Static magnetic fields induce blood flow decrease and platelet adherence in tumour microvessels. *Cancer Biology & Therapy*, Vol. 7, No. 6, (June 2008), pp. 814-819, ISSN 1555-8576

Tabrah, F., Hoffmeier, M., Gilbert, F. Jr., Batkin, S., & Bassett, C. A. (1990). Bone density changes in osteoporosis-prone women exposed to pulsed electromagnetic fields (PEMFs). *J Bone Miner Res*, Vol. 5, No. 5, (May 1990), pp. 437-42, ISSN 1523-4681

Tao, Q., & Henderson, A. (1999). EMF induces differentiation in HL-60 cells. *J Cell Biochem* Vol. 73, No. 2, (May 1999), pp. 212-217, ISSN 1097-4644

Tofani, S., Barone, D., Cintorino, M., de Santi, M. M, Ferrara, A., Orlassino, R., Ossola, P., Peroglio, F., Rolfo, K., & Ronchetto, F. (2001). Static and elf magnetic fields induce tumour growth inhibition and apoptosis. *Bioelectromagnetics*, Vol. 22, No. 6 (September 2001), pp. 419-428, ISSN 1521-186X

Tofani, S., Barone, D., Peano, S., Ossola, P., Ronchetto, F., & Cintorino, M. (2002a). Anticancer activity by magnetic fields: inhibition of metastatic spread and growth in a breast cancer model. *IEEE Transactions on plasma science*, Vol. 30, No. 4, (August 2002), pp. 1552-1557, ISSN 0093-3813

Tofani, S., Cintorino, M., Barone, D., Berardelli, M., De Santi, M. M., Ferrara, A., Orlassino, R., Ossola, P., Rolfo, K., Ronchetto, F., Tripodi, S. A., & Tosi, P. (2002b). Increased mouse survival, tumour growth, inhibition and decreased immunoreactive p53 after exposure to magnetic fields. *Bioelectromagnetics*, Vol. 23, No. 3, (April 2002), pp. 230-238. ISSN 1521-186X

Tofani, S., Barone, D., Berardelli, M., Berno, E., Cintorino, M., Foglia, L., Ossola, P., Ronchetto, F., Toso, E., & Eandi, M. (2003). Static and ELF magnetic fields enhance the *in vivo* anti-tumour efficacy of ciplatin against Lewis lung carcinoma, but not of cyclophosphamide against B16 melanotic melanoma. *Pharmacological Research*, Vol. 48, No. 1, (July 2003), pp. 83–90, ISSN 1096-1186

Traitcheva, N., Angelova, P., Radeva, M., & Berg, H. (2003). ELF fields and photooxidation yielding lethal effects on cancer cells. *Bioelectromagnetics*, Vol. 24, No. 2, (February 2003), pp. 148-150, ISSN 1521-186X

Tuffet, S., de Seze, R., Moreau, J. M., & Veyret, B. (1993). Effects of a strong pulsed magnetic field on the proliferation of tumour cells *in vitro*. *Bioelectrochemistry and Bioenergetics*, Vol. 30, (March 1993), pp. 151-160. ISSN 1567-5394

Verginadis, I., Simos, Y., Hadjikakou, S., Havelas, K., Evangelou, A, & Karkabounas, S. (2008). Mimesis of the anticancer effects of a Sn complex through its resonant electromagnetic frequencies, *Proceedings of 7th Tumour Markers Targeting Therapy Congress*, Athens, Greece

Verginadis, I. I., Karkabounas, S., Simos, Y., Kontargiris, E., Hadjikakou, S. K., Batistatou, A., Evangelou, A., & Charalabopoulos, K. (2011a). Anticancer and cytotoxic effects of a triorganotin compound with 2-mercapto-nicotinic acid in malignant cell lines and tumour bearing Wistar rats. *Eur J Pharm Sci*, Vol. 42, No. 3, (February 2011), pp. 253-261, ISSN 0928-0987

Verginadis, I. I., Simos, Y. V., Velalopoulou, A. P., Vadalouca, A. N., Kalfakakou, V. P., Karkabounas, S. Ch., & Evangelou, A. M. (2011b). Analgesic effect of the electromagnetic resonant frequencies derived from the NMR spectrum of morphine. Accepted for publication in *Electromagnetic Biology and Medicine*, ISSN 1536-8386

Walker, J. L., Kryscio, R., & Smith, J. (2007). Electromagnetic field treatment of nerve crush injury in a rat model: effect of signal configuration on functional recovery. *Bioelectromagnetics*, Vol. 28, No.4, (May 2007), pp. 256-263, ISSN 1521-186X

Walleczek, J., & Budinger, T. F. (1992). Pulsed magnetic field effects on calcium signalling in lymphocytes: dependence on cell status and field intensity. *FEBS Lett*, Vol. 314, No. 3, (December 1992), pp. 351-355, ISSN 1873-3468

Wang, Z., Yang, P., Xu, H., Qian, A., Hu, L., & Shang, P. (2009). Inhibitory effects of a gradient static magnetic field on normal angiogenesis. *Bioelectromagnetics*, Vol. 30, No. 6, (September 2009), pp. 446-453, ISSN 1521-186X

Wang, Z., Che, P. L., Du, J., Ha, B., & Yarema, K. J. (2010). Static magnetic field exposure reproduces cellular effects of the Parkinson's disease drug candidate ZM241385. *PLoS One*, Vol. 5, No. 11, (November 2010), pp. e13883, ISSN 1932-6203

World Health Organization (WHO), fact sheet No 297, October 2011, (www.who.int/mediacentre/factsheets/fs297/en/)

Williams, C. D., Markov, M. S., Hardman, W. E., & Cameron, I. L. (2001). Therapeutic electromagnetic field effects on angiogenesis and tumour growth. *Anticancer Research*, Vol. 21, No. 6A, (November-December 2001), pp. 3887-3892, ISSN 1791-7530

Yamaguchi, S., Ogiue-Ikeda, M., Sekino, M. & Ueno, S. (2006). Effects of pulsed magnetic stimulation on tumour development and immune functions in mice *Bioelectromagnetics*, Vol. 27, No. 1, (January 2006), pp. 64-72, ISSN 1521-186X

Yao, C., Mi, Y., Hu, X., Li, C., Sun, C., Tang, J., Wu, X. (2008). Experiment and mechanism research of SKOV3 cancer cell apoptosis induced by nanosecond pulsed electric field, *Proceedings of 30th Annual International IEEE EMBS Conference*, Vancouver, British Columbia, Canada, August 2008

Zhang, X., Zhang, H., Zheng, C., Li, C., Zhang, X., & Xiong, W. (2002). Extremely low frequency (ELF) pulsed-gradient magnetic fields inhibit malignant tumour growth at different biological levels. *Cell Biology International*, Vol. 26, No. 7, (2002), pp. 599–603, ISSN 1095-8355

11

Characterization of the Information Leakage of Cryptographic Devices by Using EM Analysis

Olivier Meynard[1], Sylvain Guilley[1], Jean-Luc Danger[1],
Yu-Ichi Hayashi[2] and Naofumi Homma[2]

[1]Telecom ParisTech
[2]Tohoku University
[1]France
[1]Japan

1. Introduction

Cryptographic modules (software or hardware implementations of cryptographic algorithms) are widely used in our daily life in different applications in order to secure digital transactions and exchanges. In particular, cryptographic hardware is essential in smartcards, identification systems, mobile phones, pay television set-top boxes, transportation services and so on. This hardware component embeds cryptographic algorithms that are deemed safer and unbreakable from a mathematical point of view, and thus increases the confidence and robustness of cryptographic functions. However, the hardware implementation is still vulnerable to physical attacks. Side Channel Analysis (SCA) is a major threat for crypto-systems as they disclose some information about the internal process and the sensitive data. An electronic circuit needs some time to produce its results (such as a ciphertext in the case of encryption) and an amount of energy to switch states at each clock period. The operating times, the power dissipation, or the electromagnetic radiations are directly modulated by the data that are processed. Consequently the cryptographic device leaks some clues about the inner secrets and an attacker can retrieve secrets computed by the cryptographic device, just by analysing these externally measurable quantities without touching the component. Thus, those unintentional physical emanations can be analysed in a view to derive some sensitive information from them. Such analyses are altogether referred to as Side-Channel Attacks.

Since the mid-90s, side channel attacks have attracted a significant attention within the cryptographic community. In 1999, Kocher *et al.* described a side channel attack suitable for smart-cards: the power line, supplied from the card reader, is spied. This kind of attack is named Simple or Differential Power Analysis (SPA or DPA). With this cryptanalysis method an attacker can successfully reconstruct the secure data (Kocher et al., 1999a), either with one single measurement (SPA) or with many of them (DPA), using a statistical analysis. Two years later, Gandolfi *et al.* introduced the principles of the EMA, *i.e.* the ElectroMagnetic Analysis (Gandolfi et al., 2001). In fact Gandolfi applied the method of the DPA to electromagnetic emanations. EMA exploits correlation between secret data and variations in the emitted electromagnetic radiation. The EM radiation becomes an important source of leakage because it can be conducted without tampering with the power supplies when the circuit under analysis is soldered on a printed circuit board (PCB). Such EM radiation has been studied as noise in the field of EMC (Electromagnetic Compatibility). Many studies on noise suppression or reduction have been conducted because noise interference can cause damage to other electronic devices in the vicinity. Some EMC-related committees have summarized the aforementioned knowledge and experiences, and have established guidelines on standardized acceptable values of EM radiation during device operations. Current electronic devices are usually designed so as to satisfy these EMC standards. However, these standards mainly aim to suppress and reduce EM radiation that disturbs other devices, but not necessarily the radiation that leaks secret information. Even if the EM radiation (*i.e.*, common-mode current) is below the value specified in the guidelines, extraction of secret key information from the radiation would remain a possibility. In fact, some previous studies (Kuhn, 2005) have demonstrated EM information leakage from electronic devices that are in compliance with the guidelines.

Addressing the above mentioned problem, this chapter investigates the possibility of EM information leakage at a distance from cryptographic devices. We first describe conventional side-channels, namely voltage drop and electromagnetic field, and then discuss how the common-mode current happens, contains the secret information and has a possibility of information leakage outside of cryptographic hardware. After briefly explaining the test device and the conventional SCA techniques, we present some measurement methods of common-mode current including those at a distance from the test device and their experimental results. Then we propose a method to investigate and characterize the EM radiation in frequency domain. With this characterization based on information theory we are in position to say which frequencies are carrying information. That helps us evaluate the possibility of EM analysis at a distance from cryptographic devices.

2. Side-channels

In this section, conventional and possible side-channels are described. As conventional side-channels, (i) voltage drop at an inserted resistor and (ii) electromagnetic fields close to the module are described. A mechanism behind EM propagation and radiation by ground bounce is also explained as a possible side-channel.

2.1 Conventional side-channels

Fig. 1 shows an overview of the conventional power-measurement method using a resistor (Kocher et al., 1999b). In this measurement, the transient current I released from the power pins of the LSI is assumed to contain information leakage. The mechanism behind the leakage based on the switching behavior of CMOS gates is discussed in (Mangard et al., 2007). The current I must be transformed into voltage units as general instruments (*e.g.*, digital oscilloscopes) which accept voltage signals. For this purpose, a small resistor is inserted in series between the pin and the PCB. The voltage observed at the resistor R is $V = RI$ according to the Ohm's law. Then the attacker can measure the voltage V which is proportional to the current I. Many studies have been using the above method due to its simplicity and reproducibility. When we consider the availability of measurement, however, the method involves manipulation (*i.e.*, insertion of a resistor) of the PCB and requires a contact to the PCB.

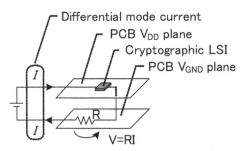

Fig. 1. Conventional side-channel: voltage drop.

The measurement method of EM field for SCAs is proposed in (Gandolfi et al., 2001), (Quisquater & Samyde, 2001) and is also being widely used. For the measurement, a magnetic field probe is placed very close to the chip. The probe transforms the magnetic field (or magnetic flux) into a voltage output based on the Electromagnetic Induction. Here, the output voltage V is

$$V = -N\frac{d\Phi}{dt},$$ (1)

where Φ represents the magnetic flux within the closed loop comprising the probe and N is the number of the loops. Since $\Phi \propto I$, it follows that $V \propto dI/dt$. A magnetic probe directly outputs V, while a current probe outputs the value proportional to I because of its loop integration. Although the output of the magnetic field probe is proportional to dI/dt but not to I, a number of experiments have shown that the attack using the magnetic probe is feasible and efficient (Peeters et al., 2007). This method also requires close access to the PCB since the electromagnetic near field decreases in amplitude in proportion to $1/r^3$, where r is the distance between the target and the probe. Therefore, signals measured from a distance suffer from the effects of external noise, resulting in a low S/N ratio.

2.2 Common-mode current

The above information leakage can be leaked via a different side-channel, which has a possibility that cryptographic modules can be attacked at distance. In the classical circuit theory, the level of the ground plane is assumed to be constantly zero. However, in reality, the ground level can change. Such transient voltage fluctuation in the ground plane is referred to as ground bounce. A transient current released from a digital circuit is a major source of ground bounce. Here, the released transient current I is transformed into a voltage fluctuation ΔV through inductance.

Fig. 2 shows an image of ground bounce. The above inductance is distributed over the PCB since conductors with finite length (*e.g.*, pins and lead lines) have parasitic inductance. When a transient current I is fed into such inductance, an electromotive force occurs due to electromagnetic induction (Sudo et al., 2004), which results in generating a voltage fluctuation ΔV in the ground plane. The fluctuation is expressed as the following equation (Sudo et al., 2004):

$$\Delta V = L_{eff}M\frac{dI}{dt},$$ (2)

where L_{eff} is the effective parasitic inductance, M is the number of simultaneous switching outputs, and dI/dt is the rate of the current change. The amount of L_{eff} depends not only on

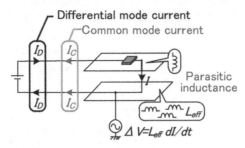

Fig. 2. Image of common-mode current.

the self-inductance within the cryptographic chip but also on the mutual inductance between the chip and the PCB. The voltage fluctuation caused by ground bounce can be modeled as an alternate voltage source as shown in Fig. 2. The model suggests that the voltage fluctuation can propagate to peripheral circuits through a common ground. Consequently, peripheral circuits, such as attached cables, are driven as antennas, which causes unintentional radiation via them (Hockanson et al., 1996).

Such radiation based on ground bounce is even more important since it generates common-mode current (Paul, 2006). If there is only differential-mode current as shown in Fig. 1, the electromagnetic fields radiated from the current pair (current with forward and reverse directions) are ideally canceled out since they are equal in amplitude and inverse in direction. The resulting radiation would be limited. In the case of common-mode current, on the other hand, the electromagnetic fields from the current pair are not cancelled out since their directions are the same. As a consequence, the common-mode current can cause strong radiation even if it is weak in amplitude. Assuming a transient current I released from a cryptographic module causes a voltage fluctuation ΔV. ΔV contains information leakage due to $\Delta V \propto dI/dt$. As a result, EMA would be possible by measuring the radiation driven by ground bounce. The mechanism also suggests that peripheral circuits interconnected to a cryptographic module can be an antenna responsible for information leakage.

3. Cryptographic devices

For the following experiments we employ a Side-channel Standard Evaluation Board (SASEBO-G) which is widely used as a uniform testing environment for evaluating the performance and security of cryptographic modules. Until now, various experiments associated with side-channel attacks are being conducted on the SASEBO boards, and many useful results are being expected to support the international standards work[1]. Fig. 3 shows the SASEBO-G used in these experiments, which employs two Xilinx FPGAs ; one FPGA is used to implement a cryptographic module in hardware or software and the other FPGA is dedicated to communicate with a host computer through either RS-232 or USB cables.

Two kinds of major cryptographic algorithms are implemented in one FPGA for the experiments: RSA crypto-system and AES (Advanced Encryption Standard) block encryption.

---

[1] http://www.rcis.aist.go.jp/special/SASEBO/index-en.html

Fig. 3. Overview of SASEBO-G.

3.1 RSA crypto-system and SPA/SEMA

The RSA crypto-system, proposed by Rivest, Shamir, and Adleman in 1977, is one of the most popular public-key ciphers. The encryption and decryption operations are given by simple modular exponentiation:

$$C = P^E \bmod N, \tag{3}$$
$$P = C^D \bmod N, \tag{4}$$

where P is the plaintext, C is the ciphertext, E and N are the public keys, and D is the secret key. Modular exponentiation is also used in other public-key ciphers such as ECC (Elliptic Curves Cryptography), and thus the following analysis technique can be widely applied to other public-key ciphers. The binary method (*aka* the square-and-multiply method) is known to be the most efficient exponentiation algorithm and is frequently used for actual applications, such as smartcards and embedded devices, because of its simplicity and low resource consumption. This algorithm performs multiplication and squaring sequentially according to the bit pattern of one exponent (E or D). There are two variations of the algorithm. The left-to-right binary method starts at the exponent's MSB and works downward. The right-to-left binary method, on the other hand, starts at the exponent's LSB and works upward. **ALGORITHM I** shows a left-to-right binary method for scanning the bits of the exponent from MSB to LSB. Each multiplication (or squaring) operation requires a large number of clock cycles due to the large length of the operand. This algorithm always performs a squaring at Line 3 regardless of the scanned bit value, but the multiplication at Line 5 is executed only if the scanned bit is equal to 1.

The basic sequence in the binary method is not changed even when major acceleration techniques such as Montgomery multiplication (Montgomery, 1985) and the Chinese Remainder Theorem (CRT) (Menezes et al., 1996) are applied to the exponentiation computation.

With this algorithm for the next experiment, we perform SPA/SEMA (SEMA is the electromagnetic counterpart to SPA) in order to distinguish between multiplication and squaring in the power/EM waveform. Fig. 4 shows an image of the SEMA on an RSA module using the left-to-right binary method. When the difference between multiplication

ALGORITHM I

MODULAR EXPONENTIATION (LEFT-TO-RIGHT BINARY METHOD) FOR A SECRET KEY OF BIT LENGTH k.

Input:	$X, N,$
	$E = (e_{k-1}, ..., e_1, e_0)_2$
Output:	$Z = X^E \bmod N$

```
1:   Z := 1;
2:   for i = k − 1 downto 0
3:       Z := Z * Z mod N;        – squaring
4:       if (eᵢ = 1) then
5:           Z := Z * X mod N;    – multiplication
6:       end if
7:   end for
```

and squaring appears as shown in this figure, the key bit pattern '10100' can be derived from the knowledge of the algorithm.

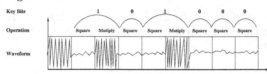

Fig. 4. Image of Simple Electromagnetic Analysis (SEMA) on RSA module.

If RSA is simply implemented with binary methods, definite vulnerabilities could exist. For example, differences between a conditional branch or an instruction sequence could be observed in the power/EM waveforms, giving strong clues about the value of the secret exponent. Even if the squaring and multiplication are performed using the same processing unit controlled by the same sequencer logic, chosen-message approaches that use specific data (Novak, 2002)–(Fouque & Valette, 2003)(Schramm et al., 2004)(Homma et al., 2008) can further enhance these differences.

One simple idea is to choose a message that has a large number of 1s (or 0s) in the bit sequence Miyamoto et al. (2008). For example, an input value of $2^{-k} \bmod N$ or R^{-1} (with $R = 2^k \bmod N$) may produce large differences between the multiplication and the squaring operations for implementations using Montgomery multiplication because R^{-1} is converted into the Montgomery domain in $Y = R^{-1}R = 1 \bmod N$ and an input of 1 is always multiplied in the modular multiplication operations. The power consumed by the multiplier for modular multiplication should be much lower than that for modular squaring that does not have an input of 1.

3.2 AES crypto-system and differential analysis

The AES is a symmetric encryption algorithm based on a SPN (Substitution Permutation Network) structure, which has a fixed block size of 128-bits and a key size of 128-bit, 192-bit, or 256-bit, that determines the number of encryption rounds (to respectively 10, 12 or 14). For our experiments, we consider the 128-bit block version that consists in 10 rounds encryption. It operates on a 4×4 array of Bytes termed the state. The AES cipher is specified as a certain number of operations per round that convert the input plain-text into cipher-text. One round of encryption consists of four operations: SubBytes, ShiftRows, MixColumns and AddRoundKey. The encryption period starts with a single AddRoundKey operation followed by nine identical encryption rounds. The final encryption round operates without any MixColumns operation. These adaptations removes cryptographically ineffective operations and allows for a decryption process that is very close to the encryption process. Simultaneously the key schedule is computed: it derives enough keys to provide for each round a different subkey.

An attacker uses an hypothetical model of the device under attack to predict its electromagnetic radiation. These predictions are correlated with the measured samples. The DPA was proposed by P. Kocher in (Kocher et al., 1999b) and is based on Hamming Weight model. For our attacks, we have chosen to use the CPA (Correlation Power Analysis). This approach is based on the Hamming distance model proposed by É. Brier *et al.* in (Brier et al., 2004) for the first time. For this CPA, the outputs of S-Box are targeted on the last round of the AES to reveal one Byte of the secret key. Given the cipher and the results of the tenth round, the attacker can predict the leakage by computing the Hamming distance model between two states of the register. An hypothesis can be made for eight bits of the key that corresponds to output of the S-Box. To perform this attack some measures of electromagnetic emanation of the device are collected. For one hypothesis of 8 bits of the subkey the attacker has to calculate $256(= 2^8)$ differential traces for all the subkey candidates. After that, the attacker has to predict the electromagnetic radiation when the bits of the register toggle. Let X_i and X_{i+1} be two consecutive values inside a register. An estimation of the radiated emanation at the time of the transition could be provide by the computation of the selected function $HD = HW(X_i \oplus X_{i+1})$ where HD is the Hamming Distance and HW the Hamming Weight. After that the attacker estimates the maximum likelihood between the theoretical predictions and the measurements by the Pearson correlation factor $\rho(W, HD)$ between the Hamming distance model and the measured power. It is defined as:

$$\rho(W, HD) \doteq \frac{\text{cov}(W, HD)}{\sigma(W) \cdot \sigma(HD)},$$

where W represents the measurement and HD the Hamming distance. We notice that $\rho(W, HD)$ follows the Cauchy-Schwarz inequality: $-1 \leq \rho(W, HD) \leq +1$. The correct key is obtained when the right key hypothesis provides the largest or smallest values of $\rho(W, HD)$. In other words, a spike is observed in the differential curve when the correct partial subkey bits have been used and where the selection function is correlated to the value of the bit being manipulated. The success of this attack depends on the number of the measured traces. Obviously this number will be changed with the distance between the component and the probe, with the selected element of the hardware implementation (S-Box) and with the leakage model used in selected function.

4. First experiments

4.1 Electromagnetic analysis based on common-mode current

Fig. 5 and Fig. 6 show a block diagram and overview of the measurement setup, respectively. The measurement system consists of the SASEBO, a digital oscilloscope, and a personal computer (PC). The SASEBO is equipped with two FPGAs, namely *FPGA1* and *FPGA2*. These two FPGAs work by using power supplied from on-board regulators and clock signal provided by an external function generator. Experiments are conducted by implementing cryptographic cores (circuits) on *FPGA1*.

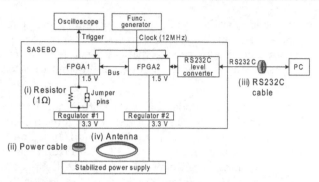

Fig. 5. Block diagram of measurement setup.

Fig. 6. Overview of four measurement methods.

Four types of measurements are conducted: (i) the voltage drop across a resistor, the current flowing in (ii) the attached power cable and (iii) the communication cable (RS232C cable), and (iv) the magnetic field around the power cable. Hereafter, these are referred to as (i) *Resistor*, (ii) *Power cable*, (iii) *RS232C cable*, and (iv) *Antenna*. It is important to emphasize that the measuring points (ii) and (iii) are not directly connected to *FPGA1*; there are circuit components (*e.g.*, voltage regulators and an RS232C level converter) between the measurement points and *FPGA1*. In addition, the locations of the measurements are about 50 cm away from the board.

Measurement instruments are summarized in Table 1. The details of the measurement methods are as follows. In measurement (i), the voltage drop across a 1-Ohm resistor inserted between the ground pin of *FPGA1* and the ground plane of the PCB is measured using a differential voltage probe. Note that the 1-Ohm resistor is short-circuited during the measurements (ii)-(iv) by using jumper pins.

In the measurements (ii) and (iii), the current flowing in cables is measured by the current probe clamping the cables as shown in Figs. 6(ii) and (iii). The measured voltage is proportional to the flowing current as described above. In measurement (ii), both lines (for

Oscilloscope	Agilent MSO6104A (with 25 MHz low-pass filter)
Voltage probe for (i)	Agilent 1130A differential (voltage) probe with SMA probe head
Current probe for (ii) and (iii)	Fischer F-2000 current probe (10 MHz – 3 GHz)
Current probe for (ii) and (iii)	AOR LA380 Wideband Active Loop Antenna (10 kHz – 500 MHz)
Pre amplifier for (ii)-(iv)	MITEQ AM-1594-9907 (+51 dB, 300 kHz – 3.0 GHz)

Table 1. Measurement instruments

$V = DD$:3.3 V and GND: 0 V) of a twisted pair cable are clamped, while the whole body of RS232C cable is clamped for measurement (iii). Since both of the current pairs are clamped, radiation due to differential-mode current is cancelled out within the probe, and thus the contribution only by the common-mode element is measured.

In the measurement (iv), a magnetic field around the power cable is measured using an antenna. We used an off-the-shelf indoor loop antenna which is used for amateur radio. We placed the antenna over the power cable at the height of about 20 cm and tuned it to maximize the measured amplitude in the range of 3–40 MHz.

In each measurement, the frequency bands of measured signals are limited up to 25 MHz by a low-pass filter equipped with an oscilloscope. This is because the measured raw traces were highly contaminated by high-frequency noise that interfered with the measurements. In addition, a trigger signal generated by the *FPGA1* is used in order to align the measured traces in time. As a probe for the trigger signal has a physical contact to the board, the measurements are not exactly contactless. However, the setup is enough to examine information leakages from the measuring points (i)–(iv). In practical scenario, an attacker would have difficulty in taking a trigger without any contact to the target, yet it is still possible. One practical way is to observe communication cables and then obtain a trigger from a specific binary sequence on them. In addition, the attacker can consult signal processing techniques to achieve precise waveform alignment (Homma et al., 2006).

4.1.1 SEMA on RSA implementation

A 1,024-bit RSA circuit using a left-to-right binary method is implemented on *FPGA1*. Modular multiplication and squaring are performed by high-radix Montgomery multiplication algorithm using a 32-bit multiplier (Cryptographic Hardware Project, n.d.). In this implementation, one Montgomery multiplication requires 4,386 cycles, and the total number of cycles for the modular exponentiation (1,024-bit RSA operation) is approximately 7 million cycles. The parameters (*i.e.*, the key and the plaintext) are embedded into the *FPGA1* in order to allow *FPGA1* to operate as a standalone module. The goal of the attack is to distinguish between multiplication and squaring in the measured trace. Since we are interested in the difference between the measurements (i)–(iv), a chosen input message of 2^{1024} is used to enhance the difference between the multiplication and squaring.

Fig. 7 shows the results of the SEMA. The traces are measured at a sampling frequency of 400 MSa/s. Each of the traces is aligned in time, in which the modular exponentiation

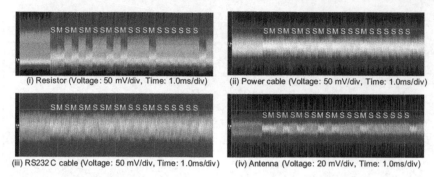

(i) Resistor (Voltage: 50 mV/div, Time: 1.0ms/div)   (ii) Power cable (Voltage: 50 mV/div, Time: 1.0ms/div)

(iii) RS232 C cable (Voltage: 50 mV/div, Time: 1.0ms/div)   (iv) Antenna (Voltage: 20 mV/div, Time: 1.0ms/div)

Fig. 7. Results of SEMAs on RSA.

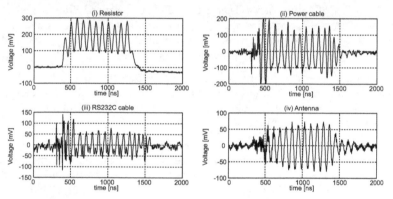

Fig. 8. Examples of measured waveforms.

starts at around 1.5 ms. Sequences of symbols 'S' and 'M' shown in the figure represent the corresponding squaring and multiplication operations, respectively.

The result of measurement (i) shows large difference between multiplication and squaring with multiplication having lower spikes compared to squaring. Although the differences are smaller than (i), they are still visible in results (ii)–(iv). The results indicate that it is possible to reveal a secret key by using any of the measurements (i)–(iv).

4.1.2 DEMA on AES implementation

In the experiments, an AES circuit, retrieved from the reference (Cryptographic Hardware Project, n.d.), supporting a 128-bit key is implemented on *FPGA1*. The circuit uses a loop architecture, where one round operation is performed every clock cycle. As a result, one encryption takes 10 clock cycles for round operations and an additional clock cycle for data I/O. *FPGA2* is configured as the control and communication circuits, and plaintexts are fed from the PC into *FPGA1* via *FPGA2*. During the encryption, the corresponding traces are captured from the four measurement points at a sampling frequency of 500 MSa/s. The measurements are repeated for 30,000 different plaintexts, and the corresponding 30,000 traces are stored for each measuring points. Examples of the measured traces are shown in Fig. 8, where the encryption process starts at around 400 ns and finishes after 11 clock cycles or 916 ns (=11× 1/12 MHz).

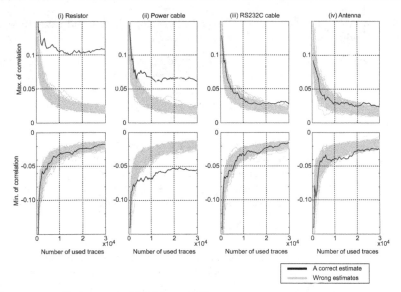

Fig. 9. Measures to disclosure in DEMA on AES.

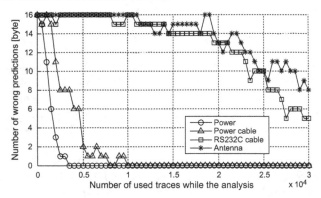

Fig. 10. Results of DEMAs on AES.

CPA (or CEMA) is applied to these traces. The 128-bit (16-Byte) register containing intermediate data is chosen as a target. The power (or EM radiation) estimates are calculated by counting changed bits of the target register in the final round of AES encryption. Here, Hamming distance model (Brier et al., 2004) is used as power model. Key candidates are searched by Byte. In other words, we generated a total of 16 power estimates corresponding to 16 Bytes of the round key. In the analysis phase, the linearity is evaluated by using Pearson's correlation coefficient.

The results are shown as error rates in Fig. 9 and Fig. 10. Fig. 9 shows the Measurement to disclosure (MTD) graph of each result. On the other hand, Fig. 10 shows error rates where the vertical axis represents the number of incorrectly predicted round-key Bytes and the horizontal axis shows the number of traces. Since the length of the secret key is 16 Bytes

($= 128$ bits), the vertical value ranges between 0 and 16, where 0 indicates the successful extraction of the whole key (*i.e.*, completion of the attack).

In the analysis with the measurement (i), the key prediction is successful when the correlation with the correct key is the highest among 2^8 candidates. On the other hand, in (ii)–(iv), the difference between maximum and minimum coefficients is used instead of the maximum coefficient since the results of (ii)–(iv) show correlation in both the positive and negative directions as shown in Fig. 8.

As shown in Fig. 10, the result for (i) goes to zero (*i.e.*, extract the whole key) fastest by 3,000 traces. In addition, the attack using the power cable (ii) also shows fast extraction. On the other hand, the attacks using the RS232C cable and the antenna require much larger number of traces. However, the error rate decreased gradually as the number of used traces increased. Therefore, we can say that all the EMAs can successfully reveal the secret keys.

4.1.3 Discussion

As shown above, the contactless SEMAs and DEMAs worked well, but the traces from (ii)–(iv) contained smaller amount of information leakage or larger noise in comparison to that from (i). Various disturbing factors between the FPGA and the measuring points have effects on the results. Such factors include the filtering effect of parasitic circuit, the compensation effect of regulators, and external noise, and so on.

For example, there is an on-board regulator between the chip's power supply pins and the power cable. Since regulators are designed to stabilize their output voltage, they feature buffering and feedback control in order to suppress voltage fluctuation. The results of (ii) indicate that the voltage fluctuation containing information leakage was able to overcome the effects of the regulator.

Measurement (iii) is also affected by such disturbing factors. In the experimental setup, the RS232C cable is connected to *FPGA1* via an RS232 level converter and *FPGA2*. First, the RS232 level converter has the same effect as the voltage regulator. In addition, *FPGA1*, *FPGA2*, and the RS232C level converter have their own separated ground planes and they are connected via noise filters (inductors). This feature is used in SASEBO in order to measure side-channel information only from *FPGA1* without those from other components. The noise filters act as low-pass filters for current through them. However, the success of measurement (iii) indicates that the voltage fluctuation from *FPGA1* can propagate to the other part of the board even if such high-frequency current is filtered out. The results suggest that the propagation of the voltage fluctuation is rather robust and should be prevented by countermeasures.

4.2 Attack at distance

Now we demonstrate that Correlation-based on ElectroMagnetic Analysis (CEMA) on a hardware-based high-performance AES module is possible from a distance as far as 50 cm (Meynard et al., 2010). The aim of this experiment is to mount a successful Correlation Power Analysis (CPA (Brier et al., 2004)) and retrieve cryptographic elements, without any additional device, such as a demodulator or a TEMPEST receiver. The device whose EM emanations are studied is cadenced by a clock running at 24 MHz. For these low frequencies, the usage of a Faraday cage is not needed as it would induce some EM reflections and could alter the measurements. Furthermore, the size of the absorber would be too large for this range of frequencies. Therefore, the material is placed on a plastic table that limits the reflection of EM radiation and avoids the conducted radiation. A plastic rod is placed perpendicularly to the board and is considered as a vertical axis to move the antenna by steps of 5 cm, as shown in the figure 11.

Fig. 11. EM measurement test bench with its antenna on a plastic rod.

We take care of keeping far away the power supply from the chip board, in order to avoid any coupling between the radiated waves and the power supply. We record the emanations on the side with the decoupling capacitors, because the signal on this part of the board has the best quality.

Firstly we check that for different distances, the curves for the same plaintext are scaled down, according to an inverse power law. as illustrated in figure 12.

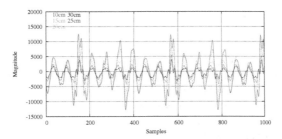

Fig. 12. Quasi-homothetic downscale of the raw curves at different distances.

In practice, the signal was amplified of 60 dB and averaged by a factor of 4096. The attack needs 51,519 measurements to break the Sbox #1, whereas only 1,000 are required at $d = 0$ cm. The correlation curve is represented in figure. 13. The correlation does not clearly stand out. We assume that the attack requires so many traces to fully disclose the key because the Hamming model is not holding anymore at this large distance. Then we propose to study the distortion of the leakage model with the distance.

It has already been noticed in the literature that the Hamming distance is not the best model in the case of very near-field analyses. For example, authors in Peeters et al. (2007) proves that under some circumstances, an ASIC can have a transition-dependent leakage. In this section,

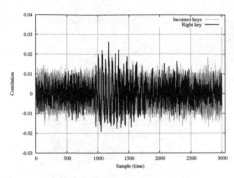

Fig. 13. CEMA on Sbox #1 at a distance $d = 50$ cm.

we show that the Hamming distance model is adequate for intermediate distance fields EM analyses, but that it distorts seriously in far-field analyses.

In near-field the leakage obeys a Hamming distance model: it is an affine function of the number of bit transitions between two consecutive states. On one hand, the Hamming distance is confirmed at $d = 0$ cm, as attested by figure. 14, at 15 cm, the model is chaotic and not consistent with an identical amount of dissipation per bit making up the analyzed Byte.

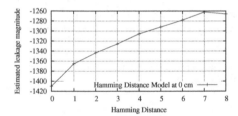
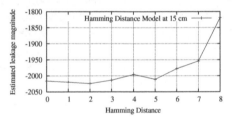

Fig. 14. Hamming distance at 0 cm and at 15 cm.

We target the Sbox #1, which exercises all the 256 transitions and propose to characterize the leakage to disclose the Sbox #1 on the tenth subkey of the AES. First we search the index t_c of the maximal correlation, that corresponds to the moment when the data are stored in the register on the last round. Then we compute for this index the average and the variance for the 256 possible Hamming distances, the key and the message being known. The figure 15 depicted the leakage model at 50 cm.

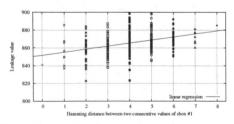
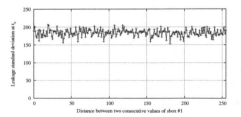

Fig. 15. Model at 50 cm for the Sbox #1.

We can observe that the standard deviation is almost independent of the Byte distances. Therefore, most of the model information is contained in the mean leakage value.

We notice that the leakage model distorts more and more with the distance. Therefore, we show that the leakage model change according to thresholds. Three regions can be identified: in near-field, the switching distance is the most suited, as initially observed in the article (Peeters et al., 2007); in medium-distance ($d \in [0,5]$ cm), the Hamming model is adequate; in long-distance ($d > 5$ cm), the model becomes less relevant.

We target in this topic situations where the difference of nature is not artificially due to a countermeasure, but naturally by the distortion into the communication channel between the leaking device and the side-channel sensor.

5. Characterization of the frequencies

EM radiations arise as a consequence of current flowing through diverse parts of the device. Each component affects the other components' emanations due to coupling. This coupling highly depends on the device geometry. Now we describe a measurement of EM radiation from a cryptographic device, whose intensity is a major suppression target in the EMC research field. We first generate an EM-field map on the entire surface of the device, and then pinpoint the points being high in EM-field intensity. Fig. 16 shows an overview of the EM measurement system in this experiment. The experimental scanner (WM7400) employs a micro EM probe whose bandwidth ranges from 1 MHz to 3 GHz, and scans the surface of the SASEBO. The probe head is arranged precisely at 2-cm distance from a target device within a tolerance of one micrometer. The system can measure the distance by the equipped laser geodesy.

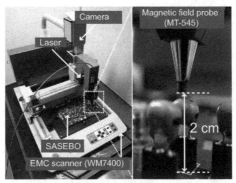

Fig. 16. EM measurement system.

Fig. 17 shows EM field maps on the entire surface of the SASEBO corresponding to the frequency bands ranging between (a) 10-100 MHz, (b) 100-200 MHz, (c) 200-300 MHz, and (d) 300-400 MHz, where the red and blue areas indicate higher and lower intensities, respectively.

The result shows that specific areas around *FPGA2* and a crystal oscillator, which is located at the upper side of *FPGA2* in Fig. 3, have higher EM-field intensities than other areas. This is because only the two components are active components on the board. We confirmed from the result that the EM-field intensity at the clock frequency is relatively higher than those of other frequencies. The phenomena of compromising signal has different origins such as

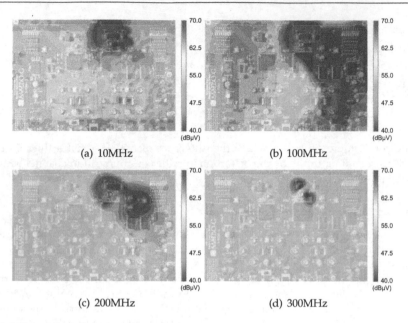

(a) 10MHz   (b) 100MHz

(c) 200MHz   (d) 300MHz

Fig. 17. Subfigure with four images

radiation emitted by the clock, crosstalk or coupling. Traditionally, we differentiate the direct emanations and the indirect or unintentional emanations. The first ones can be considered at a very short distance and requires the use of special filters to minimize interference with baseband noise. The direct emanations come from short bursts of current and are observable over a wide frequency band. On contrary, indirect emanations are present in high frequencies. According to Agrawal (Agrawal et al., 2002) these emanations are caused by electromagnetic and electrical coupling between components in close proximity. Often ignored by circuits designers, these emanations are produced by a modulation. The source of the modulation carrier can be the clock signal or other sources, including communication related signals. Li *et al.* provide in (Li et al., 2005) a model to explain such kind of modulation.

Therefore it is sometimes easier to extract information from signals unintentionally modulated at high frequencies, which are not necessarily related to the clock frequency, than baseband signals also referred to as direct emanations. The characterization of the frequencies that modulate the leakage is a scientific challenge, since as of today no relevant tool allows to distinguish which frequencies actually contain the sensitive information. For this reason, we propose a methodology in the following based on information theory.

5.1 Characterization of the EM channel in frequency domain

For the same bit sequence as in Figure 4, we obtained the EM trace illustrated on Figure 18. No difference appears between a square and multiply, even when messages are chosen to improve the result. We have even tried to improve the analysis using pattern matching techniques but without any satisfactory results in terms of contrast. First of all, the noise effect is decreased if the frequency band is reduced. Secondly, the leaked information is properly digitized whereas the strong carrier without relevant information is removed. Therefore it appears

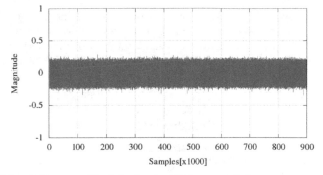

Fig. 18. Direct EM radiations emitted during an RSA computation.

strategic to find right modulated carrier. A straightforward technique consists in using a spectral analysis in order to detect the strong carrier frequencies. Another possible technique consists in scanning the frequency range of the wide-band receiver, but such demodulation process is time-consuming and one may omit some significant compromising signal. Another technique based on the STFT (Short Time Fourier Transform) has been proposed in (Vuagnoux & Pasini, 2009), but it consumes a huge amount of time as well as memory resources. We propose therefore a method based on information theory to characterize the leakage. After this characterization we are able to select the frequencies and their associated optimal bandwidth. The useful information is contained in these ranges of frequencies. Therefore, with a receiver tuned on the right frequency, we can retrieve the compromising signal. To provide this characterization, we propose an approach based on information theory. This method can be managed as follows: First we gather a large number of measurements, by knowing the key *i.e.* the operations that are computed by the chip. These EM measurements from the antenna are noisy, distorted and the operations are not distinguishable. For this step, we chose a time window where only one operation of square and one operation of multiply are performed as shown on Fig. 19. After the measurements are cut according to the operation performed. The number of samples is equal in each part of the signal, Each section of the signal is equal in term of number of samples, we get as much parts of EM signal for the multiplication as for the squaring, and we obtain two sets of measurements with the same number of traces.

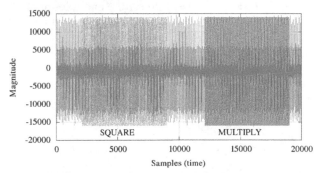

Fig. 19. EM measurement split into Square and Multiply parts.

Then, for each set, we compute: the FFT (*Fast Fourier Transform*) of every observation O_f; the mean spectrum related to each operation; and the mean of all the observations. Therefore we obtain a specific spectral signature for each operation of the modular exponentiation algorithm. Finally we compute the Mutual Information value for each frequency. Thus we attribute a specific spectral signature to each operation of the modular exponentiation algorithm. In few words, we follow the processing shown in ALGORITHM 3.

ALGORITHM 3

Input:	$O = (O_0, ..., O_{n-1}, O_n)$ Observation in time domain,
	$S = (S_0, ..., S_{n-1}, S_n)$ Secret (Operation)
Output:	Result of Mutual Information in frequency domain

1 : **for** $i = 0$ **to** n
2 :   Sort O_i Observation according to the Secret S_i;
3 :   Compute the FFT of each Observation O_i;
4 : **endfor**
5 : Compute the mean $(\mu_{Square}, \mu_{Multiply})$
      and the variance $(\sigma_{Square}, \sigma_{Multiply})$
6 : Compute the Mutual Information in frequency domain.

As a distinguisher we take for instance an information theory viewpoint to retrieve the relevant frequencies and to bring a mathematical proof that the information is not necessarily carried by the clock frequency. In 2008, Gierlichs introduced in (Gierlichs et al., 2008) the Mutual Information Analysis. This tool is traditionally used to evaluate the dependencies between a leakage model and observations (or *Measurements*). We use MIA like in previous chapter but in this case we compute for each frequency the Mutual Information (MI) $I(O_f; Operation)$ between Observations O_f and *Operation* that corresponds to the operations performed by the device (Meynard et al., 2011). Thereby, if $I(O_f; Operation)$ is close to zero for one frequency f, we can say that this frequency does not carry significant information. On the contrary, if $I(O_f; Operation)$ is high, the computed operation and the frequency are bound. As a consequence if we filter the EM signal around this frequency, we can retrieve the operations and the secret key using the SEMA.

The MI is computed as:

$$I(O_f; Operation) = H(O_f) - H(O_f|Operation), \tag{5}$$

where $H(O_f)$ and $H(O_f|Operation)$ are respectively the entropies of all the observations in the frequency domain and of the observations knowing the operations. Both these entropies can be obtained according to:

$$H(O_f) = -\int_{-\infty}^{+\infty} \Pr(O_f) \log_2 \Pr(O_f),$$

$$H(O_f|Operation) = \sum_{j \in \{Multiply, Square\}} \Pr(j) H((O_f|j)).$$

$$\text{with } H(O_f|j) = -\int_{-\infty}^{+\infty} \Pr(O_f|j) \log_2 \Pr(O_f|j),$$

where $\Pr(O_f)$ denotes the probability law of observations at frequency f. Moreover we consider that the computed operations are equi-probable events, for our time windowing therefore $\forall j \in Operation$, $\Pr(j) = \frac{1}{2}$. And the distribution is assumed to be normal $\sim N(\mu, \sigma^2)$

of mean μ and variance σ^2, given by:

$$\Pr(O_f) = \frac{1}{\sqrt{2\pi\sigma^2}} \exp\left(-\frac{(O_f - \mu)^2}{2\sigma^2}\right).$$

We call it a parametric model. We approximate this model by a parametric estimation, and we use the differential entropy defined like a 1-dimensional normal random variable O_f of mean μ and standard deviation σ as the analytical expression: $H(O_f) = \log_2(\sigma\sqrt{2\pi e})$. From this value, the Mutual Information defined in Eqn. (5) can be derived, by computing for each operation the differential entropy:

$$I(O_f; Operation) = H(O_f)$$
$$- \frac{1}{2}(H(f|Multiply) + H(f|Square)),$$

that can be simplified as:

$$I(O_f; Operation) = \frac{1}{2}\log_2 \frac{\sigma_{O_f}^2}{\sigma_{O_f, Multiply}\sigma_{O_f, Square}}. \tag{6}$$

The figure 20 represents the result of Eqn. (6). From this graph, we notice that the information might be contained in a range of frequency between 5.0 and 60.0 MHz with the presence of a large lobe spread over these frequencies.

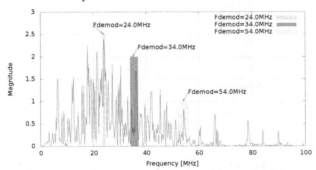

Fig. 20. Result of MIA in frequency domain.

This method provides a result with a quantity expressed in bit, that allows us to interpret easily the leakage frequencies regarding the level of compromising signal. Consequently, we are now able to fairly compare the level of compromising signal carried by different frequencies. Such Mutual Information metric allows to quantify the level of protection against SEMA attacks. Moreover it is worthwhile to underline that Mutual Information considers the non-linear dependencies that occur during the computation, such as cross-talk that occurs during the computation. The maximum in Magnitude is obtained for the frequencies around 24.0 MHz, that corresponds to the clock frequency of the component. We decide to pick up three ranges of frequencies corresponding to three peaks in Fig. 20:

• around 24.0 MHz,

- around 34.0 MHz,
- around 54.0 MHz.

and study the results of the demodulation at these frequencies to prove the efficiency of our approach.

In (Agrawal et al., 2002) Agrawal used a demodulator to measure EM emanation from an SSL accelerator. We apply a similar technique to the FPGA implementation which consumes far less power than the SSL accelerator. The EM radiation is expected to be much weaker than the previous one. We focus on a range of frequencies between 0.0 and 100.0 MHz and demodulate at the frequencies exhibited by the previous methods at 24.0 MHz, 34.0 MHz and 54.0 MHz. We employ the demodulation technique to investigate unintentional (or indirect)emanation. Each time, the demodulated signal shows a peculiarity that allows to distinguish clearly the two distinct operations.

The unintentional emanation described by Agrawal is the result of modulation or intermodulation between a carrier signal and the sensitive signal. In particular, the ubiquitous clock signal can be one of the most important sources of carrier signals. This assumption is confirmed by our results on figure 20. We tune the receiver to the clock frequency (*i.e.*, 24MHz) with a resolution bandwidth of 1MHz. Figure 21 shows one single demodulated EM waveform at 24 MHz. Indeed, the receiver improves the differences between the two operations dramatically as shown in Fig. 21.

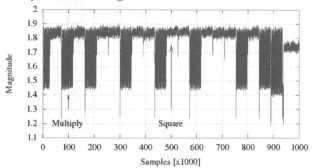

Fig. 21. One Single Demodulated EM waveform at 24 MHz.

We can obtain similar results by tuning the frequency of the receiver to the harmonics of the clock frequency. Moreover we can enlarge the distance between the FPGA and the probe despite a significant lose of S/N ratio. In order to obtain more powerful signals, we used an increased resolution bandwidth and then performed the same SEMA attacks successfully at least 5cm and more distance.

With the method developed previously we can focus on different frequencies that are not necessarily related to the clock harmonics. To measure such emanation, the probe must be placed close to the FPGA. Then an eavesdropper has to tune the receiver at every frequency of the spectrum. Interestingly, we found that the best results were not always obtained by demodulating the raw signal at the harmonics of the clock frequency.

Figures 22 and 22 show the single demodulated EM waveform at 34 and 54 MHz, which have been identified by the peaks obtained on our MI analysis on figure 20. The same sequence is replayed by changing only the demodulation frequency.

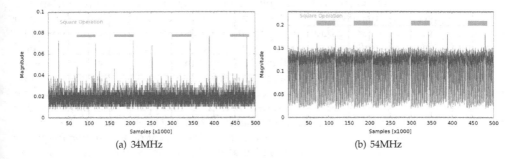

(a) 34MHz   (b) 54MHz

Fig. 22. One single Demodulated EM waveform at 34 MHz and at 54 MHz.

If we compare the figures 21 and 22 we notice that sharp peaks appear at the beginning of every square operation. These peaks are not present before a multiply operation and thus we can easily distinguish the square from the multiply operations. We obtained the same phenomena for the demodulation at 54 MHz on figure 22. Moreover it is important to notice that the magnitude of the compromising signal decreases when the frequency of demodulation increases. The magnitude of the compromising signal follows the trend obtained in the previous section. These results confirm the results obtained during the characterization as shown on the table 2.

Frequency	MI [bit]	Magnitude
24.0 MHz	2.5	0.5
34.0 MHz	1.7	0.03
54.0 MHz	1.0	0.02

Table 2. Comparison between the results.

6. Conclusions

In this chapter, we present firstly leakage mechanism behind Electromagnetic Analysis at near and far distances with contactless probe on FPGA implementations of cryptographic algorithms. The measurements are conducted with different techniques, the probe are attached to a power cable or on communication link, free space around the power cable and at distance from the electronic board. We show that an attacker from these measurements and by computing SEMA and DEMA is in position to retrieve the secret key. Different types of leakage radiation have been highlighted, such as Indirect and Direct emanation. Then we investigate a relationship between the intensity of EM radiation and the geometry of the board. In order to evaluate EM information leakage, we performed simple electromagnetic analysis (SEMA) experiments on a cryptographic device with an RSA module. We first measured EM radiations over the entire surface of a device including over the module, and then evaluated which spots and frequencies are available for EM information leakage. On the studied implementation the raw EM measurements show no obvious leakage. The result suggested that the signal (information)-to-noise ratio should be suppressed for achieving circuit and system security assuming that EM radiation can be interpreted as a signal encoding secret information.

In order to distinguish square and multiply operations in the SEMAs, we introduce a method to detect and characterize a crypto-system in frequency domain, *i.e* a distinguisher of frequencies that are carrying information. In addition we show that our method provides exploitable results and allows us to retrieve the leakages frequencies for unintentional emanations. The method proposed based on the mutual information analysis in frequency domain allows to extract the leakage frequencies of the signal related to the square and multiply operations. By following this method we are able to pinpoint the frequencies that are leaking more information and their bandwidth. Thanks to this tool we demonstrate that we are in position to give a quick diagnostic about the EM leakage of a device. Therefore an attacker is able to perform SEMAs. The method of choosing a right demodulation frequency is crucial; and thanks to our characterization based on the MI, information leaked through indirect EM emanations can be detected and observed with one single demodulated EM waveform.

7. Acknowledgment

This research was supported by the Strategic International Cooperative Program SPACES (Security evaluation of Physically Attacked Cryptoprocessors in Embedded Systems), funded by the ANR (french national research agency) and the JST (Japan Science and Technology agency).

8. References

Agrawal, D., Archambeault, B., Rao, J. R. & Rohatgi, P. (2002). The EM Side–Channel(s), *in* B. S. Kaliski Jr., C. K. Koç & C. Paar (eds), *CHES*, Vol. 2523 of *LNCS*, Springer, pp. 29–45.

Brier, É., Clavier, C. & Olivier, F. (2004). Correlation Power Analysis with a Leakage Model, *Cryptographic Hardware and Embedded Systems - CHES 2004: 6th International Workshop Cambridge, MA, USA, August 11-13, 2004. Proceedings*, Vol. 3156 of *Lecture Notes in Computer Science*, Springer, pp. 16–29.

Cryptographic Hardware Project, T. U. (n.d.).
    URL: *Website: http://www.aoki.ecei.tohoku.ac.jp/crypto/*

Fouque, P.-A. & Valette, F. (2003). The doubling attack - *hy upwards is better than downwards*, *CHES*, pp. 269–280.

Gandolfi, K., Mourtel, C. & Olivier, F. (2001). Electromagnetic Analysis: Concrete Results, *in* Ç. K. Koç, D. Naccache & C. Paar (eds), *Cryptographic Hardware and Embedded Systems - CHES 2001, Third International Workshop, Paris, France, May 14-16, 2001, Proceedings*, Vol. 2162 of *Lecture Notes in Computer Science*, Springer, pp. 251–261.
    URL: *http://link.springer.de/link/service/series/0558/bibs/2162/21620251.htm*

Gierlichs, B., Batina, L., Tuyls, P. & Preneel, B. (2008). Mutual information analysis, *CHES*, pp. 426–442.

Hockanson, D. M., Drewniak, J. L., Hubing, T. H., Doren, T. P. V. & Wilhelm, M. J. (1996). Investigation of fundamental EMI source mechanisms driving common-mode radiation from printed circuit boards with attached cables, *IEEE Transactions on Electromagnetic Compatibility* 38(4): 557–566.

Homma, N., Miyamoto, A., Aoki, T., Satoh, A. & Shamir, A. (2008). Collision-based power analysis of modular exponentiation using chosen-message pairs, *Cryptographic Hardware and Embedded Systems - CHES 2008, 10th International Workshop, Washington,*

D.C., USA, August 10-13, 2008. Proceedings, Vol. 5154 of *Lecture Notes in Computer Science*, Springer, pp. 15–29.

Homma, N., Nagashima, S., Imai, Y., Aoki, T. & Satoh, A. (2006). High-resolution side-channel attack using phase-based waveform matching, *Cryptographic Hardware and Embedded Systems - CHES 2006, 8th International Workshop, Yokohama, Japan, October 10-13, 2006, Proceedings*, Vol. 4249 of *Lecture Notes in Computer Science*, Springer, pp. 187–200.

Kocher, P. C., Jaffe, J. & Jun, B. (1999a). Differential Power Analysis, *Advances in Cryptology - CRYPTO '99, 19th Annual International Cryptology Conference, Santa Barbara, California, USA, August 15-19, 1999, Proceedings*, Vol. 1666 of *Lecture Notes in Computer Science*, Springer, pp. 388–397.
URL: *http://www.cryptography.com/resources/whitepapers/DPA.pdf*

Kocher, P. C., Jaffe, J. & Jun, B. (1999b). Differential Power Analysis, *in* M. Wiener (ed.), *Advances in Cryptology – Crypto '99*, Vol. 1666 of *LNCS*, Springer-Verlag, pp. 388 – 397.

Kuhn, M. G. (2005). Security Limits for Compromising Emanations, *in* J. R. Rao & B. Sunar (eds), *Cryptographic Hardware and Embedded Systems – CHES 2005*, Vol. 3659 of *LNCS*, Springer, pp. 265–279.

Li, H., Markettos, A. T. & Moore, S. (2005). Security evaluation against electromagnetic analysis at design time, *in* J. R. Rao & B. Sunar (eds), *Cryptographic Hardware and Embedded Systems – CHES 2005*, Vol. 3659 of *LNCS*, Springer, pp. 280–292.

Mangard, S., Oswald, M. E. & Popp, T. (2007). *Power Analysis Attacks - Revealing the Secrets of Smart Cards*, Springer-Verlag New York, Inc.

Menezes, A., van Oorschot, P. C. & Vanstone, S. A. (1996). *Handbook of Applied Cryptography*, CRC Press.

Meynard, O., Guilley, S., Danger, J.-L. & Sauvage, L. (2010). Far Correlation-based EMA with a precharacterized leakage model, *DATE'10*, IEEE Computer Society, pp. 977–980. Dresden, Germany.

Meynard, O., Réal, D., Guilley, S., Danger, J.-L. & Homma, N. (2011). Enhancement of Simple Electro-Magnetic Attacks by Pre-characterization in Frequency Domain and Demodulation Techniques, *DATE*, IEEE Computer Society. Grenoble, France.

Miyamoto, A., Homma, N., Aoki, T. & Satoh, A. (2008). Chosen-message spa attacks against fpga-based rsa hardware implementations, *FPL*, pp. 35–40.

Montgomery, P. L. (1985). Modular multiplication without trial division, *Math. Comp.*, vol. 44, no. 170, pp. 519–521, 1985.

Novak, R. (2002). Spa-based adaptive chosen-ciphertext attack on rsa implementation, *Public Key Cryptography*, pp. 252–262.

Paul, C. R. (2006). *Introduction to Electromagnetic Compatibility*, Wiley Interscience. ISBN-10: 0471549274.

Peeters, E., Standaert, F.-X. & Quisquater, J.-J. (2007). Power and electromagnetic analysis: Improved model, consequences and comparisons, *Integration, The VLSI Journal, special issue on "Embedded Cryptographic Hardware"* 40(1): 52–60.
URL: *http://dx.doi.org/10.1016/j.vlsi.2005.12.013*

Quisquater, J.-J. & Samyde, D. (2001). Electromagnetic analysis (EMA): Measures and counter-measures for smart cards, *in* I. Attali & T. P. Jensen (eds), *Smart Card Programming and Security, International Conference on Research in Smart Cards, E-smart 2001, Cannes, France, September 19-21, 2001, Proceedings*, Vol. 2140 of *Lecture Notes in*

Computer Science, Springer, pp. 200–210.

URL: *http://link.springer.de/link/service/series/0558/bibs/2140/21400200.htm*

Schramm, K., Leander, G., Felke, P. & Paar, C. (2004). A collision-attack on aes: Combining side channel- and differential-attack, *CHES*, pp. 163–175.

Sudo, T., Sasaki, H., Masuda, N. & Drewniak, J. L. (2004). Electromagnetic Interference (EMI) of system-on-package (SOP), *IEEE Transactions on Advanced Packaging* 27(2): 304–314.

Vuagnoux, M. & Pasini, S. (2009). Compromising Electromagnetic Emanations of Wired and Wireless Keyboards, *Proceedings of the 18th USENIX Security Symposium*, USENIX Association.

URL: *http://www.usenix.org/events/sec09/*

12

Biophysical Properties of Liquid Water Exposed to EM Radio Frequency Radiation

Valery Shalatonin

Belarusian State University of Informatics & Radioelectronics

Belarus

1. Introduction

Water is the most abundant substance on the surface of the earth and is the main constituent of all living organisms. The human body is about 65 percent water by weight, with some tissues such as the brain and the lung containing nearly 80 percent. Without water life would probably never have developed on our planet (Mottl et al., 2007). Increasing evidence indicates that the water has unique electromagnetic and biophysical peculiarities (Ball, 2001; Voeikov & Giudice, 2009). Since all life is based on water, all molecules in the living organisms interact with water. The water of body may not only be a carrier for nutrition and energy, but also a source and carrier for regulating electromagnetic information (Pan, 2003). Human and animal beings are bioelectrical systems and they are regulated by internal electromagnetic (EM) signals, which form an endogenous EM field. Environmental exposures to the artificial EMFs can interact with fundamental biological processes in living organisms. It is supposed that the environmental exposures to natural and artificial EM fields may interact with biological EM signals through intracellular and extracellular water. In some cases, this may lead to disease. During the past twenty years, the growing use of mobile phones (MP) has aroused great concern regarding the health effects of exposure to the EMR (Kundi et al., 2004, Khurana et al., 2009). Dual-band phones can cover GSM networks in pairs such as 900 and 1800 MHz frequencies (Europe, Asia, Australia, and Brazil) or 850 and 1900 MHz (North America and Brazil). Today's public exposure limits for telecommunications are based on the presumption that heating of tissue is the only concern when living organisms are exposed to EM radiation. In the last few decades, it has been established that bioeffects occur at non-thermal or low-intensity exposure levels thousands of times below the levels that state agencies say should keep the public safe. As reviewed in (Genuis, 2008), there are several hundred studies that support the existence of low-intensity, non-thermal effects of the MP radiation on biological systems. The consequences are mostly adverse: DNA single-and double-strand damage, changes in gene transcription, changes in protein folding, heat shock protein generation, production of free radicals, and effects on the immune system. In addition to mobile phones, new communicating systems are in use and developments of higher frequency applications are to come.

At present the question how such a low-energy of EMR could influence the functional activity of cell and organism still remains unanswered. Numerous hypotheses on molecular mechanisms of the specific biological effect of EMF have been proposed, but none have provided a reliable and exhaustive explanation of the experimental findings. The absence of such a mechanism cannot be taken as proof that health effects of environmental electric and magnetic fields are impossible. As water is the main medium where the major part of biochemical reactions is taking place, it is supposed that the environmental exposures to natural and artificial EM fields may change metabolic activity of cells and organisms using body's water as a primary receptor of the EM field (Shalatonin, 2008, 2009). Increased knowledge of the mechanisms underlying electromagnetic information storage, amplification and transduction by water may give us a fundamentally new comprehension of the processes operating in the water within biological systems (Del Giudice et al., 2010).

This work presents simple and sensitive biophysical experiments and its results to develop our notion that a possible reason of the long-term physical changes in water is conditioned by internal structures of the oxygen and hydrogen nuclei and the quantum properties of the EM radiation (V. Shalatonin, 2009, 2010, 2011; Shalatonin & Mishchenko, 2010).

2. Objects and methods

The main objects of these experiments have been wheat grains and drinking bottled artesian water. The grains were chosen to have a middle quality in order to have possibility to increase or to decrease its developmental properties. The water for their watering was preliminarily being exposed to the pulse-modulated EM radiation from GSM MP type Nokia, which was connected to a personal computer through an interface cable. The MP parameters were controlled via the nokia service program WINTESLA in order to change power, carrier frequency and other parameters of the EM radiation. The experimental setup is shown in Fig. 1. It consists of an appropriate glass with water which is located 5 - 6 cm away from the MP antenna. A dual-band phone, which can cover GSM-900 and GSM-1800 networks, was used. GSM-900 uses 890–915 MHz to send information from the mobile phone to a base station, providing 124 RF channels (channel numbers 1 to 124) spaced at 200 kHz. Frequency band of GSM-1800 is 1710-1785 MHz, providing 374 channels (channel numbers 512 to 885). The GSM standard employs the time division multiple access (TDMA) technique with eight time slots. This means that the transmitter is only ever switched on for an eighth of the time. Eight GSM phone users can share a pair of 200 kHz wide-band channels, because each user is given access only to a single 576 µs time-slot in a 4.6 ms frame, which is repeated 217 times a second. This 217-Hz cycle of power pulses is in the range of the normal bioelectrical functions both in and between cells, so it may induce low frequency power surges, causing health problems. The transmission average power in the phone is limited to a maximum of 2.5 watts in GSM-900 and 1 watt in GSM-1800. The typical radio spectra of the emitted EM radiation are shown in Fig. 2. They were measured by using a spectrum analyzer Tektronix Y400 NetTek Analyzer. The exposure time was 1.0-1.5 hour at an average power density of 55 µW/cm^2 (GSM-900) and 1.7 µW/cm^2 (GSM-1800). The power density was measured by a power meter P3-41 shown in Fig.1. Its antenna while measuring the power density at the same position as a water sample was installed. Pondus hydrogenii (pH) values of the water samples were determined by a pH meter (HI 9341, HANNA) with a precision of 0.06 pH units. A radiofrequency generator G4-76A was also used in our experiments.

Fig. 1. Photo of the experimental model

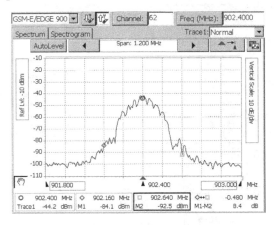

a

b

Fig. 2. Power spectra of the MP electromagnetic radiation: a – GSM-900; b – GSM-1800

3. Results and discussion

3.1 Biological properties of the water exposed to the GSM-900 MP radiation

Two samples of water, one of which was exposed to mobile phone EM radiation and the second one was the same but with regular water were used for watering the equal amount of wheat grains placed in two equal cups (200 grains in each cup). The quantity of the grains in each cup and its watering were equal throughout the experiment. New exposed water samples were being prepared in the same way once a day during the experiment. It should be noted that a first watering was being curried out in approximately 30 minutes after the exposure. The analysis of the quantity of the sprouted grains and the level of their development showed the significant differences for the sprouts in different cups. In Fig. 3 and Fig. 4 the most typical results and photos are given.

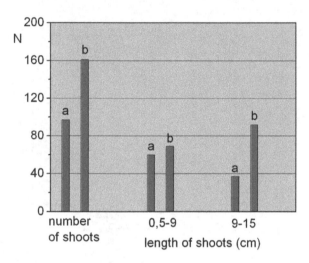

Fig. 3. Number and length of the sprouts for two groups of wheat grains are shown. The MP operated at a carrier frequency of 902. 4 MHz (62nd uplink channel of GSM-900). The experiment was conducted from March 12th to March 20th, 2010

a   b

Fig. 4. Photos of the sprouts: a – watering with the ordinary water; b – watering with the exposed water. The 8th day of the experiment

The level of development and quantity of the appeared shoots were significantly better for the grains watering with the exposed water. It can be seen in Fig. 3. The differences indicate that the EM radiation induced the stable changes in a physical state of the exposed water. The exposed water became more active and increased the grains vitality. A similar changes but not so expressive were observed in subsequent experiments, curried out for some other channels of the GSM-900 MP radiation.

It should be noted that in our experiments we wanted not only to find new results, connected with the ability of the exposed water to play an essential role in vital processes of wheat or other grains. Best of all we would like to get new data related to our approach for explaining the ability of water to store the EM or other kind of information. Therefore, let us explain the next step in the direction of our experiments (Shalatonin, 2011a).

At present most of proposed ideas devoted to this problem are being based on the molecular level of water and closely connected with the water clustering. But these results can not explain how the non-thermal magnetic and EM fields could essentially affect the hydrogen bonds in water during clustering. According to Ben-Jacob (2010) "... the detailed molecular-level properties of water are very important. Yet, it is becoming evident that they are not sufficient to provide a complete theory of water; we are missing some essential water aspects that cannot be accounted for by the molecular level investigations irrespective of how detailed and sophisticated they will be". We suppose that we are missing not only some essential water aspects but we are also missing essential properties of EM radiation, without which it is impossible to explain the phenomenon of water. Especially those related to the endogenous EM field from living beings.

Jerman et al., (1996) examined the possibility that ultra weak supposedly EM emission from living beings (bio-field) changes the structure of water. The results showed that normal seeds given water exposed to dying spruce seedlings reacted with significant slowing of germination and had a tendency to grow more slowly than controls (watered by ordinary water). This line of experiments demonstrates indirect evidence for some form of endogenous ultra weak EM emission from living beings. The authors pay attention that such emission alters water in some as yet unknown way, and that organisms can influence each other through indirect non-chemical and perhaps electromagnetic alteration of water. Slovenian scientist Detela (2002) considers that the bio-field is a subtle material structure which is permeating biological cells of living beings. It is assumed that bio-field is a three-dimensional web woven of vibrating electric and magnetic fields. Such structure of the bio-field is in close correspondence with the molecular structure of living organisms and interacts with discrete atoms and molecules in living cells. Therefore this field can regulate many processes in living organisms. Its structure must remain temporally stable and its energy must not be dissipated. In a simplified way, it can be imagined the bio-field as a "cloud" that is asymptotically approaching a zero-value in all regions that are distant from the cloud centre. Such a field pattern is a type of steady wave packet and could be described by using modified Maxwell equations. In agreement with this theory, the presence of an unusually sharp border of the electric field in the space around living organisms has been detected (Shalatonin, 2007). The main objects for measurements were some of the flower plants including roses and carnations. The sections of the field ("cloud"?) surrounding flowers are shown in Fig. 5 and in Fig.6. The author concluded that possibly in the space around living beings there is a region of the bio- field (inside of the detected border), and the region of the bio-radiation, which extends beyond the detected border. It reminds to some extent the EM field structure of an ordinary antenna. Indeed, it is well known that every EM antenna has three areas: near, intermediate, and distant. The near EM field does not participate in a process of radiation but in the distant area the magnetic and electric components of the EM field are in phase and the EM radiation is created. Coming back to the paper of Jerman et al. (1996) one can assume that just this bio-field could be a source of the information for water samples located near dying spruce seedlings.

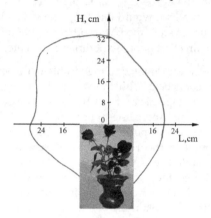

Fig. 5. The boundary of abrupt change of the electric field, surrounding the three red roses

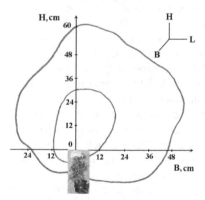

Fig. 6. The boundary of abrupt change of the electric field, surrounding the tagetes erecta:
—— in the morning;
—— in the evening

The obtained results, which we have just discussed, show that the EM field in our opinion is probably not an original field, which is responsible for the exchange information between water and bio-objects and between different bio-objects. Therefore, we support a notion (not a new one) that every kind of radiation including the EM one is with a still unexplored component. We suppose that this biologically active physical component arising due to the EM radiation is possibly the same kind of radiation as the radiation from bio-objects. It is possibly that just this radiation is responsible for the appearance of long term changes in water. In order to simplify referring to this kind of radiation it is convenient to call it as the active informational radiation (AIR) and then some particles, which correspond to this radiation, could be called as airons.

Now, we still have no real possibility to measure parameters of this radiation. Therefore, it would be useful to find something working like the presence indicator. We tried to find some materials, which, being placed between a mobile phone and a water sample could change (modify) the above discussed component of the EM radiation and this, in turn, could lead to a significant change in biological properties of the water, exposed by this changed AIR. In other words, we tried to define the presence of some factor of the EM radiation (that is the AIR), which is responsible for the water long term activation. It is important that the well known measurable EM radiation should not be shielded by these materials. Therefore, in our experiments metallic shields were not used. When choosing the suitable material, we have learned that several layers of an ordinary polyethylene film (PF) are able to modify or to shield such kind of radiation (Veinik, 1991). The polyethylene ($-CH_2-CH_2-$)n is an unusual solid in that it solidifies in long, kinked chains consisting of individual CH_2 units. It is essential that a polyethylene is a low dielectric loss polymer and so can not shield and even considerably change an intensity of the EM radiation. It was supposed that the PF placed between the MP and a water sample could change (modify) the AIR and the biophysical properties of the water samples influenced by this radiation. The experiment was conducted from March 23rd to March 31st, 2010. The MP operated as earlier at a carrier frequency of 902, 4 MHz (62nd uplink channel of GSM-900).

Let us describe this experiment in details. In Fig. 7 we can see our experimental setup. The exposed water was used for watering of wheat grains placed in two cups (200 grains for each cup). As a control group, watering by ordinary water, a third cup filled with the same amount of the wheat grains was used. New water samples for watering were being prepared in the same way once a day during the experiment. The obtained results are presented in Fig. 8, 9. We can see that the water exposed to the MP radiation has radically increased its biological activity (compare with the control group). The number of shoots has increased from 68 to 115 and their development was better. But the most surprising results have been obtained when the water samples were surrounded by the four-layered polyethylene film. The modified water influenced the number of the shoots and their growth negatively. The experiments were repeated, with slight variations several times and always led to practically the same results, but the level of growth suppression of the sprouts and its number were different. Fig. 10 shows photos, obtained during the similar experiment which was conducted from August 04th to August 12th, 2011. It can be seen that the germination of the sprouts is in accordance with the above-described experiment. The experimental results obtained at a frequency of 914, 2 MHz (121 channel) are shown in (Shalatonin & Mishchenko, 2010b). In should be noted that the shielded and exposed water was absorbed by the seeds much worse in comparison with other groups of the seeds. Therefore, this phenomenon could be considered as a visible reason of the sprouts bad development. Thus, it has been established that the EM radiation from the MP induced stable changes in the biophysical properties of the exposed water. It makes the water more active and changes dramatically the grains biological properties. The obtained results raise a lot of questions. For example, how the biological activity of the activated water depends on parameters of the MP radiation (frequency value, level of power, duration of the exposition, parameters of the film and so on) and what physical properties of the water were changed due to exposure to this radiation.

Fig. 7. The process of exposing two water samples to the MP radiation. One of the glasses is wrapped by thin (40μm) four-layered PF. The exposure duration is 1.5 hour.

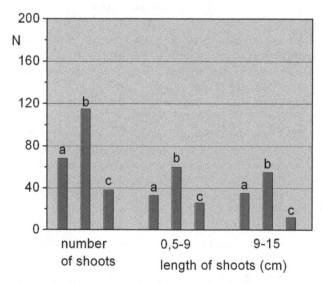

Fig. 8. Number and length of the sprouts for two groups of wheat grains. The MP operated at a carrier frequency of 902, 4 MHz (62nd uplink channel of GSM-900). The experiment was conducted from March 23rd to March 31st, 2010.

Fig. 9. Photos of the sprouts: a − watering with ordinary water; b − watering with exposed water; c − watering with shielded and exposed water. The shielded glass was wrapped by thin (40μm) four-layered PF. The exposure duration is 1.5 hour. It is the 8th day of the experiment. The experiment was conducted from March 23rd to March 31st, 2010

Fig. 10. Photos of the sprouts: a – watering with ordinary water (120 sprouts); b – watering with exposed water (131 sprouts); c – watering with shielded and exposed water (56 sprouts). The shielded glass was wrapped by thin (150μm) eight-layered PF. The exposure duration is 1.0 hour. It is the 8th day of the experiment. The experiment was conducted from August 04th to August 12th, 2011

3.2 Biological properties of the water exposed to the GSM-1800 MP radiation

Let us now describe some results of our experiments when using the same MP, which operated in the GSM-1800 band. We used practically the same procedures and in Fig. 11 the obtained results are given. It should be has in mind that the power density of the MP EM radiation is approximately 30 times less than the same value measured when using the GSM-900 operation of the MP. The diagrams in Fig. 11 show that there is a difference between the control cup and the cups watered by the exposed water. This indicates, in line with our results for the GSM-900 standard that the EM radiation exposure induces stable changes in physical state of the exposed water. It makes the water more active and changes the grains biological properties. The analysis of the quantity of the sprouted grains and the level of their germination showed the significant differences for sprouts in the different cups. The level of germination and quantity of the shoots usually were better in grains, that were watered with the exposed water preliminary wrapped in the PF in comparison with the only exposed water. The same but more evident differences were observed earlier when using a carrier frequency of 1779.6 MHz (859 channel) (Shalatonin, 2011a). But the most important that the results obtained when using the GSM-1800 radiation showed the opposite relationships in comparison with the similar experiments when using the GSM-900 mobile phone radiation. This conclusion is in accordance with the measurement of pH. The pH level of the water exposed to the GSM-1800 MP radiation is usually getting larger than this value of the water exposed to the GSM-900 MP radiation. The typical changes of the pH of the exposed water samples (GSM-900 and GSM-1800) and of the ordinary water are shown in Fig. 12. It needs also to take into account that in this experiment the water samples

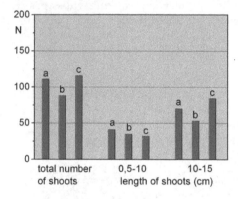 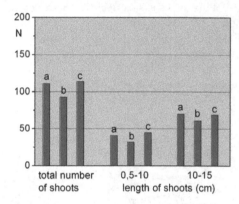

Fig. 11. Number and length of the shoots for different groups of wheat grains (on the left – f = 1784.0 MHz, channel 881; on the right – f = 1784.8 MHz, channel 883): a – watered with ordinary water; b – watered with exposed water; c – watered with shielded and exposed water. The shielded glass was wrapped by thin (40μm) four-layered PF. The exposure duration was 1.0 hour. The experiment was conducted from September 12th to September 20th, 2011

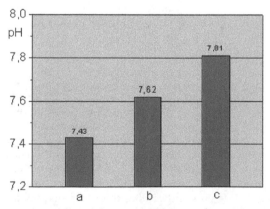

Fig. 12. Results of the measurement of the water pH: a – ordinary unexposed water; b – water exposed to the GSM MP radiation at f = 902.4 MHz; c - water exposed to the GSM MP radiation at f = 1779.6 MHz. The exposure duration is 1.0 hour.

received essentially different doses of the EM radiation. The matter is that as was noted above the power density of the applied dual-band phone when operating in the GSM-1800 standard was lesser than in the GSM-900 standard. Therefore, it can be supposed that the influence of the GSM 1800 phone radiation on the water is larger and this still remains unexplained. The experiment was carried out on August 2nd, 2011. The volume of every water sample was 450 mL.

3.3 A possible approach to explain the obtained results

Let us now consider the specific properties of the PF, and a possible biophysical mechanism, which are responsible for the obtained results. First of all it should be noted that polyethylene, like water, has a very high hydrogen concentration. It has been counted that a cubic meter of water contains 2.7×10^{29} protons (Ellis, 2011). In 1934 Enrico Fermi with colleagues were the first scientists who used hydrogenous materials including water to enhance the neutron induced radioactivity (De Gregorio, 2006). It turned out that water can be used as a suitable material in the path of the neutrons in order to slow fast neutrons down to thermal energies. Hydrogen rich substances are quite efficient at doing this as neutrons will lose more energy per collision with light atoms than with more massive substances. For other types of radiation, e.g. alpha particles, beta particles, or gamma rays, material of a high atomic number and with high density make for good shielding; frequently lead is used. However, this approach will not work with neutrons. The neutron's lack of total electric charge makes it difficult to steer or accelerate them. Charged particles can be accelerated or decelerated by electric or magnetic fields. These methods have little effect on neutrons beyond a small effect of an inhomogeneous magnetic field because of the neutron's magnetic moment. The main energy-loss mechanism occurs when they strike nuclei. It is often noted in the literature that the particles interact like billiard balls, the most efficient slowing-down occurs when the bodies that are struck in an elastic collision have the same mass as the moving bodies; hence the most efficient neutron moderator is hydrogen and some other light elements.

Since humans are mostly water, if they are standing in the way of a beam of neutrons, they will have a strong moderating effect. The slowing of the neutrons will cause damage and will induce other nuclear reactions. For example, if a thermal neutron is captured by hydrogen, a gamma ray will be released. Some substances inside of a human body will become radioactive as a result of exposure, causing the release of radiation even after the source of neutrons has been removed.

It is supposed that the energy-loss mechanism like above described could occur during a propagation of the AIR across the PF. According to (Veinik, 1991; Detela, 2001) and other authors the AIR consists of the charged particles having extremely small diameter ($\sim 1.6 \times 10^{-32}$ m) and mass ($\sim 1.2 \times 10^{-3} \times m_e$, where m_e – is the mass of an electron). It is not surprising that the AIR, modified by the PF can change its physical and, in turn, biological properties. But it should take into account that according to the abovementioned "theory" of the billiard balls, it needs to have inside of the PF some particles similar in size to that of the AIR.

A new model of the proton and neutron, that was recently experimentally discovered and our recent experimental results (Shalatonin, 2009) give a possibility to develop and justify an appropriate approach to solving the problem. It was recently shown (Miller, 2008; Islam, 2010) that a neutron has a negative charge both in its inner core and its outer edge, with a positive charge sandwiched in between to make the particle electrically neutral (Fig. 13). It means in other words that the neutrons have a three-layered charge structure. The number within the every layer is the non-integer value of the layer charge (in units of |e|). Until recently physicists have long known that neutrons are made up of three quarks (subatomic particles) of two different types – one "up" quark with an electric charge of +2/3 and two "down" quarks, each with a charge of -1/3. It is plausible that the protons could also have the similar three-layer structure that is shown in Fig. 14. The discovery changes scientific understanding of how neutrons interact with negatively charged electrons and positively charged protons. We wanted to add that in our opinion this discovery keeps us thinking if we need a notion on such particles as quarks. Ellis (2011) writes that "scattering experiments with nucleons cannot liberate free quarks. …No evidence for internal structure within quarks, or electrons, has yet been found. If quarks and leptons are discovered someday to be composite objects, bound states of some not-yet-known more fundamental constituents, then the length scale on which this binding occurs must be at least three orders of magnitude smaller than the femtometer scale of nucleons (1 fm = 10^{-15} m)". We suppose that protons and neutrons consist of electrically charged particles having the same nature as the airons. It is a very important peculiarity, because only in such a case nucleons can form lots of different wave patterns. That, in turn, may give rise to the origin of the effect of the water memory.

New findings may help understanding of how the AIR, inherent in the EMR, interacts with the protons of the PF. A physical parameter of the exposed water, which is closely related to the long-term changes in water was recently measured (Shalatonin, 2009). It was experimentally shown that multilayer standing–wave magnetic patterns (SWMP) appear around water samples exposed to the EM radiation. Fig. 15 shows a section of the magnetic three-layer structure (1, 2, and 3) obtained by using a one frequency EM radiation of 386, 4 MHz, which is very close, in our opinion, to a resonance excitation of the water protons. The quantity of the layers when using this frequency does not depend on the intensity of the EM radiation and the volume of this structure is unusually extensive. The last peculiarity as is well seen by comparing this pattern with the pattern shown in Fig. 16, is related in our opinion to the absorption resonance of the water sample protons. The SWMP remains fairly stable during a day but during several days their field energy is gradually dissipated. The patterns obtained for non-resonant frequencies had the same basic outlines but the number of the closed curves and its dimensions depended strongly on the EM radiation parameters. It is reasonable to suppose that the three-layer wave patterns of the nuclei in its resonant state may readily manifest themselves into space around the water samples. They reflect the internal charge structure of the nucleus described above. In other words, water has the macroscopic quantum properties that could give rise to the long-term physical effects. It is obvious that these magnetic structures are the result of the collective and coherent dynamics involving many nuclei. The only particular frequencies of the external EM sinusoidal radiation related to the resonant states of the hydrogen and the oxygen nuclei could lead to the resonant states of water. It should be noted that in according with the above described approach a real resonant frequency of the proton or neutron is the frequency of the AIR, which is transmitted together with the EM radiation. Its value is very high and at present there is no technical possibility to measure it. It was still nothing said of the particles in the PF, which, as was supposed earlier (the "theory" of billiard balls) should provide the energy-loss mechanism of the AIR during its passing through the PF. We assume that every proton or neutron has external field patterns that could consist of the particles similar to that in the AIR. Therefore, it could be plausible to assume that the macroscopic SWMPs surrounding water samples are the AIR formed (at least partly) from the summarized individual AIR of the nuclei.

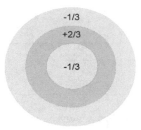

Fig. 13. A new model of neutron with charge layers. It has a negative fractional charge both in its inner core and its outer edge, with a positive fractional charge sandwiched in between to make the particle electrically neutral

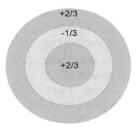

Fig. 14. A new model of proton with charge layers. It has a positive fractional charge both in its inner core and its outer edge, with a negative fractional charge sandwiched in between to make the particle electrically positive

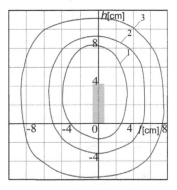

Fig. 15. Section of the three-layer standing wave magnetic pattern of the water sample following the exposure to the sinusoidal resonant EM radiation ($f_e \approx 386.4$ MHz). It is a resonance excitation frequency of water protons

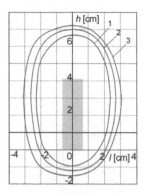

Fig. 16. Section of the three-layer standing wave magnetic pattern of the water sample following the exposure to the sinusoidal resonant EM radiation ($f_a \approx 381.5$ MHz). It is a resonance absorption frequency of water protons

Thus, our interpretation of the transition of the electromagnetically exposed water from the ordinary state to the long-range magnetically ordered state is mainly related to the inherent properties of the hydrogen protons and very probably the protons and neutrons of oxygen in molecules of water. New physical state of the water changes its biological properties. The exposed water can activate (mainly the GSM-900 MP radiation) or suppress (mainly the GSM-1800 MP radiation) the development of the cereals and possibly other bio-objects. The above interpretation is perhaps not so surprising. It is long known that almost all of an atom's matter is located in the nucleus. Atomic nuclei are thus unimaginably dense compared to chemical elements or chemical compounds. The density of a nucleus is more than 14 orders of magnitude greater than that of water. Therefore, without doubt a nuclear matter can play key role in the processes under consideration.

There is at least one more experimentally proven pattern of the manifestation of the behavior of atomic nuclei into the phase behavior of bulk liquid water (GSI Helmholtzzentrum für Schwerionenforschung GmbH, see References). The forces between individual components of the nucleus - the nucleons - vary according to distance in a manner remarkably similar to those between molecules in a liquid. At very short distances, the binding forces repel; at medium nucleon distances, they attract. In fact, in many ways, atomic nuclei behave very much like drops of liquid. Water, like all of the other matter surrounding us, is solid, liquid, or gaseous depending on its temperature and pressure. In the same way, nuclear matter – the charged protons and electrically neutral neutrons forming the nucleus of an atom – can assume various states. It turned out that even at a relatively slight increase in energy the nuclear matter undergoes a phase transition from the normal, liquid-like state to a nucleon gas. The experiments established that a caloric curve of nuclear matter (temperature-energy diagram) exhibits behavior analogous to the temperature curve of boiling water. This phenomenon was experimentally proved by the world's leading research laboratories.

In according with our notion not only water but theoretically all matter including solid materials, even stones, under certain conditions can manifest itself in a variety of effects associated with memory. For example, various methods of activation of various substances are applied in homeopathy. Jerman at el. (2004) shown that biologically effective information from two chosen substances (herbicide glyphosate and pharmaceutical substance diazepam) can be non-chemically imprinted into a polyacryl based compound material to be polymerized by a high voltage electric field. Under suitable conditions the stored information can be reproduced, evoking specific effects, biological or physical, without any chemical contact with the original substance. The positive results were obtained by electrophotography and by volunteers.

Our recent experiments proved that the above discussed magnetic patterns can arise not only around pure water but also around solid hygroscopic materials having different water content. It turned out that the water which is dissolved in a wood sample keeps nevertheless its property to perceive and accumulate the non-thermal EM (AI) radiation (Fig. 17 and 18).

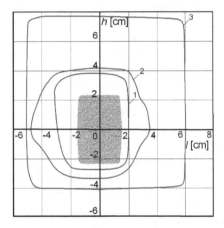

Fig. 17. Section of the SWMP of the first piece of wood following the exposure to the sinusoidal resonant EM radiation ($f_e \approx 386.4$ MHz). The exposure duration is 0.5 hour. The water content is 52%

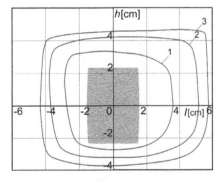

Fig. 18. Section of the SWMP of the second piece of wood following the exposure to the sinusoidal resonant EM radiation ($f_e \approx 386.4$ MHz). The exposure duration is 0.5 hour. The water content is 18%

The experiment was as follows. Two dried up equal conifer's pieces were saturated with water until the required moisture content and then exposed to the sinusoidal EM radiation at the resonant frequency of 386, 4 MHz. The prepared pieces of the wood were used for measuring the SWMPs. The obtained SWMPs are mainly similar but their volumes are different. It is seen that the larger water content leads to the bigger magnetic patterns.

3.4 Electromagnetic radiation, water and memory of living beings

As was noted at the beginning of this article, water is the main constituent of all living organisms. Many its properties are related to the storing and transferring of the vital information. And so a working hypothesis is that memory might be in particular susceptible to the extent of hydration. It is proved by recent experiments. Benton & Burgess (2009) investigated the cognitive functioning of 40 children (mean of 8 years and 7 months) twice, once after drinking 300 mL of water and on another day when no water was provided. Memory was assessed by the recall of 15 previously presented objects. Recall was significantly better on the occasions when water had been consumed. The ability to sustain attention was not significantly influenced by whether water had been drunk. Bar-David et al. (2005) found that memory but not four other measures of cognition was disrupted by dehydration. In according with our notions a basic phenomenon of the storing and transferring of the information in water is the interaction of the AIR (or the aeronic field), inherent in the informative EM radiation, with the protons and neutrons of the water (exactly speaking with the aeronic field of the nucleons). Therefore, it can be supposed that living beings exposed to the EM radiation could change their memory condition. Based on our experiments with wheat sprouts it can be assumed that the non-thermal EM radiation from the GSM-900 MP hardly may make memory worse and at the same time it is highly probable that the memory may get worse after exposing to the GSM-1800 MP radiation. Some studies using exposed rats at the GSM-900 radiation confirm our assumption (Dubreuil, 2003; Sienkiewicz, 2000 and others). There were no effects seen at a low SAR level but only some effects were found on exploratory activity at a high SAR level (3.5 W/kg). The recent study (Ntzouni at el., 2011) was conducted in order to investigate whether short-term memory is affected by ordinary GSM-1800 MP exposure. The authors concluded that an acute exposure did not affect mice memory but the chronic exposure had an impact on the recognition memory in a statistically significant manner. On the contrary, recently the first evidence has been reported that exposure directly associated with cell phone use (GSM 900 MP, SAR value, 0.25 W/kg) provides cognitive benefits i.e. improvement in transgenic Alzheimer's mice performance after long term (8 months) EMF exposure (Arendash at el., 2010). Wiholm et al. (2009) studied spatial behavior and learning (a virtual Morris water-maze) in subjects with (N=23) and without (N=19) symptoms related to mobile phone use. The design was both double-blind and crossover, and the exposure (884 MHz, SAR value, 1.4 W/kg) lasted for 2.5 hours. Spatial performance was measured before and after the exposure and the order of sessions was counterbalanced. The authors claim that the symptomatic group improved their performance during RF exposure. The authors themselves state that there is a need for replication. Luria et al. (2009) studied GSM (915 MHz, modulation 217 Hz, 0.25 W mean) effects in crossover, single blinded design on 48 subjects performing spatial working memory task, but found no effects after correction for multiple comparisons.

4. Conclusion

The experimental results presented here and previous studies show that non-thermal influences of various kinds, including EM radiation can considerably change the biophysical properties of water. We have demonstrated that wheaten grains can alter significantly their germination and development when watering with water, preliminarily exposed to the non-thermal mobile phone radiation. The biological response depends on parameters of the EM radiation especially on its carrier frequency and exposure duration. The main quality of the biological information (stimulating or depressive) preserved in water is mostly predetermined by choice of the GSM standard (GSM-900 or GSM-1800 mobile phone radiation). Based on conventional theory it is very difficult to interpret satisfactorily the revealed in this study and some other properties of water. Our results related to the polyethylene film were explained on the base of the well-known interaction of the polyethylene with fast neutrons. We suppose that the physical mechanism of the appearance of stable physical changes in the exposed water is conditioned by the presence of the biologically active field component that is inherent in the ordinary EM radiation. In order to simplify referring to this kind of radiation it is convenient to call it as the active informational radiation (AIR) and then some particles, which correspond to this radiation, could be called as airons. Polyethylene film due to a high concentration of hydrogen (H atoms) scatters and slows airons. It is supposed that the polyethylene action is similar to its action to the fast neutrons passage. Therefore, the polyethylene and other hydrogenous polymers could be used to study the physical properties and parameters of the AIR. As it turned out, the changes in the physical properties of the AIR lead to significant changes in biological properties of the exposed water. In further work it would be important for checking our assumptions to use a pure hydrogen gas instead of the polyethylene. The hydrogen is one of the main compounds of water and of all organic matter. Since humans are mostly water, it is plausible to assume that the body fluids could be the primary receptor of the aironic field. And so, it is not surprising that biophysical properties of the exposed water were being changed when placing the polyethylene film between a water sample and a source of the EM radiation (mobile phone). The next assumption is that the mechanism underlying the changes in the biophysical properties of the exposed water is related to the nuclear properties of matter and water in particular. The multi-layer standing-wave magnetic patterns, which appear around the water samples, preliminarily exposed to the EM radiation most likely go hand in hand with the recently discovered complex three-layered charge structure of protons and neutrons. It is supposed that these patterns are the result of the manifestation of the collective and coherent microscopic phenomena involving many nuclei in molecules of the exposed water. In other words, exposed water has macroscopic magnetic properties that correspond to the microscopic properties of the charge structures of its nuclei. The EM radiation (the AIR) interacts with water at a nuclear level: airon-driven nuclear reactions lead to the long-term changes of specific charge configurations of nuclei that, in turn, affect the biological properties of water. The investigation of the resonant properties of the protons and neutrons related to the AIR are important for possible technological and biomedical applications (Shalatonin, 2011a). Especially it is related to the hydrogen which possesses certain beneficial characteristics and widely used as a source of energy. The obtained results could provide a practical approach to the problem of the harmful EM radiation shielding. Finally, it should be noted that these results should be treated with some caution until other studies will confirm them.

5. Acknowledgment

The author thanks Dr. Valery Mishchenko for encouragement throughout preparation of this paper. This work is partly supported by the Ministry of Education of Belarus.

6. References

Arendash, G., Sanchez-Ramosc, J., Morie, T., Mamcarz, M., Linc, X., Runfeldt, M., Wang, L., Zhang, G., Savad, V., Tang, J. & Cao, C. (2010). Electromagnetic field treatment protects against and reverses cognitive impairment in Alzheimer's disease mice, *J Alzheimer's Dis*, Vol.19, No.1, (January 2010), pp. 191–210, ISSN 1387-2877

Ball, P. (2001). *Life's Matrix: A Biography of Water* (first published 1999), University of California Press, ISBN 0520230086, USA, California

Bar-David, Y., Urkin, J. & Kozminsky, E. (2005). The effect of voluntary dehydration on cognitive functions of elementary school children. *Acta Paediatrica*, Vol.94, No.11, (November 2005), pp. 1667–1673, ISSN 1651-2227

Benton, D. & Burgess, N. (2009). The effect of the consumption of water on the memory and attention of children. *Appetite*, Vol.53, No.1, (August 2009), pp. 143-146, ISSN 0195-6663

Ben-Jacob, E. (2010). Towards A Systemic View of Water as the Fabric of Life. *Water: A Multidisciplinary Research J*, Vol.2, Suppl.1, (January 2010), pp. 8-10, ISSN 2155-8434

Del Giudice, E., Spinetti, P. & Tedeschi, A. (2010). Water Dynamics at the Root of Metamorphosis in Living Organisms. *Water,* Vol.2, No.3, (September 2010), pp. 566-586, ISSN 2073-4441

De Gregorio, A. (2006). Radioactivity induced by neutrons: Enrico Fermi and a thermodynamic approach to radiative capture. *American Journal of Physics*, Vol.74, Iss.7, (July 2006), pp. 614- 620, ISSN 0002-9505

Detela, A. (2002) *Magnetic knots*, Littera Picta, Ljubljana, Slovenia, in Slovenian

Dubreuil, D., Jay, T. & Edeline, J. (2003). Head-only exposure to GSM 900 - MHz electromagnetic fields does not alter rat's memory in spatial and non-spatial tasks, *Behav Brain Res*, Vol.145, No.1-2, (October 2003), pp. 51-61, ISSN 0166-4328

Ellis, S. (2011). Chapter 4. Known particles, In: *Particles and Symmetries. Based on notes by Laurence G. Yaffe University of Washington*, pp. 46-47, 2011.09.10, available from: < http://courses.washington.edu/partsym/ch04_11.pdf >

Genuis, S. (2008). Fielding a current idea: exploring the public health impact of electromagnetic radiation. *Public Health*, Vol.122, No.2, pp. 113–124, ISSN 0033-3506

GSI Helmholtzzentrum für Schwerionenforschung GmbH, Last update: 11. Nov. 2010 by C.Bisignano, The Stuff that Nuclei Are Made of, In: The Wonderful World of Atoms and Nuclei, 2011.07.12, Available from: < http://www.gsi.de/portrait/Broschueren/Wunderland/05_e.html >

Islam, M., Kašpar, J., Luddy, R. & Prokudin, A. (2010). Proton-Proton Elastic Scattering at LHC and Proton Structure. *Proceedings of the 13th International Conference on Elastic and Diffractive Scattering*, pp. 48-54, ISBN 978-92-9083-342-0, Switzerland, Geneva, June 29th-July 3rd, 2009 (see also http://cerncourier.com/cws/article/cern/41014)

Jerman, I., Berden, M. & Ruzic, R. (1996). Biological influence of ultra weak supposedly EM radiation from organisms mediated through water. *Electro & Magnetobiology*, Vol.15, No.3, pp. 229-244, ISSN 1061-9526

Jerman, I., Ružič, R. & Škarja, M. (2004). Effects of elecrtomagnetic information inprinted into solid material, In: *BION, Institute for Bioelectromagnetics and New Biology; Ljubljana, Slovenia,* 2011.08.09, Available from: <http://www.bion.si/gradiva-objave/Kerrock_2stolpca.pdf >

Khurana, V., Teo, C., Kundi, M., Hardell, L. & Carlberg, M. (2009). Cell phones and brain tumors: a review including the long-term epidemiologic data. *Surgical Neurology,* Vol.72, No.3 (September 2009), pp. 205-215

Kundi, M., Mild, K., Hardell, L. & Mattsson, M. (2004). Mobile telephones and cancer — a review of epidemiological evidence. *J Toxicol Environ Health B Crit Rev,* Vol.7, No.5, (Sep-Oct 2004), pp. 351-384, ISSN 1093-7404

Luria, R., Eliyahu, I., Hareuveny, R., Margaliot, M. & Meiran, N. (2009). Cognitive effects of radiation emitted by cellular phones: the influence of exposure side and time. *Bioelectromagnetics,* Vol.30, No.3, (January 2009), pp. 198-204

Miller, G. & Arrington, J. (2008). Neutron negative central charge density: An inclusive-exclusive connection. *Phys. Rev. C,* Vol.78, No.3, (September 2008), pp. 032201.1-032201.4, ISSN 0556-2813, (Featured in Phys. Rev. Focus: http://focus.aps.org/story/v22/st11)

Mottl, M., Glazer, B., Kaiser, R. & Meech, K. (2007). Water and astrobiology. *Chemie der Erde/Geochemistry,* Vol. 67, No.4 , (December 2007), pp. 253-282, ISSN 0009-2819

Ntzouni, M., Stamatakis, A., Stylianopoulou, F. & Margaritis, L. (2011) Short-term memory in mice is affected by mobile phone radiation. *Pathophysiology,* Vol.18, No.3 (June 2011), pp. 193-199, ISSN 0928-4680

Pan, J., Zhu, K., Zhou, M. & Wang, Z. (2003). Low Resonant frequency storage and transfer in structured water cluster, *Proceedings of the IEEE International Conference on Systems, Man, and Cybernetics,* pp. 5034-5039, USA, Washington, DC, October 5-8, 2003

Shalatonin, V. (2007). A study of the endogenous electromagnetic field into the space around the flower plants. *Proceedings of the Joint 32nd Int. Conf. on Infrared and Millimetre Waves and 15th Int. Conf. on Terahertz Electronics,* ISBN 978-1-4244-1438-3, pp. 293-294, Cardiff, UK, September 2-9, 2007

Shalatonin, V. (2008). Mobile phones and health: key role of human body fluids in bioeffects of non-thermal EM radiation. *Proceedings of the 18th Int. Crimean Conference "Microwave & Telecommunication Technology",* ISBN 978-966-335-166-7, Vol.2, pp. 848-849, Sevastopol, Ukraine, September 8-12, 2008.

Shalatonin, V. (2009). Standing-wave magnetic patterns of water exposed to UHF sinusoidal electromagnetic radiation. *Proc. of the 19th Int. Crimean Conference "Microwave & Telecommunication Technology",* ISBN 978-1-4244-4796-1, Vol.2, pp. 893-894, Sevastopol, Ukraine, September 14-18, 2009

Shalatonin, V. & Mishchenko, V. (2010a). Biological properties of water activated by GSM mobile phone radiation. *Proceedings of the 20th International Crimean Conference "Microwave & Telecommunication Technology",* ISBN 978-1-4244-7184-3, pp. 1159-1160, Sevastopol, Ukraine, September 13-17, 2010

Shalatonin, V. & Mishchenko, V. (2010b). Biological Properties of Water Exposed to Mobile Phone Radiation. *Proceedings of the First European Conference on Moisture Measurement,* pp. 353-359, Weimar, Germany, October 05-07, 2010

Shalatonin, V. (2011a). Biological properties of water exposed to the GSM-1800 EM radiation, passed through a polyethylene film. *Proceedings of the 21st International Crimean Conference "Microwave & Telecommunication Technology"*, ISBN 978-966-335-351-7, pp. 1013-1014, Sevastopol, Ukraine, September 12-16, 2011.

Shalatonin, V. (2011b). Mixed proton/EM radio frequency radiation as a possibility to improve the quality of the treatment. *Book of abstracts of the Workshop "Physical and biological basic of hadron radiotherapy"*, pp. 55, Krakow, Poland, September 2-3, 2011

Sienkiewicz, Z., Blackwell, R., Haylock, R., Saunders, R. & Cobb, B. (2000). Low-level exposure to pulsed 900 MHz microwave radiation does not cause deficits in the performance of a spatial learning task in mice, *Bioelectromagnetics*, Vol.21, No.3, (April 2000), pp. 151–158

Voeikov, V. & Del Giudice, E. (2009). Water Respiration - The Basis of the Living State. *Water: A Multidisciplinary Research J*, Vol.1, (July 2009), pp. 52 – 75, ISSN 2155-8434

Veinik, A. (1991). *Thermodynamics of real Processes*, Science and Techniques Publ., Belarus, Minsk, in Russian

Wiholm, C., Lowden, A., Kuster, N., Hillert, L., Arnetz, B., Akerstedt, T. & Moffat, S. (2009). Mobile phone exposure and spatial memory. *Bioelectromagnetics*, Vol.30, No.1, (January 2009), pp. 59-65

Permissions

The contributors of this book come from diverse backgrounds, making this book a truly international effort. This book will bring forth new frontiers with its revolutionizing research information and detailed analysis of the nascent developments around the world.

We would like to thank Prof. Saad Osman Bashir, for lending his expertise to make the book truly unique. He has played a crucial role in the development of this book. Without his invaluable contribution this book wouldn't have been possible. He has made vital efforts to compile up to date information on the varied aspects of this subject to make this book a valuable addition to the collection of many professionals and students.

This book was conceptualized with the vision of imparting up-to-date information and advanced data in this field. To ensure the same, a matchless editorial board was set up. Every individual on the board went through rigorous rounds of assessment to prove their worth. After which they invested a large part of their time researching and compiling the most relevant data for our readers. Conferences and sessions were held from time to time between the editorial board and the contributing authors to present the data in the most comprehensible form. The editorial team has worked tirelessly to provide valuable and valid information to help people across the globe.

Every chapter published in this book has been scrutinized by our experts. Their significance has been extensively debated. The topics covered herein carry significant findings which will fuel the growth of the discipline. They may even be implemented as practical applications or may be referred to as a beginning point for another development. Chapters in this book were first published by InTech; hereby published with permission under the Creative Commons Attribution License or equivalent.

The editorial board has been involved in producing this book since its inception. They have spent rigorous hours researching and exploring the diverse topics which have resulted in the successful publishing of this book. They have passed on their knowledge of decades through this book. To expedite this challenging task, the publisher supported the team at every step. A small team of assistant editors was also appointed to further simplify the editing procedure and attain best results for the readers.

Our editorial team has been hand-picked from every corner of the world. Their multi-ethnicity adds dynamic inputs to the discussions which result in innovative outcomes. These outcomes are then further discussed with the researchers and contributors who give their valuable feedback and opinion regarding the same. The feedback is then collaborated with the researches and they are edited in a comprehensive manner to aid the understanding of the subject.

Apart from the editorial board, the designing team has also invested a significant amount of their time in understanding the subject and creating the most relevant covers. They scrutinized every image to scout for the most suitable representation of the subject and create an appropriate cover for the book.

The publishing team has been involved in this book since its early stages. They were actively engaged in every process, be it collecting the data, connecting with the contributors or procuring relevant information. The team has been an ardent support to the editorial, designing and production team. Their endless efforts to recruit the best for this project, has resulted in the accomplishment of this book. They are a veteran in the field of academics and their pool of knowledge is as vast as their experience in printing. Their expertise and guidance has proved useful at every step. Their uncompromising quality standards have made this book an exceptional effort. Their encouragement from time to time has been an inspiration for everyone.

The publisher and the editorial board hope that this book will prove to be a valuable piece of knowledge for researchers, students, practitioners and scholars across the globe.

List of Contributors

A. Draux
Institut National des Sciences Appliquées (INSA) de Rouen, Département de Génie Mathématique, Place Emile Blondel, BP08, 76131, Mont-Saint-Aignan Cédex, France

G. Gouesbet
Laboratoire d'Electromagnétisme des Systèmes Particulaires (LESP) Unité Mixte de Recherche (UMR) 6614 du Centre National de la Recherche Scientifique (CNRS) COmplexe de Recherche Interprofessionnel en Aérothermochimie (CORIA) Université de Rouen et Institut National des Sciences Appliquées (INSA) de Rouen BP12, Avenue de l'Université, Technopôle du Madrillet, 76801, Saint-Etienne-du Rouvray France

Jan Olof Jonson
Stockholm University, Sweden

X. Li
California Polytechnic State University, USA

D.M. Pierce and H.F. Arnoldus
Mississippi State University, USA

Mei Xiaochun
Institute of Innovative Physics, Department of Physics, Fuzhou University, Fuzhou, China

J. W. Maluf and F. F. Faria
Instituto de Física, Universidade de Brasília, Brasília, DF, Brazil

V.V. Aristov
Institute of Microelectronics Technology and High Purity Materials RAS, Russia

Nikolay Smolyakov
National Research Center "Kurchatov Institute" Moscow, Russia

Igor Protsenko and Alexander Uskov
Lebedev Physical Institute, Plasmonics ltd., Russia

Hidajet Salkic and Amir Softic
PE Elektroprivreda BiH, Bosnia and Herzegovina

Adnan Muharemovic
Energoinvest d.d. Sarajevo, Bosnia and Herzegovina

Irfan Turkovic
University of Sarajevo, Faculty of Electrical Engineering, Bosnia and Herzegovina

Mario Klaric
Dalekovod d.d. Zagreb, Croatia

I. Verginadis, A. Velalopoulou, I. Karagounis, Y. Simos, D. Peschos, S. Karkabounas and A. Evangelou
Laboratory of Physiology, Faculty of Medicine, University of Ioannina, Ioannina, Greece

Olivier Meynard, Sylvain Guilley and Jean-Luc Danger
Telecom Paris Tech, France

Yu-Ichi Hayashi and Naofumi Homma
Tohoku University, Japan

Valery Shalatonin
Belarusian State University of Informatics & Radioelectronics, Belarus